What are the health consequences of a transition from an active 'hunter–gatherer' lifestyle to that of sedentary modern living? In this book, the impact of 'modernization' is assessed in various populations in the circumpolar regions. The hazards of living in polar regions, and the adaptations shown culturally, behaviourally and physically by the indigenous peoples are examined and the effect of changes in habitual activity, diet, and general lifestyle due to more urban living patterns on the body composition, pulmonary function and susceptibility to disease is discussed. The implications of this switch are important not only for all those concerned about the survival of indigenous communities around the world, but for all of us living increasingly sedentary, urban lives.

Anthropologists, physiologists and those interested in population fitness will find this a comprehensive and valuable volume.

Cambridge Studies in Biological Anthropology

The health consequences of 'modernization'

Cambridge Studies in Biological Anthropology

Series Editors

G.W. Lasker
Department of Anatomy, Wayne State University,
Detroit, Michigan, USA

C.G.N. Mascie-Taylor
Department of Biological Anthropology,
University of Cambridge

D.F. Roberts
Department of Human Genetics,
University of Newcastle-upon-Tyne

R.A. Foley
Department of Biological Anthropology,
University of Cambridge

Also in the series

G.W. Lasker *Surnames and Genetic Structure*
C.G.N. Mascie-Taylor and G.W. Lasker (editors) *Biological Aspects of Human Migration*
Barry Bogin *Patterns of Human Growth*
Julius A. Kieser *Human Adult Odontometrics – The Study of Variation in Adult Tooth Size*
J.E. Lindsay Carter and Barbara Honeyman Heath *Somatotyping – Development and Applications*
Roy J. Shephard *Body Composition in Biological Anthropology*
Ashley H. Robins *Biological Perspectives on Human Pigmentation*
C.G.N. Mascie-Taylor and G.W. Lasker (editors) *Applications of Biological Anthropology to Human Affairs*
Alex F. Roche *Growth, Maturation, and Body Composition – The Fels Longitudinal Study 1929–1991*
Eric J. Devor (editor) *Molecular Applications in Biological Anthropology*
Kenneth M. Weiss *The Genetic Causes of Human Disease – Principles and Evolutionary Approaches*
G.W. Lasker and C.G.N. Mascie-Taylor *Research Strategies in Biological Anthropology – Field and survey studies*
S.J. Ulijaszek and C.G.N. Mascie-Taylor *Anthropometry: the individual and the population*

The health consequences of 'modernization': evidence from circumpolar peoples

ROY J. SHEPHARD

School of Physical and Health Education, and Dept. of Preventive Medicine and Biostatistics, Faculty of Medicine, University of Toronto, and Health Studies Programme, Brock University, St. Catharines, Ontario

ANDRIS RODE

School of Physical and Health Education, University of Toronto

CAMBRIDGE
UNIVERSITY PRESS

Published by the Press Syndicate of the University of Cambridge
The Pitt Building, Trumpington Street, Cambridge CB2 1RP
40 West 20th Street, New York, NY 10011-4211, USA
10 Stamford Road, Oakleigh, Melbourne 3166, Australia

First published 1996

Printed in Great Britain at the University Press, Cambridge

A catalogue record for this book is available from the British Library

Library of Congress cataloguing in publication data

Shephard, Roy J.
  The health consequences of modernization : evidence from circumpolar peoples /
by Roy J. Shephard and Andris Rode.
  p.   cm. – (Cambridge studies in biological anthropology ; v. 17)
  Includes bibliographical references.
  ISBN 0 521 47401 9 (hc)
  1. Arctic peoples – Anthropometry. 2. Arctic peoples – Health and hygiene.
3. Arctic peoples – Social conditions. 4. Physical anthropology – Arctic regions.
5. Human physiology – Arctic regions. 6. Human biology – Arctic regions. 7. Arctic
regions – Social conditions. I. Rode, A. (Andris), 1940– . II. Title. III. Series.
GN57.A73S545 1996
306,4′61′0911–dc20   95-9112   CIP
ISBN 0 521 47401 9 hardback

Contents

Preface

Many 'modern' industrial countries have experienced major epidemics of disorders such as atherosclerotic heart disease, hypertension, diabetes, and various forms of cancer during the past few decades. Among other hypotheses, a departure from the traditional human lifestyle has been suggested as a primary factor precipitating such epidemics. After many centuries of successful human adaptation to the diet and physical demands of life as a hunter–gatherer or a low technology agriculturalist, there has been a dramatic and rapid concentration of the world's population in large conurbations where there is little possibility to sustain either a traditional diet or historic patterns of physical activity. The energy demands of city life have decreased progressively as developed countries have experienced widespread use of cars, automation of factories, mechanization of the home, and easy access to sedentary forms of entertainment such as television, video-films and computer games. A combination of store purchases and increasing affluence has shifted food consumption from traditional 'country' foods such as grains and vegetables to refined carbohydrates, saturated animal fat and protein. Medical advances have increased human lifespan, but at the same time advertising and other social pressures have encouraged a large fraction of 'modern' populations to become addicted to tobacco and other drugs.

Given the wide discrepancy between our current lifestyle and that appropriate to the evolutionary roots of our human species, adverse health consequences are hardly surprising. Unfortunately, technology was not available to document the physical condition of city-dwellers before the disturbance of their lifestyle began, and our knowledge of the effect of 'modernization' upon current health is correspondingly incomplete. However, the studies of indigenous populations that were carried out by the International Biological Programme Human Adaptability Project (IBP-HAP) in the mid and late 1960s provide a unique benchmark, allowing us to trace within these populations both the social processes of acculturation to urban life and the resulting consequences for health. The pace of change has varied from settlement to settlement, but in many instances the populations concerned have moved from a near neolithic technology to the lifestyle, social habits and behaviours of the late-twentieth-century city-dweller over the course of two or three decades.

One set of traditional communities that the International Biological Programme (IBP) studied in particular depth were the various small settlements that had adapted successfully to the environmental challenges of the circumpolar habitat. When these populations were first examined, some of them had been affected but little by modern innovations. However, over the last 20–30 years, almost all of these groups have adopted a pattern of life that in many respects mirrors that of the 'modern' city-dweller. Sequential study of these settlements has thus allowed the biological anthropologist to document, albeit on an accelerated scale, the likely health consequences of a transition from a traditional to an urban, post-industrial type of civilization.

Our story moves broadly over the panorama of the circumpolar peoples, as they have undergone this process of 'acculturation', although it draws particularly upon the experience of our own research group in Igloolik (North-West Territories of Canada, soon to be the independent territory of Nunavut) and Volochanka (Northern Siberia). We begin with a brief account of the origins, traditional social customs, nutrition, and lifestyle of the several distinct indigenous circumpolar groups, noting early research findings on their respective levels of health and fitness. The concept, scope and methodology of the International Biological Programme is then examined, with particular reference to the human adaptability project and its studies of circumpolar communities. Subsequent chapters look at changes in the factors which affect the ability of the circumpolar peoples to adapt to what remains a very challenging habitat: altered social structures and behaviour patterns, secular trends in diet, metabolism and body composition, changes in physical fitness and cold tolerance, alterations of lung function, smoking habits and the prevalence of respiratory disease, patterns of growth and development, and current health status. The concluding chapter examines broader social implications of the data. It is argued that the findings have relevance for indigenous populations in many parts of the world (where, in some instances, the 'white' city-dweller's worst mistakes of lifestyle still can be avoided). There are also useful lessons for health professionals in major conurbations, as they devise appropriate tactics to counter the sedentary habits imposed by our affluent society. Finally, the data have practical importance to a growing segment of city-dwellers who visit the far north during the winter months, as this geographic region becomes integrated into our global economy.

In presenting this monograph, it is important that we acknowledge the contribution of the indigenous populations concerned. They have not only given freely of their time, but have been willing to share with us and other investigators their profound understanding of the north and its challenges. We also much appreciate the financial and logistic support of Health and Welfare Canada, Indian and Northern Affairs, Canada, and more recently of Canadian Tire Acceptance Ltd. Without the continued generous help of these organizations, our component of this research could not have been completed.

Roy J. Shephard and Andris Rode
Toronto, 1995

1 The circumpolar habitat and its peoples: traditional lifestyle and early research findings

The circumpolar habitat

Boundaries

The circumpolar region is defined by climate and culture, rather than by a specific latitude such as the Arctic circle. Geographic and climatic markers include the tree line, the zone of perennially frozen ground that even in summer only thaws superficially (the permafrost), and the mean July isotherm of 10 °C (Bone, 1992; Burch, 1986; Damas, 1984; Kimble & Good, 1954; Péwé, 1966; Wenzel, 1991). These various boundaries have shifted markedly over the past 10–20 millenia (Lamb, 1965). During the last Pleistocene period of glaciation, ice covered much of North America, Europe and Asia.

The cultural criterion of the circumpolar habitat, at least in North America, has traditionally been the region exploited by a people that early investigators, ourselves included, sometimes termed Eskimos (Damas, 1984). 'Inuk' ('Inuuk,' two, three or more 'Inuit') is the descriptor currently preferred, at least by the Canadian segment of this population. Also, some Amerindian groups can be found living within the geographic boundaries of the circumpolar region even in North America.

Climate

Throughout the circumpolar territories, the temperature remains below freezing for much of the year, and in some of the colder settlements of the eastern Canadian arctic mean daily air temperatures as low as −50 to −60 °C are recorded during the winter months. The average windspeed is about 5 m/s (18 km/h), but windspeeds of 10 m/s (36 km/h) and higher are encountered on some days, and then the windchill is particularly severe (Landsberg, 1970). In contrast, the 24 hour sunshine of late June and early July can occasionally bring mean daily temperatures as high as 25 °C.

In part because the low winter temperatures give a very low absolute humidity, precipitation is light in most of the circumpolar region. The typical winter snow cover ranges from 0.1 to 0.4 m. Nevertheless, in the early part of winter, high winds lead to an almost continual movement of the snow, with drifting and frequent 'white-outs', blizzards when blowing snow reduces visibility almost to zero. In the islands of the high arctic, pack ice offers a further hazard to shipping for most of the year, but at lower latitudes the ice melts in late June, and here the coastal settlements can be supplied by boat for several of the summer months (Bone, 1992; Stager & McSkimming, 1984).

The local climate is influenced by the extent of snow and ice cover. The snow and ice reflect incident solar radiation and thus reduce potential heating of the ground surface by as much as 80%. Other modulating factors include the proximity of open water, air and water currents and altitude. There are some quite high mountain ranges in the circumpolar territories; air temperatures decrease by 6–7 °C per 1000 m of altitude, and wind exposure is also greater in mountainous regions. In consequence of these several variables, the climate at a given latitude is much colder in the east than in the western Canadian arctic (Fig. 1.1). Likewise, the tempering influence of the Gulf Stream gives ice-free water along the coastline of south-western Greenland and Scandinavia for much of the year.

Geography and economic resources

The climatic and geographic characteristics of the region have had important economic implications for humans who have wished to colonize the circumpolar habitat.

For several months of the year, all outdoor work must be performed under conditions where it is difficult for humans to sustain heat balance. There is partial or total darkness, and visibility is further restricted by blowing snow. Once the tundra is firmly frozen, with light snow cover, travel by dog or reindeer sled or snowmobile is generally possible, but in the more southerly wooded areas or taiga, accumulations of soft snow impede winter travel by all means except snowshoes. During the summer months, much of the region becomes a treacherous swamp of permafrost that can only be explored by boat, plane, or hover-craft.

The soil is thin, poor and dry over much of the arctic. In the more southerly settlements, sparse herbs, moss, lichens, grasses, sages and even small willow shrubs emerge from the snow during the spring, but in the more northerly regions a terrain of gravel and rock is devoid of vegetation. The main potential sources of local ('country') food are hunting, trapping and fishing, although some communities have also learned the arts of

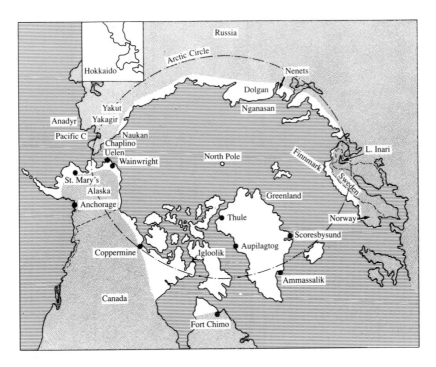

Fig. 1.1. Circumpolar regions, as defined by treeline. Unshaded land mass is devoid of trees. Arctic polar projection.

herding caribou (reindeer), foraging for berries and growing root vegetables. Many areas of the tundra such as Northern Québec still have large herds of wild caribou, but in other locations a sparse fauna and flora have caused animals and their pursuers to range over vast distances. Available species (Freeman, 1984) include various large mammals (seals, whales, walruses, bears and musk-oxen), smaller mammals that are prized mainly for their furs (wolves, arctic foxes and hares), birds (ptarmigan, ducks and geese) and fish (particularly arctic char and lake trout). A combination of local ground conditions and seasonal variations in the quality of the pelts has favoured the hunting of different species at different times of the year (Shephard, 1978; Fig. 1.2), to the extent that some Inuit communities have named the various months in terms of the hunting opportunities that they offer.

Most parts of the arctic give little scope for the growing of vegetables, even during that short period of the year when there are 24 hours of daylight, but some communities have had success in cultivated root crops, particularly potatoes. Some groups also gather berries, roots and mushrooms

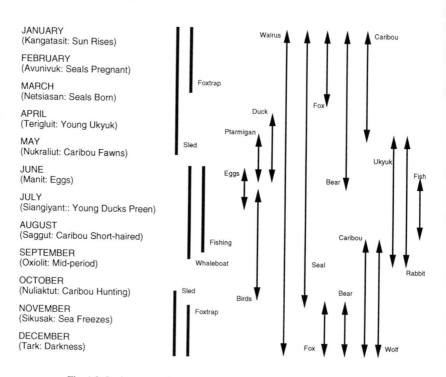

JANUARY
(Kangatasit: Sun Rises)

FEBRUARY
(Avunivuk: Seals Pregnant)

MARCH
(Netsiasan: Seals Born)

APRIL
(Terigluit: Young Ukyuk)

MAY
(Nukraliut: Caribou Fawns)

JUNE
(Manit: Eggs)

JULY
(Siangiyant:: Young Ducks Preen)

AUGUST
(Saggut: Caribou Short-haired)

SEPTEMBER
(Oxiolit: Mid-period)

OCTOBER
(Nuliaktut: Caribou Hunting)

NOVEMBER
(Sikusak: Sea Freezes)

DECEMBER
(Tark: Darkness)

Fig. 1.2. Inuit names of months of year, with hunting equipment used (solid bars) and game sought (arrows). Source: Godin & Shephard (1973a).

by foraging during the summer months. Nevertheless, until recently it has been necessary for the indigenous population in most parts of the arctic to obtain much of their normal vitamin C requirement from sources other than typical vegetables, including plankton found in the stomachs of raw fish, raw meat, blood ('blood soup') and the stomach contents of the walrus and the caribou.

The local fauna contributed to the circumpolar economy in other important ways. The skin, bones, intestines and membranes of animals such as the caribou and the seal once provided the indigenous peoples with raw materials for the fabrication of clothing, footwear and tents, plus a variety of domestic and hunting implements. As contacts with 'white civilization' developed, a trade in animal pelts progressively provided local residents with a cash or barter income for the purchase of both manufactured goods (including modern hunting weapons, domestic implements, twine and thread) and food (as during the failure of the Canadian caribou migrations in the period 1915–1924).

Communities living above the tree-line could not derive heat from the combustion of timber (except in coastal settlements, where there was the occasional and unpredictable bonus of finding driftwood or wreckage from shipping). Adaptation to the harsh winter environment thus depended to some extent upon physiological mechanisms, but was based largely upon the development of clothing that provided good insulation during outdoor activities (Renbourn, 1972), and the design of very effective snow shelters (the igloo and the quarngmaq or sod-house). The snow shelters allowed the accumulation of heat from both body metabolism and traditional oil lamps. During the summer months, a change in prime hunting locations and a substantial increase of ambient temperatures led families to adopt differing tactics for both their clothing and their overnight accommodation (for example, the eastern arctic Inuit established temporary tent camps near the floe edge to facilitate the hunting of sea mammals).

The 'cold war' of the 1950s led to the establishment of a chain of radar stations across the North American arctic (the 'DEW' line), with a substantial influx of 'white' immigrants. Airstrips developed at some of these stations were adequate to accommodate large cargo planes and jet aircraft. A more detailed exploration of mineral resources now became possible. Some parts of the arctic proved to have extensive reserves of oil (for example, the Alaskan north shore) and/or minerals (for example, the lead, zinc and silver deposits at Nanisivik, near Arctic Bay, and the uranium found at Baker Lake). Economic exploitation of such resources by southern entrepreneurs has progressed relatively slowly, in part because of the need to resolve the land claims of the indigenous populations, and in part because of a continuing search for industrial technologies that can withstand the rigours of the arctic winter.

The circumpolar peoples and their origins

Before extensive contacts with 'white civilization' had developed (1750–1800), the total number of inhabitants of the circumpolar belt of the northern hemisphere was thought to be about 48 000 people (Milan, 1980).

Rychkov & Sheremet'eva (1980) identified 16 groups indigenous to the arctic. Some of these populations currently inhabit Central Siberia (the Evenki, nGanasan and Dolgans) and North-Eastern Asia (the Yu'pik Inuit, Chukchi and other smaller indigenous groups of the Chukotka region, Vahtkin, 1992). Larger numbers of the indigenous circumpolar peoples are found in North America and in Greenland (Amerindians, Aleut and Inuit). Other circumpolar groups which have been studied by the International Biological Programme include the Ainu living in the northern part of Hokkaido (Kodama, 1970) and the Lapps of northern Norway (Finnmark), Sweden, Finland and Russia.

An earlier monograph in the present series (Mascie-Taylor & Lasker, 1988) examined the origins of these various populations. A review of geological, archaeological, linguistic, cultural, anthropometric, dental, dermatoglyphic and genetic evidence suggested that many of these peoples had a common ancestry (Crawford & Duggirala, 1992; Dahlberg, 1980; Eriksson et al., 1980; Fagan, 1987; Kirk & Szathmary, 1985; Laughlin & Harper, 1979, 1988; Turner, 1989). Early authors had argued for a migration from the Jenisej River and the Taimir peninsula in central Siberia, the region currently populated by the Dolgans and the nGanasan (Larsen & Rainey, 1948; Sollas, 1924). More recent investigators have suggested that the nGanasan are a distinctive population (Ferrell et al., 1981), and the postulated 'cultural cradle' of the North American native peoples has been moved eastward, to around the mouth of the Anadyr River, much closer to the Bering Strait (Giddings, 1960; Spuhler, 1979).

The current hypothesis is that forerunners of the North American indigenous populations crossed the Beringia land bridge. This land-mass linked Siberia with Alaska from 25 000 to 14 000 years before the present day (BP), during the last (Wisconsin) glaciation of the Pleistocene era (Hopkins, 1967). There has been speculation that genetic changes emerging late in the Pleistocene epoch may have given the migrant Inuit/Northern Amerindian stock an unusual ability to adapt to the very severe cold of the circumpolar regions, and thus to exploit the migratory opportunity offered by the land bridge.

Until recently, linguistic (Greenberg et al., 1986) and genetic (Schanfield et al., 1990; Shields et al., 1992; Williams et al., 1985) evidence was interpreted to suggest that several successive waves of migrants, arriving in North America between 20 000–15 000 years BP, developed through divergent evolutionary pathways into North Amerindians, Aleut and North American and Greenlandic Inuit.

There are continuing attempts to deduce inter-relationships between the various populations and sub-populations from such gene markers as blood groups and platelet antigens (ABO, Rh, Duffy, Diego and MNS), HLA (histocompatibility) markers, immunoglobulins (Gm1 and Km systems), proteins such as haptoglobin and transferrin, enzymes such as glucose-6-phosphate dehydrogenase, cholinesterase and acid phosphatase, DNA polymorphisms and miscellaneous markers such as phenylthiocar-bamide testing and isoniazid inactivation (Dossetor et al., 1973; McAlpine et al., 1974; Roychoudhury & Nei, 1988) (Fig. 1.3). Critics of the resulting dendrograms have argued that even with the propensity for in-breeding that is an inevitable consequence of life in isolated settlements, intermingling of supposedly distinct populations has continued over many centuries, making the precise characterization of racial origins almost impossible.

Despite such objections, Laughlin & Harper (1988) compared genetic diversity within and between population groups, and by relating this information to [14]carbon-dating studies they established time-estimates for a 'genetic clock'. They argued strongly for the arrival of a single, small group of perhaps 300 people in North America about 15 000 BP. Others (for example, Weiss, 1988) have maintained that genetic evidence is insufficient to decide whether there were one or several waves of immigrants.

Contrary to some early hypotheses, it is now considered that much of the Bering land bridge was a frigid polar desert (Ritchie & Cwynar, 1982). In consequence, it is hypothesized that the small band of early settlers remained scattered along its southern coastline for several thousand years. Carbon-dating of currently known archaeological sites in the interior of Alaska has revealed no human remains dating from earlier than 11 000– 12 000 BP. Nevertheless, it is dangerous to conclude from this that the interior of Alaska remained uninhabited for several millenia. Given the vast area of the state, it is quite conceivable that fossils indicating an earlier settlement of the interior are yet to be discovered.

The time clock analysis suggested to Laughlin & Harper (1988) that the Athapascan Amerindians diverged from the Bering Sea coastal population around 15 000 BP. The Aleut/Inuit divergence was set around 9000 BP, and the Yu'pik/Inupiaq divergence around 5000 BP. Currently, North American Inuit have greater genetic similarity to Athapaskan Amerindians than to Aleuts, or St. Lawrence Island or Yu'pik Inuit (Szathmary & Ossenberg, 1978; Nazarova, 1989). This may reflect the inland migration of a small sub-sample of the original coastal colonists, with subsequent survival under extremely isolated conditions (Laughlin & Harper, 1988).

Current ethnic groupings

Ainu

The Ainu currently live in Hokkaido, Sakhalin and the Kuril Islands in Eastern Asia. Weiss (1988) has argued that they were one of the original groups inhabiting eastern Asia, but that they were displaced by the expansion of 'Chinese' people. The Ainu share some genetic characteristics with the Aleuts, Inuit and Amerindians (Omoto, 1973b; Weiss, 1988), but

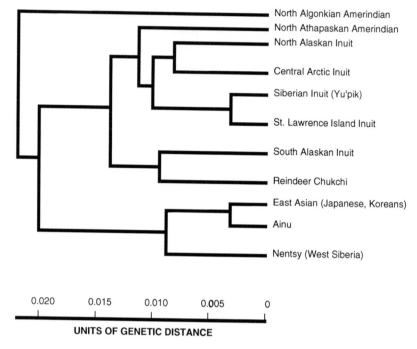

North Algonkian Amerindian
North Athapaskan Amerindian
North Alaskan Inuit

Central Arctic Inuit

Siberian Inuit (Yu'pik)

St. Lawrence Island Inuit

South Alaskan Inuit

Reindeer Chukchi

East Asian (Japanese, Koreans)

Ainu

Nentsy (West Siberia)

0.020   0.015   0.010   0.005   0

UNITS OF GENETIC DISTANCE

Fig. 1.3 (a). Dendrogram showing affinities between the various circumpolar peoples based on genetic data. Source: Adapted from Szathmary (1981).

they are physically, linguistically and genetically distinct from most of the circumpolar peoples discussed above. Indeed, they seem more closely related to the Japanese and Polynesian groups (Omoto & Misawa, 1974; Weiss, 1988). In recent years, there has been much inter-marriage between the Ainu and the Japanese. At the time of the IBP project, the amount of genetic admixture was already estimated at 40–50% (Omoto & Harada, 1972).

Siberian populations

Chuchki

The Chuchki have been of particular interest to Soviet/Russian scientists. This population has quite a close genetic relationship to the Aleuts and the South Alaskan Inuit. About 12 000 Chuchki currently live in the Eastern part of the former Soviet Union, along with smaller numbers of Yu'pik Inuit, Evenki, Yakuts, Yukaghir, Koryak, and Chuvan.

Present-day settlements of the Chukotki and Yu'pik Inuit are found on

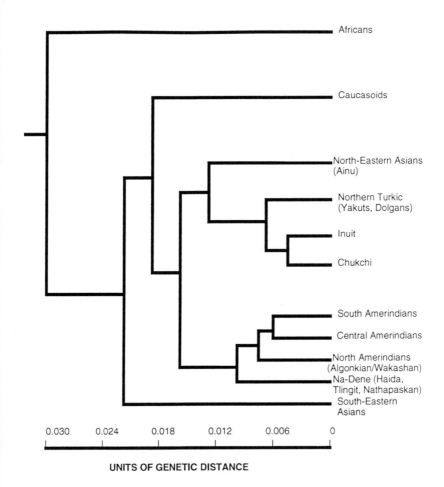

Fig. 1.3 (b). Dendrogram showing affinities between the various circumpolar peoples based on a combination of genetic, archaeological and linguistic evidence. Source: Adapted from Cavalli-Sforza *et al.* (1988).

the Chukotka Peninsula and around the Anadyr River basin (Bering Sea). The Koryak are located mainly on the Kamchatka Peninsula (Pacific Coast), and the Yakuts along the Yana River basin, on the north-eastern arctic coast of Siberia. Inter-marriage has blurred divisions between these various populations (Milan, 1980).

The habitat ranges from the arctic tundra to transitional zones. As with other indigenous groups in the former Soviet Union, the Stalinist government of the 1930s forcibly replaced the traditional Chuchki lifestyle of hunting and fishing by wage employment as reindeer herders on collective farms.

Evenki

The Evenki are an indigenous group currently living in Central and Eastern Siberia, but having some genetic similarities to the Chuchki. Settlements are now concentrated along the Stony Tunguska River, some 750 km north of Krasnoyarsk. Rychkov & Sheremet'eva (1980) estimated that at the time of the IBP study, the total population of Evenki amounted to some 2100 people, although Hannigan (1991) more recently classed 30 000 Siberians as Evenki. Like most of the arctic, the Evenki habitat is subject to permafrost. Much of their territory is forested, but the higher lands are arctic tundra.

During the Stalinist era, the bulk of the population was moved to 16 collective settlements, each of some 200 people (Rychkov & Sheremet'eva, 1980). Their principal occupations are currently reindeer herding, hunting and fishing, although some are now finding employment in forestry, mining and urban occupations.

Somewhat to the north of the Evenki territory, a group of some 1100 Keto people occupy the fishing village of Sulami (Hannigan, 1991).

nGanasans and Dolgans

The nGanasan (population 680) and Dolgans (population 4080) are two ethnically and genetically distinct small groups of people who are thought to have migrated from central or eastern Mongolia to the Taimir region of central Siberia north of Norilsk (Rychkov & Sheremet'eva, 1980). The first arrivals were the nGanasan. They were followed around 1600 by Russian migrants, and around 1700 by the Dolgans and Yakuts from the southeast. Other native groups further to the west include the Entsy, Nentsy, Selkups, Khants, Komi and Lapps.

The Dolgans are currently the dominant aboriginal culture within a social system that has been imposed by the Russian migrants. The habitat lies at the boundary of two eco-systems: forest and tundra. Traditionally, the nGanasan and Dolgans exploited the resources of the tundra, following the migration patterns of the reindeer as far as the arctic north shore during the warmer summer months, setting up hunting camps at the river and lake crossings of the reindeer during the fall, and retreating to the forests to escape from the heavy blizzards of the Siberian winter. The primary occupations were reindeer herding, hunting, trapping and fishing. The nGanasan used reindeer sledges, and the Dolgans exploited the reindeer for its milk and for riding.

The bulk of the indigenous population of this region was settled on cooperative reindeer farms in the Stalinist era. Our laboratory has made a particular study of one such community, at Volochanka (71° N, 94° E).

Lapps

The genetic origin of the Lapps is unknown, although the frequency of some blood groupings suggests a distant kinship with the Inuit. It is thought that the Lapps once colonized a large part of the Fenno-Scandia, but that they were forced to retreat progressively northwards with the arrival of the Finns around 300 AD (Eriksson *et al.*, 1980).

By 1970, Lewin estimated there was a world-wide total of some 35 000 Lapps (Lewin, 1971). The majority lived north of the Arctic Circle, with some 20 000 in the Finnmark region of Northern Norway, 10 000 in Sweden, and smaller numbers in Finland and Russia. Von Bonsdorff *et al.* (1974) noted that in 1972, some 3800 Lapps were living in Finland; they accounted for 18% of the total population of northern Finland. Many of these people had been displaced westward in 1944, with the annexation of a part of Finland by the former Soviet Union.

About 30% of Lapps are currently characterized as Skolt Lapps (a group who lived in relative isolation until after World War II). The remaining 70% are classed as Mountain and Fisher Inari Lapps. Many of these latter groups have inter-married with Finns during the present century.

In recent times, the majority of Lapp communities have based their economy upon a combination of reindeer herding and coastal fishing. In a few regions, the local climate has also permitted the growing of potatoes. Both Norwegian and Finnish investigators studied the health and fitness of Lapp populations during the IBP-HA project.

North American and Greenlandic populations

Aleuts

The Aleuts are currently found in the Aleutian islands and along the west coast of Alaska. Like the Inuit, they share their physical appearance and genetic characteristics with the indigenous populations of Mongolia.

Many features of language and culture are shared with the Inuit, although Laughlin & Harper (1988) have argued from genetic data that the two populations diverged from a common ancestry about 9000 years ago. In addition to the game pursued by the Inuit, sources of 'country' food available to the Aleuts include sea otters and salmon.

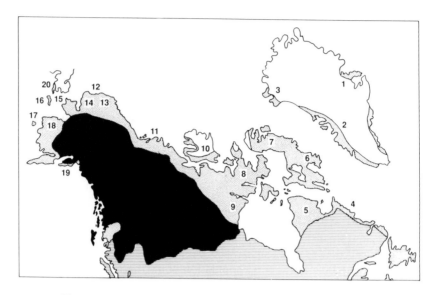

Fig. 1.4. Traditional territories of the 20 tribe-like groupings of Inuit, and the two groups of Amerindians colonizing the arctic habitat (Nadene shown in black, Algonkian Wakashan in horizontal shading). 1. East Greenland Inuit. 2. West Greenland Inuit. 3. Polar Inuit. 4. Labrador Coastal Inuit. 5. Northern Québec Inuit. 6. Baffinland Inuit. 7. Iglulik Inuit. 8. Netsilik Inuit. 9. Caribou Inuit. 10. Coppermine Inuit. 11. Mackenzie Delta Inuit. 12. North Alaskan Inuit. 13. Interior Northern Alaskan Inuit. 14. Kotzebue Sound Inuit. 15. Bering Strait Inuit. 16. St. Lawrence Island Inuit. 17. Nunivak Inuit. 18. Southwestern Mainland Alaskan Inuit. 19. Pacific Inuit. 20. Siberian (Yu'pik) Inuit.

Amerindians

During the period of recorded history, two major linguistic and tribal groupings have exploited hunting and fishing in those regions of North America immediately to the south of the traditional Inuit habitat: the Nadene and the Algonkian-Wakashan Amerindians (Fig. 1.4). Because the climate of the Amerindian habitat is a little less extreme than that of the Inuit, the resources of 'country' food available to more northern settlements are supplemented by variable reserves of moose, deer, and fresh-water fish.

Probably because the Amerindian tribes have faced a somewhat less harsh climate than the Inuit, they have shown correspondingly fewer technological initiatives in combatting cold weather. The snow in the interior of northern Canada is deeper and less packed than that on the arctic shoreline. The traditional winter transport of the Amerindian has thus been by snowshoes and toboggan rather than by dog-sled. During the summer months, their favoured method of travelling has been by canoe.

Inuit

The adoption of a written language and regular communication between the various regions of the arctic are quite recent developments. In part for these reasons, names other than Inuit have been adopted by people who share the same ethnic background, including Inuvalialiut (the people of the MacKenzie Delta), Inupiat (Northern Alaska), Yu'pik (Central Alaska) and Kalaallit (Greenland). However, for the purpose of the present monograph, we have chosen to use the common name Inuk/Inuuk/Inuit, adding where necessary information on the geographic location of particular Inuit settlements.

The small size and considerable mobility of traditional Inuit social units have precluded the description of clear 'tribes', although some 20 tribe-like groups have been identified (Damas, 1984; Fig. 1.4). The Inuit seem to have spread progressively eastward from the Bering Strait, first reaching Greenland and the Canadian arctic about 4000 BP. In the Eastern part of Canada, four distinct cultural waves have been described (Melgaerd, 1960; Crowe, 1969). The first three were the Sarqaq (4000–2900 BP), Dorset (2800 BP–700 BP), and Thule (900–200 BP) people. The Dorset people lived in houses built of sod and stone, and apparently did not use either the dog-sleds or the kayaks and umiaks (larger, open, skin-covered boats) typical of the Thule whale hunters (Taylor, 1963). Some evidence of the Dorset culture still persists around Hudson's Bay (Hughes & Milan, 1980), possibly through a renewal of contacts between Amerindians and Inuit. During the past two centuries, the Thule culture has been supplanted in its turn by a new wave of migrants, comprising the Iglulik, Netsilik and Coppermine Inuit (Fig. 1.4). Among other innovations, the newcomers have introduced a more complicated religious system, with a greater number of taboos (Glassford, 1970a).

The total number of Inuit exploiting the circumpolar habitat has varied widely over the past two centuries. Large segments of the population were killed periodically by contacts with unfamiliar 'white' diseases such as measles and tuberculosis (Chapter 8; Keenleyside, 1990). In the last 30 years, the establishment of a chain of modern nursing stations across the arctic has been associated with a rapid growth of the indigenous population. Nevertheless, the precise number of Inuit in various parts of the arctic is still hard to determine, since population censuses usually rely upon a self-ascription of ethnicity.

Danish IBP investigators made extensive study of the Greenlandic west-coast settlement of Upernavik. There has been much inter-marriage between Greenlandic Inuit and Danes throughout the past century. By

1988, Greenland claimed to have a population of some 45 000 indigenous Inuit (Bjerregaard, 1991); however, the number of genetically 'pure' Inuit was probably substantially less than this.

Canada had 34 000 self-reported Inuit in 1986. About two thirds of this group lived in the North West Territories (Norris, 1990b; Statistics Canada, 1989). Medical officers from Indian and Northern Affairs, Canada, have offered anecdotal comments on the health of many Inuit settlements over the past four to five decades. The detailed observations of our own laboratory have been concentrated at Igloolik (69°40′ N, 81° W), a community that is almost entirely Inuit, and that has grown from 500 to some 1150 people over the past 20 years. Auger and associates (1976, 1980) have obtained additional data from Kuujjuaq (formerly Fort Chimo) in northern Québec, a mixed settlement of some 600 Inuit and Cree Amerindians.

The US (US Congress, Office of Technology Assessment, 1986) reported a population of 42 000 Inuit (Eskimos) and 14 000 Aleuts in 1980, although only a small fraction of these people were living in the circumpolar regions of Alaska. IBP investigators made a particularly detailed study of the North Alaskan communities of Wainwright (population about 330, mainly Inuit), Point Barrow (population about 2500, 'white', Inuit and mixed parentage), and Point Hope.

The Chukotka region of Siberia still has some 1000 people who are identified as Yu'pik Inuit (Young, 1994), although there has been much inter-mingling with the Chuchki. The Yu'pik Inuit have been studied by Soviet/Russian scientists, both during the IBP and subsequently.

Lifestyle and adaptive tactics

Social structures and behaviour

Detailed knowledge of social structures and behaviour covers no more than a few centuries, although inferences about earlier times have been drawn from findings at a few archaeological sites. Probably because of the limited economic base provided by a very low density of game, most traditional circumpolar settlements have been small in size and temporary in nature. A complete community often comprised only 80–100 people, amounting to no more than a few extended families. In transitional climatic zones, the development of reindeer herding allowed the construction of semipermanent settlements, but for many of the more northern arctic populations life was essentially nomadic, whole communities migrating seasonally to areas that favoured particular types of hunting. For instance, the Inuit took their spears to the river mouths in the spring season, when the char were returning to the sea, and tents were moved to the floe edge in late June, when the skins of the young seals were at their prime. Likewise, the nGanasan established camps at river-crossings used by the caribou in the fall, when the skins of these animals were at their best for the making of winter clothing, but as winter set in, they retreated to the protection of the forests.

Marriage patterns

The limited choice of marriage partners has led to some in-breeding, particularly in the smaller and more isolated settlements.

At the time of the IBP surveys, Milan (1980) rated communities on a scale extending from 0 (no in-breeding) to 0.25 (complete in-breeding). Reported coefficients ranged from 0.001 for Igloolik, NWT and 0.004 for Wainwright, Alaska, to 0.045 for the Nellim-Pasvik Skolt Lapps in Finland. However, there was no clear evidence that this degree of in-breeding had caused the emergence of any recessive diseases or disorders.

Traditional religions

The traditional religious cultures of most circumpolar groups were strongly linked to nature, with a profound respect for the earth, the sea and their respective harvests. Thus, for the Huron Amerindians, the deity was perceived as Gitchi Manitou, a mighty hunter, and in Inuit mythology Sedna, the sea goddess, was mother of the many creatures found in the oceans (Boas, 1964). The Inuit believed that all animals had spirits, and certain taboos were developed based on this belief. For instance, the mingling of meat from land and sea mammals was prohibited. If the members of a settlement broke this particular taboo, it was held that the animals would avoid them during subsequent hunting excursions.

In recent years, elements of rival cultures have been imported and sometimes assimilated. For example, the beliefs of the present generation of the Ainu not only show a respect for the forces of nature, but also include an element of ancestor worship that seems to have been borrowed from their Japanese neighbours. Likewise, the kaitvitjvik (winter festivities) of the MacKenzie Inuit now incorporate some concepts drawn from the southern Canadian Christmas season.

Shelter and transportation

Traditional forms of shelter have shown much regional variation. The Canadian Inuit once favoured a tent for summer, and an igloo and qarngmat for winter protection. However, by 1960, Hart *et al.* (1962) found that the igloo had become no more than a temporary overnight shelter. More permanent dwellings were usually fashioned from canvas, burlap and moss, set on a wooden frame. At this same period, the inland Inuit of Northern Alaska lived in dome-shaped tents that were covered with thick caribou skins (Irving, 1972). In both cases, body heat and the burning of animal fat kept interior temperatures at a comfortable level throughout the coldest days of winter.

Traditional methods of transportation such as walking in snow-shoes, or the use of a dog or reindeer sled, were relatively slow. In consequence, adult males were absent from the main settlement for periods of several weeks during extended hunting trips.

Pelto (1962) speculated that those colonizing the circumpolar habitat had a special personality characteristic ('social uninvolvement') that helped them to complete difficult tasks in lonely situations. Certainly, those living in the arctic had a need for inventiveness in finding their way through unfamiliar territory, coupled with patience, logical thinking, attention to detail and a capacity for innovation when repairing damaged equipment at an isolated site (Forsius, 1980). Survival depended on the effective management and skilled interpretation of detailed information gleaned from the physical terrain, snow conditions, weather, game, and the performance of personal equipment such as sleds and harnesses.

Most traditional circumpolar societies were cooperative in spirit. Behaviour was concentrated upon meeting the challenge of the hostile environment rather than on out-performing peers. The capture of game with primitive weapons often required close inter-personal cooperation. Nevertheless, previous cultures were transmuted if not physically subdued over the centuries (for instance, the transition from Dorset through Thule to Iglulik traditions in the Canadian arctic). In some circumpolar communities, there has been persistent friction and overt or repressed hostility between successive waves of immigrants (for example, between the nGanasan and the Dolgans in the Taimir Peninsula of Siberia).

Acculturating influences

Despite the vast distances that separated individual settlements, there was probably some communication between different circumpolar groups, even before the modern era. During the last one to four centuries, much stronger acculturating influences have been brought to the arctic by missionaries, explorers, traders, representatives of colonizing governments and other immigrants from 'developed' societies.

For several hundred years, missionaries have endeavoured to convert many of the circumpolar populations to their particular branches of the Christian church (Catholic, Orthodox, Anglican or Pentecostal), with varying degrees of success. In some settlements, the establishment of two or more rival Christian congregations has had an unfortunate, disruptive effect upon community life. In the Stalinist era of the Soviet Union, similar systematic attempts were made to convert indigenous peoples to a Marxist/Leninist faith.

Both in the former Soviet Union and in North America, the use of indigenous languages and customs was discouraged by church or party and local government officials until about 20 years ago. In Canada and the US, community life was further disrupted by sending sick adults to centrally located sanatoria and young children to distant residential schools (Chapter 9).

Russia

Regions such as the Taimir tundra were settled by land-hungry Russian peasants in the early seventeenth century. As early as 1618, the Cossacks were reportedly imposing a fur tax on the nGanasans (then described as the Pyasina Samoyeds). Further major upheavals occurred in the Soviet Union of the 1930s. Enforced collectivization of all indigenous peoples was accompanied by the establishment of a rigid, centrally controlled state economy, with a stringent rationing of most commercial goods and the imposition of fur and meat quotas to be met by each of the circumpolar settlements.

Canada

The eastern arctic was visited by several English mariners in the late sixteenth century, in the course of their search for a North-West Passage. Martin Frobisher made three voyages to the eastern arctic in 1576–1578, discovering Frobisher Bay (Iqualuit) and the Hudson Strait that separates Baffin Island from the mainland of Canada. In 1585, John Davis rediscovered Greenland and the Strait that separates it from Baffin Island. The ill-fated Henry Hudson pressed further into the Strait and James Bay in 1610, but continuation of the voyage under adverse conditions provoked a mutiny, and Hudson was forced to abandon his vessel to meet an unknown end.

By the eighteenth century a brisk fur trade had been established by the Hudson's Bay Company. In the nineteenth century, north-shore Inuit settlements were visited increasingly by the crews of whaling vessels, and in 1860 Charles Francis Hall wintered on Baffin Island. The visits of English and Scandinavian merchants brought timber, metals, manufactured goods and European genes to the Arctic territories.

The twentieth century was marked by a continued development of the Hudson Bay Company monopoly on 'market' food. This persisted until the mid-1960s. Then the federal government appointed 'Cooperative Development Officers' and began to encourage the establishment of cooperative stores in each of the major arctic settlements.

Beginning in the 1940s, the health, education, welfare and policing of the circumpolar peoples were also the subject of increasing federal intervention. At the behest of government, the previously nomadic indigenous populations were concentrated into relatively large permanent settlements, and compulsory schooling was introduced. For example, in the eastern arctic, the Canadian government began to encourage the Iglulik Inuit (who to this time had been widely scattered over the Northern Foxe basin) to move to a central location at Igloolik. The introduction of firearms reduced the need for cooperative hunting (Glassford, 1970a), and as permanent settlements such as Igloolik developed, the population showed a rapid transition from the barter of furs and skins to an economy based upon federal employment and subsidies (Table 1.1).

Since the 1970s, there has been a conscious effort to revive the traditional Inuit culture. This culminated with the passage in 1993 of an act of parliament that will see the establishment of an autonomous territory (Nunavut) in the Eastern Arctic in 1999.

Alaska

Early colonization of Alaska was by Russians, with a further strong influence from US immigrants subsequent to the Alaska purchase. The process of acculturation to the lifestyle of 'developed' society took hold much earlier in Alaska than in Canada. For instance, permanent settlement of the Wainwright area began as early as 1904, with the founding of a small village school (Milan, 1970). From the period of World War II, the construction of military bases and the exploitation of north shore oil has led to extensive contacts with 'modern' US society.

Table 1.1. *Progressive transition of Canadian Inuit to wage economy*

	Sources of income (%)			
Community	Hunting	Wages	Crafts	Subsidies
Baker Lake (1951/52)	56	6	0	38
(1961/62)	3	46	3	48
Lake Harbour (1963/64)	69	12	6	14
(1966/67)	28	38	17	18

Based on data of Freeman (1971a).

Scandinavia

As long as 400 years ago, Denmark began to exert an analogous colonizing influence upon the Inuit living in south-western Greenland. In north-western Greenland, around Thule, the Inuit have undergone much less acculturation and inter-marriage (Eriksson *et al.*, 1980).

Most circumpolar Lapp groups have had extensive contact with more southern populations, and in recent years extensive inter-marriage has made it difficult to distinguish those who are of indigenous ancestry.

Patterns of physical activity

Measurement problems

There have been very few studies of activity patterns and energy demands imposed by the traditional lifestyle of circumpolar communities. Reasons for the lack of information include a tremendous seasonal variation in hunting patterns (Fig. 1.2), problems inherent in documenting the behaviour of a nomadic people, and constraints peculiar to the arctic habitat.

Often, an observer must live with a small segment of the population through at least an entire year in order to obtain even a limited sample of data. Furthermore, the extreme cold encountered on many hunting trips causes a rapid failure of the traditional tools of the ergonomist, ranging from a freezing of ball-point pens, mouthpieces and valve boxes, to an impaired function of electric motors on ECG tape-recorders, and both motors and shutters on movie cameras. Even note-taking by pencil is hampered because the observer is wearing thick mittens and hand coordination is impaired by severe local cooling of the extremities.

Unless laboratory staff are accustomed to operating their equipment in the arctic environment, subjects may be asked to attend to the physical needs of the observers and their apparatus quite frequently, thereby biasing both activity patterns and daily energy expenditures. If representative, accurate and valid data are finally accumulated on a small group of subjects, it still may be difficult to generalize the findings to other settlements or other ethnic groups. Because of regional difference in fauna, flora and geography, the isolation of individual settlements, and varying degrees of acculturation to the lifestyle of the dominant society, hunting, trapping and fishing techniques vary widely from one community to another. Further, activity patterns vary from one season to another, and from one year to another.

Nevertheless, information on activity patterns has considerable theoretical interest, both as a potential explanation of the unusual fitness of circumpolar peoples, and also as a means of assessing the ecological balance between the energy requirements of the local population and the potential energy yield of hunting, trapping and fishing in a specific habitat (Godin & Shephard, 1973a; Harrison, 1982; Kemp, 1971).

Studies of food intake

In some of the more permanent circumpolar communities, it has been possible to estimate the total daily energy expenditures of the indigenous residents by examining their food intakes.

If careful dietary records are maintained over a period of 1–2 weeks, the estimated energy value of the food consumed often matches expenditure to within 10–15% (Pollitt & Amante, 1984). However, even in an urban society, dietary data commonly under-estimate total daily energy expenditures by 10–20%, mainly because subjects forget to record their snacks. In the circumpolar environment, there are additional problems of communication between subject and observer, and the energy content of some 'country' food items is unclear. Questionnaire data are thus highly suspect. A highly committed nutritionist can obtain better data by living with a small group of villagers for a long period, weighing all foods accurately, but even this approach cannot be applied to subjects who are often absent from the settlement, taking long hunting trips or herding reindeer across the distant tundra.

Heart rate data

One option, developed by Spurr *et al.* (1988), is to use tape-recorded heart rate data to divide the day's activities into three categories of energy expenditure: sleep, resting activities (below a heart rate breakpoint, assigned a fixed energy cost as determined from representative metabolic measurements on the local population) and vigorous activities (each assigned a specific energy cost, based on the observed heart rate and a heart rate/oxygen consumption calibration curve obtained in the base laboratory). Katzmarzyk *et al.* (1994) applied this method to reindeer herders in central Siberia.

A major weakness in most heart rate estimates is that the calibration curve is obtained when the subject performs a large muscle, rhythmic activity such as cycle ergometry or stepping. However, many daily tasks involve isometric contractions of small muscle groups (where the heart rate is much higher in relation to oxygen consumption). Change of environmental temperature and emotional disturbances also alter the relationship between heart rate and oxygen consumption, and in the study of Katzmarzyk *et al.* (1994), female subjects with heart rates as high as 106 beats/min. were classed as engaged in 'resting' activities.

Diary records

One commonly used method of estimating daily energy expenditures in the urban environment is for a technician to observe, or for a subject to record all of the activities that are performed during a typical day on a minute by minute basis. The energy cost of the various items thus identified is obtained by reference to standard tables such as those of Durnin & Passmore (1967).

Secular trends to a decrease in the energy cost of common daily tasks are a problem even when this method is applied to city-dwellers. Plainly, it is inappropriate to adopt standard energy costs for circumpolar populations if they use very different technologies than the city-dweller in order to accomplish a given task. Irrespective of the availability of 'modern' technology, the performance of the circumpolar group is often impeded by the need to wear heavy winter clothing, and energy costs are further increased if the task involves moving across rough, snow-covered surfaces. However, it is possible to use the principle proposed by Durnin & Passmore (1967), developing population-specific tables that indicate the energy cost of undertaking a variety of common daily tasks in the arctic environment.

After some preliminary experimentation in a laboratory cold chamber, we were able to insulate and pre-heat Kofranyi–Michaelis respirometers sufficiently to obtain measurements of oxygen consumption during a number of hunting and fishing trips in the Igloolik region (Godin & Shephard, 1973a). Well-greased 100 ml syringes were used to bring samples of expired air back to the base laboratory for electronic gas analysis. In a few instances, the nature of the activity that the subject was undertaking

Table 1.2. *Energy cost of eight different types of traditional hunting. Based on data of Godin & Shephard (1973a), adjusted to standard body mass of 65 kg*

Type of hunting	24-h energy expenditure (mJ)
Fishing: Summer	18.6
Ice	16.9
Caribou: Summer	16.2
Winter	16.1
Walrus	15.5
Seal: Ice-hole	14.5
Boat	14.4
Floe edge	10.6
Average (8 hunts)	15.4

Source: Godin & Shephard (1973a).

(for example, skin chewing) precluded use of a mouthpiece, and in such situations the oxygen consumption was estimated from electrocardiographic recording of heart rates, using a heart rate/oxygen consumption graph for that individual, as established in the base laboratory. In other situations, the observer's store of syringes became exhausted. Predictions of oxygen consumption were then made from the individual's respiratory minute volume and a graph of ventilation versus oxygen consumption, as established in the base laboratory. Detailed diary records of the type, intensity and duration of daily activities were kept throughout a variety of hunting trips. Combination of such descriptive records with on-site measurements and predictions of the oxygen costs of the tasks observed allowed us to estimate the total daily energy expenditure during eight different types of hunting (Table 1.2).

Activities to be considered

Our observations in the Igloolik region gave some indication of the range of activities that must be considered when estimating the average daily energy consumption of populations exploiting the circumpolar habitat. Walrus were available for much of the year, particularly when an east wind drove pack-ice toward the shore line. The condition of the seal pelts was optimal for harvesting in the spring, but some seal were caught during about 10 months of each year. Caribou were also hunted throughout the year, but were a particularly useful source of meat in the winter months, when the animals could be pursued easily over the frozen tundra, and the carcasses could be transported by sled back to the base camp. Polar bear were encountered rather infrequently, but because of the high market value of the pelts, they took precedence over other types of hunting once they had been sighted. During the winter months, traps for smaller mammals were set over a wide area of Baffin Island. The average hunter would cover 80–100 km during a 2–3 week trip, visiting 15–16 traps, and commonly would find one or two foxes.

Environmental effects

During the winter months, protective garments added several kilograms to the clothed weight of the Igloolik Inuit when they were working out of doors. Tests with and without winter clothing (Godin & Shephard, 1973a; Pandolf *et al.*, 1976; Romet *et al.*, 1986) suggested that there was an increase in the oxygen cost of movement that was almost directly proportional to the added weight of clothing at any given environmental temperature. However, under very cold conditions the oxygen cost of a given task was also increased as the resultant of an augmented basal metabolism, shivering, various forms of non-shivering thermogenesis (Shephard, 1993a), an increase of muscle viscosity, anaerobic activity (at least until the subject had warmed up), a lowering of metabolic activity in inactive tissues (the Q_{10} effect), and increased energy costs imposed by movement across snow or ice-covered surfaces.

Traditional activity patterns

Bearing in mind the limitations of the experimental data, what conclusions may we draw regarding traditional activity patterns in the various circumpolar communities?

Ikai *et al.* (1971) suggested that the Ainu were relatively sedentary. Their main activities were said to be wood carving and merchandising. The data of Koishi *et al.* (1975), collected between 1967 and 1972, showed that the Ainu had a relatively low total energy expenditure (10.2 MJ/day in the men, 7.5 MJ/day in the women). Nevertheless, the subjects examined were very light, and if the observed values had been adjusted to the standard body masses proposed by Durnin & Passmore (1967) for 'developed' societies (65 and 55 kg, for men and women, respectively), energy expenditures would have risen to 14.8 and 9.0 MJ/day.

Andersen *et al.* (1962) noted that Norwegian Lapps supplemented their primary activity of reindeer herding by ptarmigan hunting, trapping and fishing, but they gave no details of daily energy expenditures in this population. Most of the observed activity was pursued on foot or on skis, with occasional use of a reindeer sleigh. Many of the Finnish Lapps had already found regular salaried employment in the forest industries at the time of the IBP studies, although some were still engaged in fishing and/or the small-scale cultivation of root vegetables. Hasunen & Pekkarinen (1975) estimated a daily food intake of only 9.3 MJ for Skolt Lapps who had maintained a traditional nomadic life, but much higher figures of 15.9 MJ/day were observed among the men who had moved into forestry. Apparently, neither of these figures was adjusted for body mass, which tends to be quite low in traditional Lapps, as in other circumpolar populations.

Rennie (pers. commun.) estimated the daily energy expenditure of Inuit living in the northern interior of Alaska, at Anaktuvuk Pass. In 1962, he reported that the hunters of the community were expending 15 MJ/day (3600 kcal/day).

The main traditional activities of West Greenlandic Inuit have been hunting and fishing. Even at the beginning of the IBP studies, Inuit living in the Upernavik region based these activities from a motor launch. When sitting on board the launch, the heart rates of the hunters remained close to resting values, but when they began tracking the seals in their kayaks, heart rates of 120–140 beats/min. were observed. Lammert (1972) estimated that the average hunter spent 25% of the total hunting time in a kayak. During this portion of the day, bouts of activity at some 50% of aerobic power came in bursts of 2–3 min. duration.

The total daily energy expenditures observed at Igloolik in 1969/70 were higher than in some circumpolar communities. Nevertheless, our estimates of the energy costs of hunting (Godin & Shephard, 1973a) agree well with observations on the central Alaskan Inuit of Anaktuvuk Pass made by Rennie in 1962. A city dweller would certainly regard an expenditure of 15.4 MJ/day as very heavy physical work. However, even in 1969/70, not all of the men in Igloolik could be classed as regular hunters. Moreover, Godin & Shephard (1973a) estimated that the typical hunter spent only about 161 days/year on trips outside the village. The remaining days were spent relaxing and in repairing equipment, so that when averaged over the entire year, the energy expenditure of the hunters dropped to a more modest 12.6 MJ/day. On days when hunting conditions were particularly favourable, the chase would continue for much of a 24-hour period.

In consequence, we may conclude that although quite large total expenditures were reached on days of traditional hunting, such expenditures were accumulated more by persistent activity than by bursts of very intensive physical activity. A casual observer might not even realise that the Inuit hunters were highly active. Nevertheless, both the frequency and the total energy cost of traditional hunting seem more than sufficient to have had a favourable effect upon health-related fitness (Chapter 5) and the prevalence of ischaemic heart disease (Chapter 8).

Diet

Traditional dietary patterns

As noted above, problems of communication between dieticians and local populations, and uncertainties regarding the composition and mode of preparation of 'country' foods have hampered dietary studies.

The traditional diet of the various circumpolar communities has shown much regional variation, depending on the nature and extent of local resources. In general, it has been high in protein and in fat, and very low in carbohydrate (Chapter 4). For example, Rennie *et al.* (1962) estimated that in the early 1960s, the central Alaskan Inuit of Anaktuvuk Pass were gaining 65% of their total food energy from protein.

In the high arctic regions, where survival of a settlement has sometimes depended almost entirely upon the local resources of land and sea, adverse weather conditions have led to episodic starvation. In recent history, one period of particular privation was associated with a failure of normal caribou migration patterns from 1915 through 1924. During the present century, most circumpolar groups have had some access to 'market' food, although Danish authorities were reluctant to make such provision for fear of creating dependency among the Inuit, and in the North-West Territories of Canada as recently as the early 1960s, a visit to the store may have required several days journey by dog sled (Hart *et al.*, 1962).

Clinical aspects of diet

In many communities, dieticians and clinicians have focused on the relative fractions of daily energy requirements obtained from 'country' and from 'market' sources and any resulting deficiencies in key nutrients, together with parasitic infections (Cameron, 1968; Freeman & Jamieson, 1976) and diseases such as botulism (Brocklehurst, 1957; Eisenberg & Bender, 1976; Hauschild & Gauvreau, 1985) that may have arisen from traditional methods of preparing foods.

Because the diet has contained little carbohydrate, a number of circumpolar groups have lacked adaptive mechanisms to metabolize refined carbohydrates if these suddenly become available. Thus, an intolerance of glucose has been noted in several samples of Canadian Inuit (Rabinowitch & Smith, 1936; Schaefer, 1968 a and b; and Schaefer *et al.*, 1972) and in Aleuts (Dippe *et al.*,

1976), but not in Ainu (Kuroshima *et al.*, 1972) or Inuit students attending a residential school at Point Hope, Alaska (Feldman *et al.*, 1975). Intolerance has also been described for lactose in Finnish Lapps (Sahi *et al.*, 1976); in Greenlandic and Alaskan Inuit (Bell *et al.*, 1973; Duncan & Scott, 1972; McNair *et al.*, 1972) and in Alaskan Indians (Bell *et al.*, 1973; Duncan & Scott, 1972), and for sucrose, encountered in 10.5% of Greenlandic and 2–3% of north coastal Alaskan Inuit (Bell *et al.*, 1973; McNair *et al.*, 1972).

Photon absorptionmetric data suggested that in the early 1970s a combination of a high protein, low phosphorus, low calcium diet and a poor tolerance of milk products gave many circumpolar residents a 10–15% deficit of bone calcium relative to 'white' standards for any given age (Martin *et al.*, 1985; Mazess & Mather, 1974, 1975).

Community energy balance studies

Comparison of community energy requirements with 'local' country resources of food energy has allowed investigators to calculate overall energy balance sheets, matching the needs of various communities against the resources available in the local habitat. This in turn has allowed ecologists to estimate the total population that could be supported from 'country' food resources in a given region of the arctic (Godin & Shephard, 1973a; Harrison, 1982; Kemp, 1971).

Some circumpolar communities adopted the diet of 'developed' societies many years ago. In 1975, Koishi *et al.* reported that the food patterns of the Ainu were already indistinguishable from those of other low income Japanese. Likewise, a survey of northern Lapp communities that was completed in 1969/70 found that the people of this region were eating similar relative proportions of protein, fat and carbohydrate to those observed in the diet of southern Finns (Hasunen & Pekkarinen, 1975; Aromaa *et al.*, 1975). In contrast, at the time of the IBP survey, Lapps living on the mountain plains of Norway had conserved a high protein intake (up to 23% of daily energy intake, mainly in the form of reindeer meat), and those living in the coastal regions still had a large intake of fish (Gassaway, 1969; Ogrim, 1970).

Bang *et al.* (1976) made detailed dietary studies on seven West Coast Greenlandic Inuit. At that period, seal, whale and fish were already being supplemented by 'market' items, but the protein intake was still sufficient to satisfy some 26% of daily energy needs. Earlier data had suggested that in 1914, Greenlandic Inuit obtained 44% of their daily energy from protein, 47% from fat, and 8% from carbohydrate.

An early study of an isolated Inuit group living in central Alaska (Rennie *et al.*, 1962) found that 56% of their dietary energy needs were met from protein. However, by 1971–1972, the adult Inuit of Wainwright were deriving less than a half, and children only a quarter of their energy needs from traditional foods of all types (Bell & Heller, 1978). Dietary acculturation was even more advanced in Point Hope, Alaska, where 'country' foods were meeting only 22% of the adults' and only 8% of children's daily energy needs.

Godin & Shephard (1973a) estimated the total energy input of the Igloolik community that was derived from hunting in the Foxe Basin. They concluded that at the time of their survey (1969/70), some 31% of the energy needs of the Igloolik community were being met from 'country' sources. However, they underlined that it was difficult to obtain complete information on either the quantities of local game and produce that were being harvested, or their respective contributions to the energy requirements of the community. There was a small wastage of edible meat through skinning. Estimates were further complicated by the provision of food to the 500 dogs then living in the village, plus some pillaging of caches by wolverine and weasel.

Fatty acid intake

In many circumpolar regions, the traditional 'country' diet has had a high fat content. However, the fat has usually come from fish and marine mammals (Chapter 4). It has thus been ingested mainly in a polyunsaturated form (Bang *et al.*, 1976). A substantial fraction of total fat has consisted of atherosclerosis-inhibiting omega-3 fatty acids (Bang *et al.*, 1980; Innis & Kuhnlein, 1987), and there has been an absence of the unnatural and unhealthy trans- and cis-isomers that are found in synthetic oils such as margarine (Booyens *et al.*, 1986).

Populations with a traditional, fish-rich diet have conserved low levels of serum cholesterol (Chapter 4, Table 4.10), examples being found in Canada (Corcoran & Rabinowitch, 1937; Nutrition Canada, 1975; Rode *et al.*, 1995; Thouez *et al.*, 1989), Chukotka (Alexseev, 1991; Gerasimova *et al.*, 1991; Nikitin *et al.*, 1981), Greenland (Bang & Dyerberg, 1972; Bang *et al.*, 1976) and Alaska (Scott *et al.*, 1958; Wilber & Levine, 1950). In contrast, Draper (1976) found cholesterol levels similar to those for developed societies in Wainwright, Alaska (4.85 ± 1.06 mM/l for males and 5.11 ± 1.32 mM/l for females and Point Hope, Alaska (5.50 ± 1.01 mM/l (males) and 5.47 mM/l (females)), with undesirably high readings in the Inuit of South Western Alaska (6.40 ± 0.93 mM. (males) and 6.74 ± 1.17 mM/l (females).

Vitamin and mineral intake

One important nutritional weakness of many traditional circumpolar diets has been a deficiency of vitamin C (Bell & Heller, 1978; Forbes, 1974; Nutrition Canada, 1973). 'Country' foods generally have provided substantial intakes of vitamins A and the B complex, but some Canadian Inuit groups have shown folacin deficiencies (Forbes, 1974).

Despite the increased need of vitamin E imposed by the high intake of polyunsaturated fatty acids, plasma levels of vitamin E were judged as adequate, at least in one early study of Alaskan Inuit (Wo & Draper, 1975).

Some US and Canadian Inuit communities had a low iron intake, and thus low haemoglobin levels at the time of the IBP study (Draper, 1980), although we found very normal haemoglobin readings in both children and adults in all of our surveys of Igloolik (Rode & Shephard, 1971, 1984a, 1992a).

Socio-cultural implications of diet

'Country' food has a mystic significance for many traditional circumpolar groups. Borre (1991) underlined the integral relationship between a continued consumption of 'country' food and the cultural identity of aboriginal peoples. For example, Canadian Inuit traditionally believed that seal blood could become an element of human blood, and that a child weaned on 'country' food would have an inherent desire to adopt a traditional, hunting-oriented lifestyle.

Physical Fitness

Modifying factors

The harsh environment of the circumpolar regions could conceivably have modified fitness in a number of ways:

1. As a prelude to the Pleistocene land-bridge migration, the inaccessibility of small arctic settlements might have encouraged the emergence and subsequent persistence of an unusual 'genetic isolate', with features of body build, metabolism and cardio-respiratory physiology peculiarly adapted to survival in an arctic habitat.
2. A diet high in fat and protein, but low in refined carbohydrate, with superimposed episodes of starvation, may have encouraged the development of a thermally efficient compact body form with thin skinfolds and internal reserves of fat, either as a genetic variant or

because the total energy intake was insufficient to realise the inherent genetic potential for linear growth.

3. The vigorous and sustained physical activity essential to both effective hunting and survival may have had either a selective or a shorter-term training effect, reducing blood lipids and augmenting the functional capacity of the oxygen transporting system.

4. Repeated cold exposure may have modified vascular and metabolic regulation, either through genetic change or maximal exploitation of existing mechanisms of acclimatization.

5. Occasional major epidemics of diseases such as measles and tuberculosis may have retarded child development and/or caused permanent impairments of cardio-respiratory function, while also improving the ultimate fitness of the community through the selective elimination of weaklings.

Traditional levels of physical fitness

Physiologists once expressed the opinion that the fitness of traditional circumpolar populations was poor. However, a review of the early literature (Table 1.3) does not altogether support this conclusion (Shephard & Rode, 1973).

Early tests were made on small and unrepresentative samples of the communities concerned, usually with a selective exclusion of the most active hunters. For example, Brown *et al.* (1963) studied six male Inuit (supposedly hunters) at Coral Harbour, and using a skinfold prediction method found an average body fat content of 37%! Contrary to this assessment, Hart *et al.* (1962) one year earlier had shown that the body fat content averaged only 13% in a more carefully selected traditional group, living up to 200 km from the neighbouring settlement of Pangnirtung (Cumberland Sound).

In early investigations, methods for the testing of fitness were often poorly standardized and open to suspicion. For example, fitness was predicted from respiratory recovery curves (Eriksson, 1958), or from performance of the Harvard step test (Rodahl, 1958). More recent investigators (Table 1.3) have either measured the maximal aerobic power directly (Andersen & Hart, 1963; Karlsson, 1970) or have predicted maximal oxygen intake from submaximal data on heart rate and oxygen consumption or work-rate (Andersen *et al.*, 1960a; Andersen, 1963; Rennie *et al.*, 1970). In some of the supposed maximal tests, the attained values for peak heart rate (173–180 beats/min.) and blood lactate (7 mM/l) have fallen below accepted maxima for young adults in 'developed' societies. This

Table 1.3. *Early data on the cardio-respiratory fitness of circumpolar communities*

Author	Number and type of subjects	Age (years)	Test criterion	Result
Eriksson (1958)	10 Arctic Inuit	16–24	Respiratory recovery curves	'Better cardio-resp. function than "white" servicemen'
Rodahl (1958)	? Arctic Inuit	?	Harvard step test	3.5 times airmen 3.5 times trained subjects 2.5 times arctic soldiers
Andersen et al. (1960a)	8 Arctic Amerindians	17–39	Cycle ergometer Max. heart rate predicted at 180 beats/min.	Aerobic power 49.1 ml/[kg·min.]
Andersen & Hart (1963)	8 Arctic Inuit	19–27	Cycle ergometer 2 min. exhaustion. Max. heart rate 173 beats/min.	44 ml/[kg·min.]
Andersen (1963)	16 Nomadic Lapps 6 Nomadic Lapps	20–40 50–60	Cycle ergometer (as in Inuit?)	53 ml/[kg·min.] 44 ml/[kg·min.]
Karlsson (1970)	31 Skolt Lapps	10–60	Cycle ergometer to lactate 7 mMol/l	52 ml/[kg·min.] at age 15–19 yr
Rennie et al. (1970)	65 Alaskan Inuit	6–64	Sub-maximum step test ?predicted heart rate	50 ml/[kg·min.] at age 15–25 yr

Source: Shephard & Rode (1973).

suggests that the effort of some of those tested was really submaximal rather than maximal. Tellingly, the authors of one early report admitted that their technique (two minutes of exhausting exercise) may have under-estimated the true aerobic power of their subjects by at least 20% (Andersen & Hart, 1963). Further, in some instances, the physical condition of the subjects who were tested was prejudiced by starvation and/or tuberculosis (Andersen & Hart, 1963).

Sometimes, these disclaimers have been overlooked in subsequent evaluation of the early reports. But if due attention is paid to such comments, it would appear that throughout the early period of physiological history, the aerobic power of the various indigenous adult males who were tested in the arctic must have been at least 50 ml/ (kg.min), a higher figure than would be anticipated in the city-dwellers of 'modern' urban society.

Adaptations to cold

Traditional circumpolar populations could have met the specific challenge of the arctic winter climate by either cultural or biological acclimatization and adaptation.

Cultural adaptations

The extent of cold exposure in any given habitat depends very much upon the individual's skills and behaviour patterns (Moran, 1981; Steegman *et al.*, 1983): exposure can be greatly attenuated by an accurate interpretation of weather conditions and local topography, the design of effective clothing (Renbourn, 1972), and the ability to build simple wind-shelters. The indigenous populations of the arctic have undoubtedly transmitted such knowledge from generation to generation, although much of this expertise is now being lost as they become acculturated to the lifestyle of urban society.

Modern technology compares poorly with traditional wisdom. For example, the Canadian military clothing provided for arctic patrols offers an insulation of some 4 CLO units. However, the indigenous people of central Siberia (with four or even six layers of long quilted garments) are much better insulated. Likewise, the double layer of caribou skin that is worn by traditional Inuit during the winter months can provide up to 11 CLO units of protection under the worst environmental conditions. At least in still air, the Inuit clothing ensures thermal equilibrium for a resting subject at temperatures of −40 °C (Burton & Edholm, 1969; Renbourn, 1972; Shephard, 1982).

Because the traditional clothing of the circumpolar populations prevents heat loss over the trunk almost totally, a much larger blood flow can be directed to the hands (Hart *et al.*, 1962), enhancing manual dexterity. Moreover, most types of traditional clothing provide flexible insulation. When a person is working, the belt can be loosened, and if necessary mittens and the outer layer of skins can be removed to avoid an accumulation of sweat within the clothing.

Influence of body form and composition

Differences of body form and composition may also reduce cold exposure (Newman, 1960; Roberts, 1953; Steegman, 1975). The rules of Bergmann (1847) and Allen (1877) attributed a heat-conserving advantage to the rounded, compact body form of traditional circumpolar populations, and the snub nose of the Mongolian and the Inuit may have reduced the likelihood of frostbite to this region of the face. Renbourn (1972) suggested that the Inuit had a relatively thick layer of subcutaneous fat in the hands and feet. Hildes *et al.* (1961) noted that heat loss from the extremities was limited further in the Inuit by the small size of their hands and feet.

Metabolic and cardiovascular adjustments

In reviewing metabolic and vascular adjustments to cold, a distinction must be drawn between acclimatization, a process of biological adjustment that is available to all ethnic groups, and adaptations which reflect the emergence of an inherited advantage specific to a given ethnic group (Folk, 1966; Shephard, 1985).

In the general population, the possibility of developing a biological adjustment to cold seems quite limited. A person may become habituated to a cold environment, so that less discomfort is experienced (Glaser & Shephard, 1963), heat production may be increased (metabolic acclimatization), or heat loss may be reduced by a restriction of peripheral blood flow (the insulative tactic). Habituation is seen most clearly as a diminished heart rate and a blood pressure response to local cold exposure of the hands during immersion in iced water (LeBlanc, 1975).

Some early reports suggested that the traditional Inuit, Alaskan Indians and Lapps were much more tolerant of local cold exposure than were 'white' city dwellers (Eagan, 1966; Elsner *et al.*, 1960; Meehan, 1955; Miller & Irving, 1962). However, this is also true of other ethnic groups, such as Gaspé fishers who work in cold water (Table 1.6). Meehan (1955) found no difference in the tolerance of painful thermal radiation between Inuit from the isolated Brooks Range of Central Alaska, Arctic Indians living in the larger settlement of Fort Yukon, and 'white' men.

General acclimatization to cold is often reflected by a reduction of shivering. Traditional Lapps seem able to sleep without visible shivering under much cooler conditions than can their 'white' counterparts (Andersen *et al.*, 1960b). However, such habituation and any associated local circulatory adjustments seem a consequence of prolonged exposure to severe cold rather than the result of an unusual inheritance.

Potential metabolic adjustments to cold include voluntary exercise, shivering, and non-shivering thermogenesis. The effectiveness of exercise as a means of maintaining or increasing core temperature depends in part on the individual's metabolic capacity (a function of peak aerobic power and available food reserves), and in part upon the ability to conserve heat that is produced within the body.

Because of a high aerobic power (above), a number of traditional circumpolar groups had an initial advantage over 'white' immigrants in terms of their ability to generate heat by exercising, although if cold exposure continued, their advantage was soon offset by their low carbohydrate diet, with a resultant limitation of glycogen reserves and difficulty in sustaining a high rate of aerobic activity. Exercise is often a rather ineffective way of warming the body, since limb movement tends to pump cold air or water beneath protective clothing and displaces the insulating layer of still air or water adjacent to the outer surface of a person's clothing. An exercise-induced accumulation of sweat within the fabric may also reduce the insulation of protective garments (Rennie *et al.*, 1962), and an associated increase of pulmonary ventilation necessarily augments respiratory heat losses (Shephard, 1982). Mainly as a consequence of accumulated experience of arctic conditions, the traditional Inuit have developed some specific techniques for dealing with these problems. In particular, traditional garments have been closed very effectively at the neck, wrist and ankles, and the hunter has known to remove the outer layers of clothing as sweating begins. There is also some evidence of an alteration in sweat gland distribution and/or function among the Inuit. Sweating is most marked over the face rather than the trunk, thereby minimizing the risk of soaking clothing when exercise is performed in the cold (Schaefer *et al.*, 1974; Shephard & Rode, 1973). Nevertheless, Rennie *et al*, (1962) suggested that the Inuit produced a larger total volume of sweat than 'white' subjects when they were exposed to an equivalent heat. The total number of sweat glands per unit of skin surface does not appear to be particularly low in the Inuit (Kuno, 1956), although it is less than average in the Ainu (Kawahata & Sakamoto, 1951; Kuno, 1956). It remains unclear whether any unusual sweat gland responses among the circumpolar populations reflect a specific genetic adaptation, since Davies (1973) has also reported small winter changes in sweat gland activity among 'white' workers living in Antarctica.

Shivering can boost metabolism to four to five times its resting level, at least for short periods (Shephard, 1985, 1993a). Meehan (1955) and Meehan *et al.*, (1954) suggested that shivering had an unusually early onset when Alaskan natives were exposed to cold. However, Adams & Covino (1958) found a parallel relationship of shivering to skin temperatures among 'whites' and Anaktuvuk Pass Inuit, both groups showing a similar increment over their resting metabolism when they had been exposed to cold air for 55 minutes. In contrast, Andersen *et al.* (1963) and Hildes (1963, 1966) both found that the increase over resting metabolism was smaller in Pangnirtung Inuit than in University of Oslo students when both groups were exposed to a similar cold stress. Shivering continued when the Inuit were exercising, until the combined metabolic cost of exercise plus shivering was enough to keep them in thermal equilibrium (Andersen *et al.*, 1963). However, the ability of the Inuit to produce heat when working in the cold did not seem to be unusual. Glycogen is the major source of fuel for shivering (Jacobs *et al.*, 1985; Vallerand & Jacobs, 1989, 1990), so that the ability of the traditional arctic inhabitant to sustain long-term shivering may have been handicapped by poor muscle glycogen reserves.

A variety of types of non-shivering thermogenesis can increase the resting metabolism of cold-exposed subjects (Shephard, 1985, 1993a). The basal metabolic rate showed some seasonal variation in traditional circumpolar groups (Brown, 1954), and unfortunately not all investigators indicated the timing of their data collection. Values for the Ainu, obtained in 1974, seemed essentially normal (Itoh, 1974), but there were many early reports of a high metabolic rate in isolated, traditional Alaskan and Canadian Inuit communities (Table 1.4, Hart *et al.*, 1962; Itoh, 1980; Milan & Evonuk, 1966; Roberts, 1952; Rodahl, 1952). When observers have had only limited contact with the community, it has sometimes been difficult to verify that subjects have not taken food, caffeine containing drinks or cigarettes prior to examination. Another factor tending to increase basal energy expenditures in many circumpolar communities has been a high protein diet, since substantial heat is liberated during the conversion of amino-acids to glucose (Shephard, 1983). Nevertheless, some groups have found a persistent elevation of basal metabolic rates in Inuit subjects even after several days of adherence to a standard hospital diet (for example, Hart *et al.*, 1962, and Milan *et al.*, 1963). Another possible distorting factor has been the expression of results relative to an estimate of body surface area that is derived from body mass, since many Inuit groups have a low percentage of body fat and a high percentage of lean tissue. However, Rennie *et al.* (1962) maintained that the basal metabolism of the Inuit remained higher than that of 'white' subjects, even if it was expressed per unit of lean body mass.

Table 1.4. *Basal metabolic rate (BMR) of Inuit circumpolar populations*

Author	Reported BMR (% of DuBois standard)	Location
Heinbecker (1928)	+33%	Cape Dorset, Baffin Island
Rabinowitch & Smith (1936)	+26%	Eastern Canadian Arctic
Levine (1937)	Eleven 'normal' Five +19 to +30%	Akulurak & Nome, Alaska
Hoygaard (1941)	+13%	Eastern Greenland
Bøllerud & Blakeley (1950)	+14 to +17%	Gambell, Alaska
Rodahl (1952)	+16% (male) + 2 +10% (female)	Various parts of Alaska
Brown (1954)	+36%	Coral Harbour, Southampton Island
Brown (1957)	+25 to +30%	Southampton Island
Adams & Covino (1958)	+38%	Anaktuvuk Pass, Alaska
Hart *et al.* (1962)	+32%	Pangnirtung, Baffin Island
Rennie *et al.* (1962)	+47%	Anuktuvuk Pass and Point Barrow, Alaska
Andersen *et al.* (1963)	+61%	Pangnirtung, Baffin Island
Hildes (1963)	+27%	Cumberland Sound, Baffin Island
Milan & Evonuk (1966)	+21%	Wainwright, Alaska

Finally, it has often been difficult to assess the possible influence of anxiety when the basal metabolic rate has been determined in an unfamiliar situation. Rodahl (1952) argued that the metabolic rate of the Inuit became essentially normal if due allowance was made for these various influences.

We collected data on basal metabolic rates at the end of the winter of 1981/82. Values for the Inuit at that time were appreciably higher than for their white peers living in the same community (Table 1.5), the discrepancy being particularly great for the older and less acculturated members of the indigenous population. Care was taken to ensure that none of the subjects had eaten, consumed any beverages or smoked any cigarettes prior to data collection, and measurements of ventilation and heart rates demonstrated that all of those tested were well relaxed. Respiratory quotients at all ages were around 0.89 in the men and 0.87 in the women, presumably reflecting a high intake of protein in the older subjects, and of carbohydrate in the younger subjects. However, there was no evidence of the low respiratory quotient values that would be expected from a high fat diet.

Itoh (1974) suggested that the output of thyroid hormones was increased during the cold winter months, thus uncoupling the normal linkage between ATP production and operation of the sodium pump. However, this response occurred equally in the Ainu and in Japanese immigrants to

Table 1.5. *Basal metabolic rate of Inuit residents of Igloolik (mean ± SD, kJ/m²/h). Comparative data are given for 4 'white' men and 2 'white' women living in the same community*

Age group (years)	Males	Females
14–19	185 ± 17	174 ± 26
20–29	183 ± 18	178 ± 20
30–39	166 ± 17	159 ± 21
40–49	181 ± 10	164 ± 33
50–59	180	156
60–69	178	

Values for 4 'white' males, 30–44 years: 154 ± 5
Values for 2 'white' females, 25 and 31 years: 144

Unpublished data, Rode and Shephard.

Hokkaido. In support of the role of thyroid hormones, Gottschalk & Riggs (1957) noted high protein-bound iodine levels in Canadian Inuit. A cold-induced release of catecholamines could also trigger futile metabolic cycles (Bodey, 1973; Shephard, 1993a). The metabolic response to injections of noradrenaline was found to be enhanced in the Ainu (Itoh *et al.*, 1970a), but a similar change has been shown in 'white' subjects after cold acclimatization (Shephard, 1993a). There have also been suggestions that cold exposure can lead to a synthesis of brown fat, which has a high intrinsic rate of metabolism (Huttunen *et al.*, 1981). None of these mechanisms of increased heat production is unique to circumpolar groups. The trigger has been merely a prolonged exposure to cold. Hong *et al.* (1986) and Rennie (1988) demonstrated that whereas a combination of vigorous physical activity and non-shivering thermogenesis pushed the energy expenditures of traditional Korean pearl divers (Ama) to very high levels, values decreased dramatically when they began to wear modern insulative diving suits, reducing their exposure to cold water and thus their cold acclimatization.

When the unacclimatized individual is exposed to cold, insulative vasoconstriction limits peripheral heat loss, but this is achieved at the expense of a considerable decrease in the temperature of the hands, and thus a loss of manual dexterity (Shephard, 1985, 1993a). Elsner *et al.* (1960) found that in Arctic Indians the time to onset of a cold-induced vasoconstriction became shorter during the winter, but several other studies showed that the cold vasodilatation associated with a major drop in limb temperature also developed earlier in circumpolar populations (Itoh, 1980; Livingstone *et al.*, 1978). Thus, during sustained exposure to severe cold, the peripheral blood flow was greater among the indigenous arctic residents. Their hands cooled more slowly and manual dexterity was enhanced relative to their unacclimatized peers. Such findings have been replicated (Table 1.6) in Ainu (Itoh *et al.*, 1969, 1970b), Lapps (Andersen *et al.*, 1960b; Krog *et al.*, 1960), Inuit (Andersen *et al.*, 1963; Brown & Page, 1952; Brown *et al.*, 1955; Covino, 1961; Eagan, 1966; Hart *et al.*, 1962; Hatcher *et al.*, 1950; Hildes, 1966; Hildes *et al.*, 1961; Milan *et al.*, 1963; Miller & Irving, 1962; Page *et al.*, 1954) and Alaskan natives (Meehan, 1955; Milan *et al.*, 1963). The response seems a matter of altered neural regulation, since the resting blood flow in a warm environment does not differ between Inuit and 'white' subjects (Krog & Wika, 1978).

Adams & Covino (1958) concluded that there were no major physiological peculiarities in the acute response of the Inuit to cold, but Wika (1971) and Livingstone *et al.* (1978) both found that the circulation to the skin of the cheek was less than anticipated in the adult male Inuit and older Skolt Lapps (presumably, two segments of the circumpolar population who had the greatest cumulative exposure to cold weather). By conserving body heat, facial vasoconstriction could be a very useful adaptation to cold, allowing thermal equilibrium despite a greater hand blood flow (Livingstone *et al.*, 1978).

A number of investigators have tested vascular reactions to immersion of the hand in cold water. Page & Brown (1953) and Page *et al.* (1954) noted a slow decrease of local blood flow when the hand of the Inuit was cooled, and unlike 'white' subjects, local cooling of the legs had little effect on hand flow. Moreover, higher rectal temperatures were maintained and less heat was lost from the arm muscles during hand immersion. Brown *et al.* (1963) attributed the conservation of core temperature to a more efficient exchange of heat between arteries and veins in the Inuit, but Hildes *et al.* (1961) pointed out that another important factor could be that the hand size of the arctic dwellers was small. Finally, Livingstone *et al.* (1978) noted further that finger temperatures recovered from immersion much more quickly in Inuit than in 'white' subjects. Irrespective of mechanisms, these changes could make an important contribution to survival in the arctic, by conserving manual dexterity in the face of extreme cold.

Health experience

Early data

At the time of first contact with Europeans, skeletal studies suggest that the average survival of the North American Indian was no more than 23 years

Table 1.6. *A comparison of local acclimatization between circumpolar groups and other populations*

Non-comparative studies			
Non-acclimated individuals			
Adams & Smith (1962)	4 exposures/day for 1 month	10M US Caucasians	Acclimation reduces pain and results in higher skin temperatures during exposure and earlier rewarming after exposure
Livingstone (1976)	2 wks; whole body; 1 wk; whole body	129M military personnel; 10M military personnel	Acclimation slows vasodilatation and lowers skin temperature (general acclimation dominates local effect).
Savourey et al. (1992b)	2 exposures/day for 2 months	5 volunteers	Acclimation increases peripheral bloodflow during exposure, no metabolic changes observed
Acclimated individuals			
Elsner et al. (1960)	Lifetime	8M Arctic Indians	Time to vasoconstriction on exposure shorter after winter
Hellstrom & Andersen (1960)	Lifetime	9M Arctic fishers	No differences in heat output of hands relative to young men not accustomed to working in cold water
Comparative studies			
Brown & Page (1952)	Lifetime; None	6M Eskimos, 5M medical students	Hand bloodflow two times higher in cold in Eskimos, skin temperature higher in students and fell more slowly in Eskimos
Krog et al. (1960)	Lifetime; None	13M Lapps, 11M Norwegian fishers, 12M controls	More rapid vasodilatation on exposure in Lapps versus controls, no difference in hand bloodflow of fishers versus controls
LeBlanc et al. (1960); LeBlanc (1962)	Throughout year; None	11M Gaspé fishers, 9M controls	Pressor response less in fishers versus controls
Meehan (1955)	Lifetime; None	9M Alaskan natives, 8M military personnel	Earlier shivering, hands and feet warmer and finger bloodflow higher in natives versus military personnel

Miller & Irving (1962)	Lifetime	7M White, 3M Arctic residents, 8M, 3F Eskimos	Cooling of hands slower and higher minimum temperature in Eskimos than in Whites; White arctic residents show intermediate response
Nelms & Soper (1962)	Career None	11M fish filleters, 9M controls	Earlier vasodilatation in hands, and skin temperatures higher in filleters versus controls
Paik (1972)	Career None	8 Ama, 8 Non-divers	Vasoconstriction on exposure greater in divers versus non-divers
Wika (1981)	Lifetime		Eskimos had lowest blood flow to face, relative to other ethnic groups

Abbreviations: M = male, F = female, wk = weeks.
Based on data accumulated by Shephard (1993a).
For details of references see Shephard (1993a).

(Johansson, 1982). However, much of the short average lifespan was attributable to perinatal mortality. Adults in circumpolar communities seem to have been basically healthy (Hart *et al.*, 1962; Fortuine, 1971), although they were commonly affected by minor problems such as lice and systemic parasites (Chapter 8; Freeman & Jamieson, 1976).

Population growth was checked by infanticide (especially the selective killing of female children, Balicki, 1967; Chapman, 1980; Schrire & Steiger, 1974; Freeman, 1971b), accidents (particularly drowning and boating accidents, with a selective loss of males, Bjerregaard, 1990), starvation, and (at times of food shortage) the carefully contemplated suicide of older members of the community (Leighton & Hughes, 1955).

Lack of immunity and limited access to medical care led to high mortality rates from a number of 'European' infectious diseases (Chapter 8), including such conditions as measles (Brody & Bridenbaugh, 1964; Peart & Nagler, 1954), anterior poliomyelitis (Adamson *et al.*, 1949; Rhodes, 1949), meningitis and respiratory infections (particularly haemophilus influenzae, Bjerregaard, 1983; Nicolle *et al.*, 1981; Ward *et al.*, 1986; Wotton *et al.*, 1981), pneumococcal pneumonia (Davidson *et al.*, 1989), gastroenteritis (Bender *et al.*, 1972; Fournelle *et al.*, 1966), hepatitis (Baikie *et al.*, 1989; Larke *et al.*, 1981, 1987; Minuk *et al.*, 1981, 1985; Skinhøj *et al.*, 1977) and tuberculosis (Grzybowski *et al.*, 1976), as the various causal organisms became imported into northern communities for the first time (Keenleyside, 1990; Marchand, 1943).

In the early 1960s, crude mortality rates for the North West Territories of Canada showed that the main causes of death were respiratory disorders (320/100 000), injuries and poisoning (210/100 000), congenital and perinatal problems (150/100 000), neoplasms (60/100 000), and infectious and parasitic conditions (50/100 000).

Lifestyle diseases

During the early 1960s, circulatory disorders, at around 40/100 000, were toward the bottom of the list of causes of death in the North West Territories. Echocardiography, autopsy reports and analyses of death certificates have all shown a low relative risk of atherosclerotic heart disease in northern populations (Chapter 8), including Alaskan natives (both Inuit and Indians, Arthaud, 1970; Lederman *et al.*, 1962; Middaugh *et al.*, 1990), Inuit from the North West Territories (Hildes & Schaefer, 1973; Schaefer *et al.*, 1980a; Young *et al.*, 1993), natives of the Chukotka region (Astakhova *et al.*, 1991) and Greenlandic Inuit (Bjerregaard & Dyerberg, 1988; Dyerberg, 1989; Hart-Hansen *et al.*, 1991).

Blood pressures have also been low in Alaskan Inuit (Colbert *et al.*, 1978; Mann *et al.*, 1962; Maynard, 1976), and in natives of the Chukotka region (Astakhova *et al.*, 1991), but in contrast the Aleuts have shown a high prevalence of hypertension (Torrey *et al.*, 1979). The mortality rate for stroke now does not differ between Inuit and non-natives in Alaska (Middaugh, 1990) and among Greenlandic Inuit the incidence of stroke is high relative to that in developed societies (Bjerregaard & Dyerberg, 1988; Kromann & Green, 1980); reasons for this paradox are explored in Chapter 8.

Early studies suggested that the circumpolar populations had been spared the diabetes epidemic which has afflicted North American Indians living in more southern communities (Chapter 8) (Mouratoff & Scott, 1973; Sagild *et al.*, 1966; Stepanova & Shubnikhov, 1991). Scott & Griffith (1957) found a prevalence of only 0.5 cases of diabetes per 1000 among Alaskan natives.

However, traditional circumpolar populations have been at high risk of a number of cancers that are rarely seen in developed societies (Chapter 8), including tumours of the nasopharynx, the salivary glands and the oesophagus (Hildes & Schaefer, 1984). Nitrosable constituents of the diet are a possible explanation of such tumours (Stich & Hornby, 1985).

Conclusions

We may conclude that, perhaps as the result of some genetic change, one or more waves of migrants were able to move from eastern Siberia to colonize the circumpolar regions. They crossed to North America at the time of the Bering Strait land bridge. The limited fauna and flora of the new habitat only allowed the formation of small and isolated settlements, the economy generally being based upon semi-nomadic hunting–gathering and fishing.

The available food sources were very high in protein and polyunsaturated fats, and very low in carbohydrates. Perhaps as a consequence of this dietary pattern, a proportion of these populations is currently intolerant of glucose, lactose and sucrose, but the blood lipid profile has remained more favourable than that currently observed in developed societies.

The physical demands of the traditional lifestyle were substantial, and (perhaps because of the training effect of regular, hard physical activity) circumpolar groups have had relatively high average levels of physical fitness. Physiological adjustments to the cold environment has not differed greatly from what might have been expected if city-dwellers had moved to the arctic; although some thermal advantage may have been gained from a compact body form and the restriction of sweating to the facial region, survival of the traditional indigenous populations has depended mainly upon behavioural adjustments to the environment.

There were few diseases in the circumpolar region until the coming of 'white' settlers. Population size was limited largely by infanticide, accidents, starvation, and the deliberate suicide of the elderly. Contacts with urban civilizations began almost four centuries ago, so that even the baseline data collected by IBP scientists have been coloured by the effects of trade and the importing of disease. However, the processes of 'acculturation' and 'modernization' have accelerated very rapidly over the past two to three decades. Succeeding chapters will examine the impact of these changes upon the health and fitness of the indigenous circumpolar peoples.

2 Concept of the International Biological Programme Human Adaptability Project, and IBP studies of circumpolar populations

This chapter examines the concepts underlying the Human Adaptability Project of the International Biological Programme (IBP-HAP), with particular reference to its studies of circumpolar populations. Critical issues for the project were an appropriate sampling of the populations of interest, an inter-laboratory standardization of methodology, and the adoption of common units for the reporting of data. Comment is offered regarding the study's reliance on cross-sectional rather than longitudinal data, and the underlying concept of human adaptability is evaluated critically.

Human Adaptability Project

The concept of biological adaptability extends back to Charles Darwin (1859). Studies conducted during the voyage of the 'Beagle' convinced Darwin that the unusual inherited characteristics of certain species gave them a competitive advantage in colonizing a particular habitat – there was a 'survival of the fittest'.

The IBP-HAP, which originated in the early 1960s (Harrison, 1979; Weiner, 1964; Worthington, 1978), adopted a Darwinian perspective. It argued that small inherited differences of body build and biological characteristics had emerged between various ethnic groups over many centuries of separation from each other, and that such differences had given certain populations a survival advantage when exploiting a specific habitat. One basic objective of the IBP-HAP was thus to trace the potential extent of human adaptation to adverse environments, determining how far differences in biological characteristics had influenced the observed patterns of colonization. It was further hypothesized that humans as a whole had made an evolutionary adaptation to the lifestyle of hunter–gatherer or subsistence farmer over many centuries, and that levels of both fitness and health would be much lower in populations which had abandoned the

43

traditional way of living. It was also suggested that in traditional indigenous communities, the importing of various diseases had acted as a selective process, restricting physical activity patterns, causing an acute decrease of fitness in infected individuals and killing the least-well-endowed members of the local population. Finally, it was proposed that a study of athletes from various populations would document the diversity of body build and biological function which could evolve successfully both within and between populations.

The organizers of the IBP-HAP recognized that already, in the 1960s, many of the traditional indigenous populations were undergoing rapid 'modernization', and that the technical innovations offered by 'developed' societies would quickly obscure any gains which had been achieved by many centuries of biological adaptation to a particular environment. It was thus proposed that baseline data be collected on isolated and traditional populations from a wide range of habitats, if possible making such measurements before the 'acculturation' process had begun. In regions where 'modernization' was already underway, an alternative tactic was suggested: examining several genetically similar populations at different stages in the course of acculturation.

As discussed in Chapter 1, a variety of genetic markers were proposed to examine how far the observed phenotypic differences between communities reflected genetic divergence, and how far they were attributable to the more immediate influence of socio-cultural factors.

IBP-HAP studies of circumpolar populations

The arctic environment has presented a severe challenge to human adaptability because of extreme climatic conditions, and at certain seasons of the year an acute shortage of 'country' foods (Chapter 1). Indeed, there is good evidence that starvation has led to failure of a number of attempts at colonization of the far north, particularly in Greenland. Many of the indigenous populations who are now living successfully in the circumpolar regions appear to have developed from a common stock, although their biological characteristics have diverged through several millenia of life in small, widely separated and relatively isolated communities (Chapter 1). Whether colonizing the tundra of the interior or the arctic coastline, they have also faced strong selective pressures, and at a first inspection such groups thus seem good candidates for testing several of the basic IBP-HAP hypotheses.

By the 1960s, some circumpolar communities already had quite a long history of contacts with 'developed' society. The primary focus of the IBP-HAP circumpolar studies was on selected Inuit settlements. US investigators concentrated their efforts on the village of Wainwright, on the Alaskan north shore. Here, a school had been opened almost 70 years earlier. Additional Alaskan data were collected from Point Hope, Point Barrow and Anaktuvuk Pass. By 1969/70, Point Barrow had grown to a relatively large town of over 2000 people, including some 100 non-Inuit. It had a military base, a research station and a long-established hospital, with a substantial airport, hotels and shops that were operated largely by migrants from the south.

The main Canadian studies were conducted on Inuit living in Igloolik, a small island settlement located near the tip of the Melville peninsula (69°40′ N, 81° W). Some IBP observers made additional observations on Inuit people living at Hall Beach, some 80 km to the south, and Auger (1974) carried out extensive anthropometric measurements at Kuujjuaq (Fort Chimo), a village in Northern Québec on the air route from Montréal to Iqualuit. At Kuujjuaq, there had been much inter-marriage between Inuit and Cree Amerindians. When the Canadian IBP-HAP team first selected Igloolik as the prime site of study, there had been limited contact between this community and 'white civilization' through a small Hudson's Bay trading post, a Belgian priest, itinerant Anglican missionaries and occasional visits of the Inuit residents to the 'DEW' line (distant early warning radar) station at Hall Beach.

Nevertheless, all of the indigenous people over the age of 40 years had been born in small camps, scattered across the Foxe Basin. By the time that the first physiological studies were undertaken, in 1969/70, the process of 'modernization' had already advanced considerably. A small diesel generator now provided the settlement with electricity. The last Inuit family who had been living in an igloo had just moved into a prefabricated government bungalow. An eight-grade school and a small nursing station had been established in the village, and depending on weather and the availability of aircraft, light planes were arriving from Iqualuit once or twice per week. Although the villagers maintained about 40–50 dog-teams, some 30 families had already purchased snowmobiles, and were using them for hunting expeditions. Given the small size of the settlement, the arrival of about 30 scientific investigators and graduate students for the IBP-HAP study was in itself a further powerful 'acculturating' influence.

The Scandinavian IBP-HAP studies were undertaken at Aupilagtoq and Kraulshavn, in the Upernavik district on the west coast of Greenland, and at Ammassalik and Scoresby Sound in Eastern Greenland. Each of these communities was relatively isolated and had reputedly maintained its traditional culture, although beginning around 1950 the Danish government had encouraged a concentration of the regional population into the main settlements. In one published physiological study from this era, residents of the Upernavik district apparently had access to a motor launch for the hunting of marine mammals (Lammert, 1972).

Other circumpolar communities that were studied under the auspices of the IBP-HA programme included Ainu, Lapps, and the various small Siberian populations noted in Chapter 1. The Ainu were living in the communities of Nükappu and Usu on the island of Hokkaido (Koishi *et al.*, 1975). Lapps were recruited from Svettjärvi (northeast of Lake Inari) and Nellim (south of Lake Inari) in Finland, from the Kola peninsula of the USSR, and from Finnmark (Norway).

During this period, studies of the various Siberian indigenous populations seem to have been limited mainly to the collection of genetic data (Chapter 1).

Other studies of circumpolar populations

The IBP-HAP stimulated several important initiatives of data collection and reporting. Beginning in 1967, an International Conference on Circumpolar Health has been held approximately every three years: Fairbanks (1967), Oulu (1971), Yellowknife (1974), Novosibirsk (1978), Copenhagen (1981), Anchorage (1984), Oulu (1987), Whitehorse (1990), Rekjavik (1993) and Anchorage (1996). This series of meetings has been stimulated by an increasing input from the indigenous populations surveyed; the reports presented have examined data on the growth, body build, nutrition and physiological function of the circumpolar peoples in the broad context of their social and clinical health.

The International Union of Biological Sciences (IUBS) has also sought (although with limited financial success) to sponsor a successor programme to the IBP-HAP, entitled 'Man and Biosphere' (Collins & Roberts, 1988). The IUBS research, focused particularly on life in the tropical biosphere, has elaborated one of the concepts inherent in the original IBP-HAP: the interdependent relationship between humans and their immediate environment (Foote & Greer-Wootten, 1966; Lee, 1969). Energy flow within any given community has thus been viewed as a measure of success in adapting to the limitations imposed by a particular habitat.

Many individual research projects involving indigenous circumpolar populations have been reported to the International Conferences on Circumpolar Health and/or have been published in the journal *Arctic Medical Research*. Our studies of the Igloolik population, which began as a cross-sectional investigation under the auspices of the IBP-HAP, have evolved into a longitudinal cohort study of child growth and development, with repeated cross-sectional evaluations of the adults as these have adopted a 'modern' lifestyle (Rode & Shephard, 1992a). Cross-cultural comparisons have also continued between the Inuit and the less acculturated nGanasan populations (Rode & Shephard, 1994b).

Population sampling

Biological scientists have been slow to recognize the importance of appropriate population sampling as a precondition to the generalization of their research findings. The handbook of methodology prepared for the IBP-HAP (Weiner & Lourie, 1969; 1981) stressed that the total sample size should be sufficient to ensure an adequate statistical power when making comparisons between populations. However, in terms of defining and selecting the population to be tested, the only admonition offered by the handbook was to ensure that sampling was 'random'.

An important first step in sampling is to define the boundaries of any given community. Boundaries are more difficult to establish for circumpolar nomads than for city dwellers of relatively fixed address, and surprisingly few IBP investigators have addressed this issue. For example, some members of the Canadian IBP-HAP team confined their observations strictly to the community of Igloolik. This settlement lies on a neatly circumscribed but irregularly-shaped island with a diameter ranging from 8–16 km. However, other investigators included data obtained on Inuit living in Hall Beach, apparently without considering how this might affect the comparability of their results. In the settlement of Hall Beach, many 'modern' social concepts were introduced with establishment of the 'DEW' Line station in the early 1950s, and one index of the biological impact is that by 1969/70, the average age of menarche at Hall Beach (13.4 ± 0.2 y) was substantially advanced relative to that observed in Igloolik (14.3 ± 0.2 y, Milan, 1980).

Even if the nature of the settlement allows the definition of a clear geographic boundary, the population living within a given region unfortunately rarely remains static. In most arctic settlements, there is substantial outward migration, as teenagers are sent to larger settlements to receive advanced schooling, young adults migrate to cities of the south in search of paid employment, and those with tuberculosis are sent to regional sanatoria. At the same time, less acculturated people continue to move into the community from small and isolated camps scattered over the tundra and along the coastline. The population mix within a settlement can thus change dramatically over a period of 10 or 20 years.

If data are to reflect the physiological status of a population on any given date, the ideal approach is to test all willing volunteers, or at least a stratified and representative sample of the community under investigation. But in practice, many factors conspire to bias subject selection. Members of indigenous populations vary widely in their knowledge of the first language of the investigators, and unless the observers are prepared to learn the local language, they will probably recruit the most 'acculturated' individuals, because these are the people who can understand and speak the language of the colonists. Hunters and itinerant reindeer herders tend to be excluded from surveys of short duration, since they are absent from their homes for several weeks at a time.

If a survey is led by a physician or a nurse, there will be a tendency for observers to recruit the 'worried well' (Criqui, 1985) and those with overt manifestations of chronic disease. Conversely, a team that is led by an exercise scientist may attract selectively people who have a healthy lifestyle and are of above average fitness (Stephens, 1989; Stephens & Craig, 1990). The magnitude of differences between results from the two types of survey was examined in Igloolik, where both a physician and an exercise scientist made studies of lung volumes during the same year (Shephard, 1978). The two surveys each included a large fraction of the total population, and in instances where the same individuals were tested by both teams of investigators, the two data sets agreed quite closely. Ignoring a few individuals with conditions such as congenital dislocation of the hip, poliomyelitis or a recent coronary attack, nursing station records showed a much higher proportion of villagers with pulmonary disease (56/105) among those who were not seen by the physiologists than among those who had been tested by this team (48/176, $\chi^2 = 33.6$, $p < 0.001$, Table 2.1). Conversely, the medical team examined fewer fit subjects than the exercise scientists. Thus, for men aged 20–59 years, there was a 4% inter-survey difference in average vital capacity and a 9% inter-survey difference in average one second forced expiratory volume (Shephard, 1978), with parallel inter-survey differences of lung volumes shown by other age and sex categories.

As in industrial and post-industrial societies, the proportion of the population who volunteered for testing by the exercise scientists decreased progressively with age (Table 2.2). It seems probable that this led to a selective elimination of older individuals with chronic disease. Unfortunately, we have no good method of assessing the impact of such differential sampling upon average test scores, although in one attempt to examine the

Table 2.1. *Nursing station records for Inuit villagers of Igloolik: a comparison between those tested and those not tested by the physiology team*

| | Tested | | Not tested | |
Health of subject	Number	%	Number	%
Normal health	178	78.6	49	44.1
Primary tuberculosis	9	4.0	12	10.8
Hilar calcification	11	4.9	22	19.8
Secondary tuberculosis	17	7.6	7	6.3
Pulmonary fibrosis	9	4.0	13	11.7
Emphysema	2	0.9	2	1.8
Disabled			5	4.5
Recent coronary attack			1	0.9

Source: Shephard (1988).

fitness of a representative national sample of Canadians, incomplete sampling apparently caused little bias in simple descriptive variables such as body mass index, resting heart rate and resting blood pressure (Stephens, 1989).

Many of the IBP circumpolar studies were based upon quite small numbers of subjects (Table 2.3). Sometimes, no information was provided regarding the percentage of the community that had been tested or any likely biases of data collection. The conclusions that can be drawn from such research are correspondingly limited.

Standardization of methodology

The various circumpolar populations are separated from each other not only by vast geographic distances, but also by differences in language and colonial jurisdiction, and by transportation routes that are oriented about a north–south rather than an east–west axis. It was thus almost inevitable that the individual communities would be tested by differing teams of investigators. Valid comparisons of data from one community to another then depend on a careful inter-laboratory standardization of methodology, with an on-going physico-chemical and biological calibration of all test procedures against common and valid external standards (Jones & Kane, 1979).

Details of IBP-HAP methodology for the measurement of health-related fitness and physiological work capacity were standardized by an international

Table 2.2. *Influence of age of subject on proportion of individuals volunteering for fitness testing. A comparison between the 1981 Canada Fitness Survey (CFS, Shephard, 1986) and the 1969/70 physiological survey of Igloolik (Shephard, 1978, 1980)*

| Age (years) | Percentage of population volunteering | | | |
| | Females | | Males | |
	CFS	Igloolik	CFS	Igloolik
10–19	74	65	80	77
20–39	57	56	68	67
40–59	40	48	48	41

working party (Shephard *et al.*, 1968), and the agreed techniques were subsequently summarized by Weiner & Lourie (1969; 1981).

A variety of factors unfortunately limited implementation of the standardized protocol:

1. Some laboratories were unwilling to abandon older techniques because they had used these procedures to collect data for many years. They thus ignored the recommended IBP-HAP protocol.
2. Other national and international organizations proposed alternative methodologies, often designed with slightly different objectives (such as clinical testing, or the evaluation of athletes). Too often, their recommendations were put forward without acknowledging the prior work of the IBP-HAP (American College of Sports Medicine, 1991; Andersen *et al.*, 1971; Erb, 1970; Kattus, 1972, Larson, 1974; McDougall *et al.*, 1983; Mellerowicz, 1966).
3. An overall lack of funding and/or national purchasing policies prevented some investigators from buying the equipment recommended by the IBP-HAP.
4. Because of the era in which the IBP-HAP recommendations were framed, the proposed methodology did not cover some important but recently introduced physiological and biochemical techniques (for instance, the fractionation of plasma lipids).

There is unfortunately no good method of adjusting findings for systematic errors that have arisen through inter-laboratory differences of methodology. Such errors must be accepted as a limitation upon the interpretation of apparent functional differences between circumpolar communities and between cohorts within a given community. It is particularly desirable that technology should remain consistent when inferences are to be made about the effects of acculturation upon a particular population. However, successive studies of circumpolar communities have often been undertaken by different teams of investigators, using differing equipment and differing protocols. The Igloolik study is unique in that the same observers used the same equipment and the same protocol in three surveys, conducted in 1969/70, 1979/80, and 1989/90. The one major constraint imposed by such an approach is that technology then becomes limited to methods that were current at inception of the study (in 1969/70).

Choice of units

Standard International Units (Ellis, 1971) in general have served exercise physiologists well, and they have been adopted increasingly on a world-wide basis. However, there remains some reluctance to express oxygen consumptions as molar rates of gas exchange, and blood pressures in kPa terms, as a full exploitation of the SI system would require.

A second source of controversy is the presentation of data as ratios, for example the calculation of aerobic power as the maximal oxygen intake per kg of body mass, or the evaluation of cardiac function in terms of the oxygen pulse (oxygen consumption/heart rate). The majority of published data on the maximal oxygen intake of indigenous populations has been presented as ratios to body mass. The logic of this approach is that the energy cost of performing many heavy physical tasks is almost directly proportional to the individual's body mass (J.R. Brown, 1966; Godin & Shephard, 1973b). Thus, the ratio of maximal oxygen intake to body mass provides a rough indication of a person's ability to perform heavy endurance work.

One immediate objection to use of a body mass ratio is that the power available for weight-supported activities (for example, paddling a kayak) depends more on absolute than on relative aerobic power. Further, the usual types of exercise that are used when measuring aerobic power (a step test, cycle ergometry, or an uphill treadmill run) evaluate the performance of the legs, but in the example cited (the paddling of a kayak), the critical factor is the absolute rate of metabolism that can be sustained when using the arms and shoulder muscles.

A second argument against relating maximal oxygen intake to body mass is that the score then depends as much upon body fat content as on aerobic power. Thus, if the relative maximal oxygen intake decreases with

Table 2.3. *The completeness of sampling in some IBP anthropometric and physiological studies of circumpolar populations. Data indicating total population, subjects (absolute number and percentage of population), and likely biasing of sample*

| Population | Total | Subjects[a] | | Likely bias | Authors |
		No.	%		
Ainu	275	21	7.6	Healthy but sedentary	Ikai et al. (1971)
Wainwright Inuit	315	87	27.6	Deficient in women	Rennie et al. (1970)
Point Barrow & Point Hope Inuit	2000+	231	11.6	Mixed parentage	Rennie et al. (1970)
Anaktuvuk Pass Inuit	?	65	?	?	Auger et al. (1980)
Fort Chimo Inuit	600	300	50	Amerindian admixture	Auger (1974)
Igloolik Inuit	335[b]	224	67	Healthy, ? fit	Rode & Shephard (1971)
Upernavik Inuit	149	15	10	Convenience sample	Lammert (1972)
Upernavik Inuit	149	84	56	?	Auger et al. (1980)
Norwegian Lapps	?	806	?	Children, hospitalized adults	Auger et al. (1980)
Skolt Lapps	1000	70	7	Convenience sample	Anderson (1963)
Skolt Lapps	1000	52	5.2	Convenience sample	Karlsson (1970)

Notes:
[a] In some instances, larger groups were sampled for genetic analyses (Eriksson et al., 1980).
[b] Subjects over the age of nine years.

acculturation, it is unclear whether the reason for the decrement in score is an accumulation of body fat or a deterioration of cardiorespiratory function. Nevertheless, some method of data standardization is necessary to inter-population comparisons, since many of the traditional circumpolar groups are much smaller than their urban peers. One alternative approach is to express biological data as a ratio to body stature. Depending on the variable that is being measured, the denominator may be stature, stature squared or stature cubed (Asmussen & Christensen, 1967; Shephard, 1982; von Döbeln, 1966). Such an approach remains less popular than use of the body mass ratio, in part because of disagreement regarding an appropriate choice of height exponent for key variables. In some of the circumpolar populations, a short limb length or accelerated shortening of the spine (Rode & Shephard, 1994h; 1992a) further complicate the interpretation of all ratios to standing height.

Some statisticians have advanced a final and more general objection to any use of ratio scaling: the implicit assumption of direct proportionality between the numerator and the denominator. From a statistical viewpoint, data sets are represented more accurately by a linear regression with a specific intercept than by a simple ratio. In the case of height and body mass, inter-population differences in average values are fairly small, and the errors arising from use of a ratio rather than a linear regression are correspondingly limited. However, the problem is more important for oxygen pulse: oxygen consumption increases in an approximately linear fashion with an increase of heart rate, but only if observations are restricted to the range 50–100% of maximal oxygen intake (Åstrand, 1960). Further, the heart rate corresponding to 50% of maximal oxygen intake is 50% of the heart rate reserve, rather than 50% of the absolute heart rate. The oxygen pulse thus provides only a very crude index of cardiac function.

Longitudinal versus cross-sectional data

The physiological characteristics of many indigenous populations are affected by both aging and progressive 'acculturation' to a 'modern' lifestyle. However, there are few instances where these processes have been examined either by true longitudinal studies, or by repeated cross-sectional data collection.

Inferences about the course of growth and aging have frequently been based upon single cross-sectional surveys, using relatively small convenience samples. Often, the available number of subjects has been insufficient to allow an adequate breakdown of the population by age and gender. Even in instances where sample size has not been a problem, the true course of growth and aging has often been masked by concomitant acculturation. In communities where lifestyle has been changing, the more recent cohorts of children have tended to be taller at any given age, and they have reached their pubertal growth spurt at an earlier age than their older peers (Rode & Shephard, 1994c). More recent cohorts of adults may also have reached a greater mature height, so that cross-sectional analyses of adult stature have exaggerated the individual's decrease of stature with aging (Shephard, 1980; Shephard *et al.*, 1984). In communities where some acculturation has occurred, this has usually had a greater effect on young than on older adults; thus, cross-sectional estimates of the aging of variables such as maximal oxygen intake have reflected the opposing influences of loss of an active lifestyle among younger adults, and a true aging of cardiorespiratory function in older members of the community.

One analytic approach that we have exploited in Igloolik is to carry out three cross-sectional surveys of the entire population at intervals of ten years. This has allowed us to group the adult data by age-group and by cohort. We have then carried out a gender-specific two-way ANOVA by age decade and cohort (Rode & Shephard, 1994a). The age decade effect in this analysis reflects the true rate of aging, the cohort effect the influence of acculturation, and the (age decade × cohort) term any interaction between the two processes.

In an urban society, patterns of child growth can be examined economically by the technique of a semi-longitudinal survey, observations being repeated after a six month or one year interval. We adopted this approach in the community of Igloolik (Rode & Shephard, 1973e; 1984b; 1994c). However, the outcome of our surveys was less satisfactory than would have been likely in an urban community. Firstly, the available number of students in any single age category was quite small, so that aberrant patterns of maturation in single children sometimes led to bizarre average data for a given age group. Secondly, the semi-longitudinal technique could not overcome the difficulty that younger and older children had been exposed to very different environments, and in consequence had matured at differing rates.

True longitudinal surveys have been difficult to organize in isolated settlements, although improvements in northern air networks are now facilitating the task. Ten years of residence in Igloolik allowed one of us to initiate a longitudinal study of all willing Igloolik schoolchildren. Observations began in the fall of 1981 (Shephard & Rode, unpublished, 1995), and measurements were repeated every six months until the children left school, typically at the age of 16–17 years.

Critique of human adaptability approach

The IBP-HAP had initially assumed that there would be substantial differences in biological characteristics between different traditional populations, and that inferences could then be drawn about the survival value of the observed differences with respect to specific habitats. However, one of the major surprises from a synthesis of IBP-HAP data on human physiological work capacity (Shephard, 1978) was the great similarity of fitness and working capacity from one region of the world to another. Any differences in average fitness scores between typical circumpolar hunter–gatherers and sedentary city-dwellers were over-shadowed by inter-individual variations; most urban populations included occasional subjects whose working capacity far exceeded average values for the circumpolar group. Some circumpolar communities had slightly higher average levels of fitness than those reported for peoples living in temperate or tropical climates, but any such advantage was of doubtful statistical significance and of little practical importance.

One important component of the IBP-HAP hypothesis was that a substantial fraction of human physiological work capacity was genetically determined. Attempts have been made to determine the inherited fraction of variation through the study of urban samples comprising either similar and dissimilar twins or whole biological families (Bouchard, 1992; Bouchard & Pérusse, 1994). However, it has been difficult to disentangle true effects of heritability from the influence of a common family environment, so that estimates have been unstable (Table 2.4). Heredity apparently has influenced a number of variables, such as aerobic power, aerobic capacity, heart size and cardiac stroke volume, together with the response of these same items to training, although in most instances the estimated contribution of inheritance has accounted for less than 30% of the overall variance. Attempts to correlate aerobic power with red cell gene markers have been unsuccessful (Chagnon *et al.*, 1984; Couture *et al.*, 1986; Shephard & Rode, 1973). More recently, similar observations have been made with respect to various muscle enzyme markers (creatine kinase, adenylate kinase-1 and phosphoglucomutase-1, Bouchard, 1992). Again, no relationship was found to the individual's initial maximal oxygen intake, but small fractions of the inter-individual difference in training response were described by the genetic markers. Further progress of the human genome project may yet uncover substantial genetic descriptors of human performance. However, many supposedly stable inherited characteristics such as adult stature have changed rapidly in response to the recent pressures of acculturation, and such observations cast severe doubt on the possible

Table 2.4. *Familial concentration and heritability of selected aspects of human physiological work capacity*

Variable	Familial concentration	Heritability Initial value	Heritability Training response
Submaximal power output	+	—	+
Aerobic power	+	+	+ +
Aerobic capacity	+ +	+ +	+ +
Heart size	+ +	+	?
Stroke volume	+ +	+ +	+ +
Muscle fibre type	+ +	+	—
Muscle oxidative potential	+ +	+	+
Lipid oxidation	+ +	+	+
Lipid mobilization	+ +	+ +	+ +

+ significant effect; + + very significant effect.
Source: based on data collected by Bouchard (1992) and Bouchard & Pérusse (1994).

evolutionary role of the characteristics concerned.

In retrospect, it seems probable that several factors have minimized the influence of biological characteristics upon the survival prospects of a given community:

1. Challenges to survival in any given habitat such as the polar ice-cap have been seasonal in nature. Characteristics that favoured survival in the depths of winter were not necessarily advantageous in the warmer days of the arctic summer.
2. Many indigenous populations have exploited two or more adjacent ecosystems – for example, the tundra and the floe edge, or the tundra and the forest. Characteristics that were adaptive for one habitat may have proven a handicap in the second ecosystem.
3. Success as a colonist has depended more upon intelligence, the transmission of local knowledge from parents and community leaders, and a capacity for technical innovation than upon great strength or a high level of aerobic power.
4. Death has commonly been caused by an accident or disease, rather than by starvation.

This is not to deny that the IBP-HAP stimulated much thought about the processes of acclimatization and genetic adaptation to challenging habitats. It also remains possible that at some period in human history the emergence of unusual biological characteristics helped the Inuit to colonize the circumpolar regions. However, it seems that in recent years, the adaptive value of any population-specific inherited characteristics has been small. Nevertheless, the benchmark data gained from the Human Adaptability Project of the International Biological Programme is proving invaluable as scientists seek to chart the effects of 'modernization' upon the health and fitness of circumpolar populations.

3   *Changes in social structure and behaviour*

'Modernization' has brought a multitude of changes in social structures and behaviour patterns to the indigenous circumpolar populations – a process that some authors have termed 'acculturation' (Berry, 1976; Forsius, 1980; Sampath, 1976). This chapter looks at objective measures of acculturation and the resulting acculturative stress or alienation, examining the extent and the social effects of cultural change in groups that, until recently, have followed a neolithic hunter–gatherer or pastoral lifestyle.

Measures of acculturation

In a study of Inuit living at Baker Lake and Lake Harbour, Freeman (1971a) measured acculturation in terms of sources of income (Table 1.1). He demonstrated a substantial change in the type of employment from 1951/52 to the late 1960s.

At the time of the 1969/70 IBP-HAP survey of Igloolik, Rode & Shephard (1973b) made a three-level classification of the occupation of adult men, based on extensive observation of the villagers and discussion with the community elders. The population was divided into those who were persisting with a traditional hunting lifestyle, those who had accepted wage-earning employment within the settlement, and an intermediate, transitional group who were accepting occasional paid work, but still persisted with some hunting. This classification, although simple, revealed substantial inter-category differences in aerobic power and skinfold thicknesses, both in summer and in winter (Table 3.1). The association between acculturation and measures of fitness became even more marked when the hunters were subdivided, based upon the frequency of their long (two to three week) hunting trips. Those who had made 10 or more such trips during the year had an average aerobic power of 62.0 ± 7.5 ml/[kg.min.], compared with an average of 54.5 ± 4.8 ml/[kg.min.] in those who made only three or four long trips per year. It was estimated that in 1969/70, 6% of adult males were very frequent hunters, 11% were less frequent hunters,

64% were in a transitional group, and 17% had become acculturated wage-earners.

By 1989/90, there were about 430 adults of employable age in Igloolik, and some 180 wage-earning jobs were available to the local Inuit population. In the sample that we tested, 36.8% of adult males were wage-earners, and 18% still claimed to be hunters. The remaining 45.4% were concentrated particularly in the 17–29-year-old age range, and the economic survival of many of these individuals depended heavily upon government welfare payments. The hunters were themselves spending much less time in their search for game than in 1969/70, and they were using snowmobiles and power launches rather than dog sleds and kayaks for most of their activities. In consequence, their physiological characteristics no longer showed any advantage of fitness over the wage-earning group (Table 3.2).

By 1989/90, almost a third of Inuit women aged 20–49 years had accepted paid employment. However, there was little difference of fitness levels between the women in this group and those with other, more traditional lifestyles.

MacArthur (1974) developed a detailed empirical index to assess the extent of acculturation during the original IBP-HAP studies of Igloolik (1969/70). The characteristics used in his index were such that it could be applied to both male and female subjects of all ages (Table 3.3). A numerical score for each villager was based on six items: years of schooling, extent of English or Danish vocabulary, type of housing occupied (including the availability of modern appliances), geographic mobility (wage versus land economy), the occupation of the head of the family (or self, if an adult) and (in the case of Igloolik) the observations reported by a Belgian Catholic priest who had long been a resident of the village. Probably because many of the participants in our study were children and adolescents, the scores on this index were unrelated to age (Table 3.3). However, a cross-sectional analysis of data for 1969/70 suggested that acculturation was associated with an increase of skinfold thicknesses in both male and female subjects, and in the male subjects an increase of the acculturation score was also negatively related to aerobic power, whether expressed as an absolute value or relative to body mass.

Berry *et al.* (1981) undertook a factor analysis of variables with a potential relationship to acculturation, and based upon this analysis he included in his index age, western education, wage employment, ownership of 'modern' goods (radio, snowmobile, etc.), language used, literacy in English, and frequency of media use (newspapers and magazines).

A final potential quantitative index of acculturation is provided by dietary composition, as determined from questionnaires, or assessed

Table 3.1. *Association between measures of fitness and acculturation of lifestyle (means ± SD). Data for adult males living in Igloolik in 1969/70 during summer (S) and winter (W) seasons, classified into traditional hunters, acculturated villagers with wage-income, and an intermediate, transitional group*

| | Number tested | Skinfold thickness[a] (mm) | | Predicted aerobic power[b] | | | | |
| | | | | l/min. STPD | | ml/[kg·min.] | | |
Lifestyle		S	W	S	W	S	W	
Traditional hunter	20	5.0 ± 0.8	6.4 ± 1.2	3.72 ± 0.52	3.75 ± 0.74	56.6 ± 5.1	56.2 ± 10.1	
Transitional group	22	6.1 ± 2.7	6.7 ± 1.8	3.64 ± 0.76	3.63 ± 0.71	54.9 ± 10.9	54.9 ± 9.2	
Acculturated	18	6.7 ± 2.8	7.9 ± 4.4	3.43 ± 0.64	3.38 ± 0.49	51.2 ± 9.6	50.1 ± 7.8	

[a] Skinfold thickness is an average of three fold thicknesses (triceps, subscapular and suprailiac).
[b] Values adjusted downwards by 8% to allow for possible over-estimation of maximal oxygen intake by the prediction method.
Source: Rode & Shephard (1973b).

Table 3.2. *Association between lifestyle and fitness levels as seen in 1989/90 survey of Igloolik Inuit. Mean $\pm$ SD for sum of three skinfolds (SF, mm) and aerobic power ($\dot{V}_{O_2max}$, ml/[kg·min.]) in subjects classified as continuing hunters (H), wage-earners (W), and other lifestyles (O)*

Age group (years)	Hunters		Wage-earners		Other lifestyles	
	SF	$\dot{V}_{O_2max}$	SF	$\dot{V}_{O_2max}$	SF	$\dot{V}_{O_2max}$
17–19	32.0	62.0	40.7	53.5	27.5	56.0
(2H/3W/26O)	± 14.1	± 11.5	± 19.7	± 10.4	± 0.1	± 9.0
20–29	40.4	46.4	29.9	51.9	32.0	50.6
(5H/28W/37O)	± 24.2	± 3.4	± 16.9	± 8.2	± 15.0	± 9.2
30–39	26.6	47.4	37.3	44.7	26.5	48.1
(5H/14W/5O)	± 17.0	± 5.1	± 19.9	± 6.4	± 14.0	± 8.3
40–49	80.5	32.4	42.3	42.3	54.5	40.0
(2H/10W/4O)	± 68.6	± 1.3	± 27.5	± 6.7	± 37.2	± 6.1
50–59	44.6	33.5	62.2	36.4	32.0	33.6
(5H/5W/0O)	± 22.6	± 6.4	± 45.0	± 6.6	± 18.7	± 6.3
60–69	34.0	35.2			46.0	31.3
(5H/0W/2O)	± 19.3	± 5.6			± 42.4	± 3.6

Source: Rode & Shephard (1992a).

Table 3.3. *Coefficients of correlation between an arbitrary index of acculturation to the lifestyle of 'developed' society (R. MacArthur) and physiological variables. Data for Igloolik Inuit tested in 1969/70*

Variable	Males ($n = 132$)		Females ($n = 93$)	
	r	p	r	p
Age	-0.02	ns	-0.02	ns
Triceps skinfold	0.26	0.004	0.34	<0.001
Subscapular fold	0.20	0.021	0.26	0.011
Suprailiac fold	0.31	<0.001	0.36	<0.001
Sum of three folds	0.29	0.001	0.34	0.001
Height	-0.02	ns	-0.02	ns
Body mass	-0.02	ns	0.12	ns
Handgrip force	-0.05	ns	-0.01	ns
Knee extension	-0.15	0.068	0.00	ns
Aerobic power (l/min.)	-0.23	0.009	-0.06	ns

Source: Shephard (1980).

objectively from blood lipid profiles (Chapter 4). Among the coastal populations of the arctic, those who have maintained traditional patterns of hunting and fishing show the omega-3 fatty acid profile characteristic of a diet rich in fish and marine mammals, whereas in more acculturated individuals the lipid profile has become much closer to that of the city-dweller.

Such observations suggest that it is possible to obtain useful quantitative indices of acculturation on circumpolar populations, although to date many observers have been content to present descriptive information. The findings at Igloolik further suggest that prior to the mechanization of hunting, there was an association between the continuation of a traditional lifestyle and a high level of physical fitness. However, the findings do not necessarily prove that hunting either caused a high maximal oxygen intake or prevented the development of obesity – it may rather be that the fitter members of the community were the most successful hunters, and therefore chose to persist with a traditional pattern of living when many of their peers were opting for the lesser physical demands of a 'modern' lifestyle.

Family structures and demographics

Family structures

The traditional structure of indigenous circumpolar populations has been that of an extended nuclear family, with relatives informally assuming responsibility for children if parents were killed by accidents or disease. For example, Leonard *et al.* (1994b) noted that 'Until the 1930s, the Evenki remained socially organized into named and extended family lineages that served as hunting units'. A typical Inuit camp comprised from two to eight such extended families (Sampath, 1976). The camp leader (who was usually a skilful hunter, and sometimes also the Shaman) tended to retain leadership responsibilities for many years, but individual members of the community who became dissatisfied with some aspect of camp life could leave (either alone or with some of their kin). The emigrants would then establish a new camp of their own or join another existing camp, rather than engage in a confrontation with their peers over the issue that had disturbed them. The social structure of a region was loose and individualistic, and it was rare for a camp leader to have any authority over adjacent camps.

In contrast with the organization of agricultural communities, traditional nomadic society showed little social stratification (Berry, 1976). Indeed, even in large and modern Inuit settlements such as Iqualuit and Inuvik, the emergence of inequalities of income, jobs, housing and aspirations has been slow to engender the stratified social classes characteristic of most 'white' societies (Honigmann & Honigmann, 1966; Mailhot, 1968).

Contact with 'developed' societies and their governments has generally brought about a progressive concentration of the population into larger regional settlements. Whether in Canada, Greenland, or Alaska, the primary motivation for population concentration has been to provide better schooling and health care to the circumpolar peoples; particularly in Siberia, the extension of political and social control has also been an important consideration. Leonard *et al.* (1994a) found that in the early 1990s, the entire population of an Evenki settlement wintered in a central village, with access to electricity and heat. However, during the summer, many of the population still dispersed across the Taiga, living in temporary shelters or Chums. Those who remained in the village during the summer included labourers, party officials, elderly pensioners and young children.

The prolonged absence of family members in residential schools, distant hospitals or paid employment in larger towns (Seitamo, 1976), a substantial prevalence of alcoholism and drug abuse, cultural and linguistic gaps between the older and younger generations and frequent instances of sexual abuse have all contributed to a progressive breakdown of traditional family structures with acculturation. Frequently, the elders of a community now complain that the younger generation lacks respect for them and for their accumulated knowledge of the circumpolar habitat.

Demographics

Many circumpolar communities were marked by a rapid surge of population, beginning in the early 1950s (Fig. 3.1). For example, the population of Scoresby Sound in Greenland increased by 500% between 1925 and 1971, 'due to the extraordinarily efficient medical care provided by the Danes' (Langaney *et al.*, 1974). The number of 'white' people living in the circumpolar regions made an appreciable contribution to this increase. Thus, in Greenland there were 473 'Caucasians' in 1945 (2.3% of the population), but by 1980 there were 9184 (22.2% of the population) (Hubert *et al.*, 1985).

During the 1960s, the population pyramid for many circumpolar settlements became similar to that of a developing country, a high proportion of villagers being under the age of 20 years. At Igloolik, the median age of the Inuit was 15 years by the year 1970. The total Inuit population of this settlement almost doubled from 533 in 1970 to 930 in 1990, but the shape and composition of the population pyramid also changed, the proportion of very young children becoming substantially smaller as the birth rate decreased (Fig. 3.2). Median ages thus increased slightly from 1970 to 1990, reaching 18 years for males and 17 years for females. The village also saw an influx of some 70 'white' migrants from southern Canada over the two decades. The newcomers were mostly

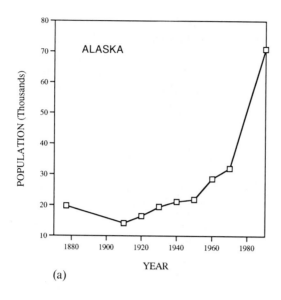

(a)

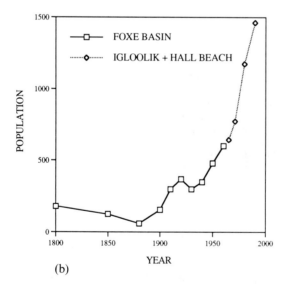

(b)

Fig. 3.1. Growth of circumpolar populations from the nineteenth century to the present day: (a) Alaskan Inuit and Aleut populations (in thousands) from US Census data, and (b) Igloolik Inuit (total population of Foxe Basin), and (in more recent years) of Igloolik and Hall Beach settlements. Based in part on data of Shephard (1978) and Milan (1980).

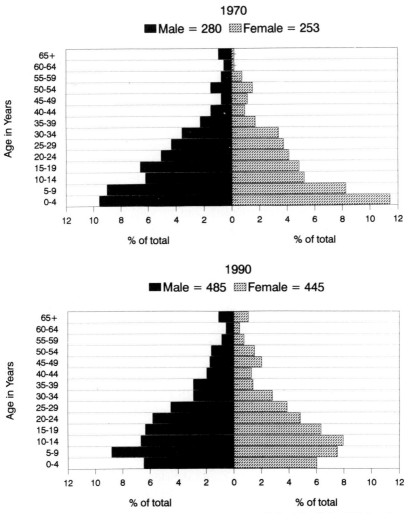

Fig. 3.2. A comparison of the population pyramid for Igloolik in 1970 (total numbers, 280 males and 243 females) and 1990 (485 males and 445 females).

temporary residents (government officials, police officers, teachers, health-care workers and their dependents).

Other circumpolar communities entered the IBP programme with equally young populations. At Wainwright, the median age for males and females was 18.6 and 13.6 years in 1970; it was 10–14 years at Kraulshavn, and around 15 years at Ammasalik (Milan, 1980).

Nevertheless, in some settlements and at some periods the potential population growth has been offset by a high rate of outward migration. The population of Wainwright actually decreased from 388 in 1940 to 225 in 1950, as young adults moved to exploit the economic opportunities associated with the larger settlement of Point Barrow. Subsequently, the numbers at Wainwright increased slowly, to reach a total of 308 in 1968 (Milan, 1980).

The territory available to the Avam nGanasan also shrank from the southern half of the Taimir peninsula to a much smaller region around Volochanka when the Soviet government established collective reindeer farms in the 1930s (Golcova, pers. commun., 1993). The increase in size of the nGanasan population (from 421 people in 1976 to 457 in 1991) was relatively small. In 1976, mixed marriages and southern migrants accounted for 13.8% of Volochanka villagers, and by 1991 this fraction had risen sharply to 31.5%, with a net *decrease* in the number of ethnically pure nGanasan. The total number of Soviet citizens identified as indigenous northern natives increased from 129 875 in 1959 to 182 000 in 1989 (Sedov, 1991). The most rapidly expanding ethnic groups were the Nenets (22 845 to 34 200), the Evens (9023 to 17 100) and the Dolgans (3925 to 8600).

Factors contributing to the rapid growth of some circumpolar populations have included governmental encouragement of migration to larger settlements and the associated arrival of nursing and/or medical care. Before World War II, some 40% of deaths among Alaskan Inuit were in children under the age of five years (Milan, 1980). Acculturation brought a dramatic decrease in infant mortality among both the Alaskan and the Canadian Inuit, due not only to improved perinatal care (much more readily available in a permanent settlement than in a seasonal camp), but also to improved maternal nutrition. Deaths from tuberculosis were also virtually eliminated among these populations with the introduction of streptomycin-based tuberculosis control programmes, and the close monitoring of those who had been in contact with 'open' cases of the disease.

Furthermore, and perhaps most importantly, the fertility of young Inuit and Skolt women increased as the tradition of prolonged (two to three years) lactation was lost (Forsius *et al.*, 1981; Langner & Steckle, 1991; Schaefer, 1981). Although advice on contraception was made available by most government health agencies during the 1960s, dissemination and acceptance of this information by circumpolar groups was relatively slow. In 1970, some 54% of women in Wainwright were practising some form of birth control (usually by means of intrauterine devices) (Milan, 1980). In Igloolik, 34% of women were using contraceptive pills by 1973 (Haraldson, 1974). In Aupilagtoq, seven of 21 women, and in Kraulshavn two of 15 women were using intrauterine devices by 1970 (Milan, 1980). However,

Golcova (pers. commun., 1993) noted that as late as 1976, only 7% of nGanasan women were availing themselves of either contraceptives or medically controlled abortions. By 1991, the figure had risen to 14%, but it was still low relative to the 67% of women from mixed marriages who were practising contraception.

Birth rates have shown parallel trends. In the relatively acculturated community of Point Barrow, Alaska, the crude birth rate peaked at almost 60/1000 population in 1964, but it had already declined to about 25/1000 by 1970. In 1970, the women at Wainwright reported an average of 7.9 pregnancies; comparable figures were 7.5 for Aupilagtoq, 7.8 for Kraulshavn and 6 for both Igloolik and Hall Beach. In the Upernavik district of Greenland, the annual fertility rate increased from 42/1000 inhabitants in 1950 to 71/1000 in 1963, thereafter declining to 25/1000 by 1974 (Green & Kromann, 1981). In Northern Québec, Inuit women had an annual birth rate of 40.1/1000 over the period 1973 to 1976, but the rate had diminished to 29.9/1000 for the period 1977–1979 (Duval & Therien, 1981). Forsius *et al.* (1981) found that Skolt women born before 1939 had an average of 6.0–6.1 births, but this dropped to 2.3 for those who were born between 1940 and 1951. The average number of live children per woman was 3.39 in the Volochanka nGanasan of 1991, but women from mixed marriages had only 2.06 children (Golcova, per. commun.). Leonard *et al.* (1994b) found four live births per woman among the Siberian Evenki.

Government and urban development

In many circumpolar communities, the initial trend was for federal governments to impose authority and laws upon previously independent peoples. However, in more recent years, several countries have moved to transfer responsibility back to indigenous populations, sometimes with an associated promise of at least regional autonomy in the near future (Chapter 9).

Concentration of the population in large settlements has (in varying degree) provided most circumpolar residents with 'modern' prefabricated housing, water and sanitation, together with easy access to commercial stores and other concomitants of a wage economy.

Government

In recent years, many circumpolar communities have received substantial cash settlements of outstanding 'land claims', and there has been an associated movement for indigenous populations to assume responsibility for their local government.

The typical course of changes in local government can be illustrated by developments in Igloolik. In 1970, all communal services in the settlement were administered by either the government of the North-West Territories or by an appropriate federal government department. The 'police force' of 1970 comprised one 'white' RCMP officer and an Inuk Special Constable. Political activity was limited largely to the fielding of candidates for the Village and the Territorial Councils. All administrative positions within the settlement were held by non-native migrants from Southern Canada, typically serving a one- or a two-year contract. The majority of these temporary sojourners spoke little Inuktittut, and had even less understanding of local culture and traditions. Tradespeople from the south handled nearly all of the repair and maintenance work that was undertaken within the settlement.

In 1976, Igloolik gained hamlet status, and a local council assumed responsibility for the management of all municipal services, including road maintenance, water and fuel delivery, and waste and sewage disposal. The administrative office of the hamlet was from the outset operated by an Inuit staff. Under the direction of the local council, these employees have been responsible for both budgeting and the daily operation of municipal services. By 1990, the police force had expanded to two 'white' officers and an Inuk Special Constable, all of whom were much more heavily committed to law enforcement tasks than had been the case in 1970. Other governmental positions had been transferred to local Inuit: job opportunities within the civil service included Wild Life Officer, Economic Development Officer, Social Service staff and Area Service officers. By 1990, the local post office, Bell Canada telephone installations, the weather station and airport services were all being handled by Inuit personnel. According to the local outreach officer, there were a total of 214 full-time paid jobs in Igloolik in 1990. Of these, more than 180 were held by Inuit (Rode & Shephard, 1992a).

Over the past two decades, the indigenous population has exploited the Canadian political system with considerable skill, and in 1992, the principle was recognized of establishing the eastern half of the North-West Territories as an autonomous region (Nunavut) under Inuit administration. This will become a reality in 1999.

There have been parallel developments in rural Alaska, but in Siberia progress has been much slower. The nGanasan currently talk of establishing autonomous regions and of restoring indigenous government. However, in the chaos following collapse of the Soviet system, power seems to have gravitated to a few powerful individuals. In Volochanka, we saw no organization comparable with the Igloolik Hamlet Council, and there seemed to be no elected organizations with responsibility for health, education or recreation of the type that are common in Canadian arctic settlements (Rode & Shephard, 1994b). In 1991, the *de facto* administrative authority for Volochanka seemed to reside almost entirely in two people – the Mayor and the Director of the local collective farm.

Private enterprise

Transition of the circumpolar communities to a wage economy has fostered the emergence of commercial enterprises. In 1970, there were two small shops in Igloolik: a Hudson's Bay Company store of some $150\,m^2$ floor area, and a very small cooperative store. The Hudson's Bay store had originally operated on a barter basis. By 1990, the Hudson's Bay Store, now renamed the Northern Stores, had expanded to a floor area of $670\,m^2$, and the coop had also built a $600\,m^2$ store. A number of the local Inuit were also operating their own businesses: there were several small general and dry goods stores, video-rental outlets, a hotel and a restaurant, a construction company, a hunting outfitter, and a taxi service (Rode & Shephard, 1992a).

Housing

Traditional Inuit dwellings have been described in Chapter 1. During the 1950s, the Canadian government began experimenting with various forms of prefabricated housing that could be shipped by coastal steamer and erected in arctic settlements. By 1970, the Igloolik settlement boasted a total of 69 houses, eight of which were occupied by teachers and government employees from Southern Canada. Of the remaining 61 dwellings, 10 had a single room, 27 had one bedroom, and 32 had three bedrooms. In the winter of 1969/70, the last Inuit family who had been living in an Igloo moved into one of these housing units. All the dwellings of that era still used a bag and bucket system for sewage disposal, and the average population density was 8.73 Inuit per housing unit.

By 1990, the housing stock had increased to 168 units, 15 of these properties being occupied by government employees. Of the remaining 153 units, 133 were managed by the Igloolik Housing Association (a predominantly Inuit group), and 20 homes were privately owned by Inuit families. All units now had central heating, hot and cold running water, a refrigerator and a stove, and all except 22 dwellings were equipped with flushing toilets. In part because families were becoming smaller, the average population density had decreased substantially to an average of 6.07 Inuit per housing unit.

Other communities in the Canadian arctic have had a similar experience (Rodgers, 1991; Sampath, 1976), although the timing of the process has differed from one settlement to another. Unfortunately, improvements in the housing stock have not always had positive social effects. Houses have tended to be allocated as they were constructed, little attempt being made to conserve traditional extended families in the process. The provision of convenient, modern homes has also robbed many males of one of their traditional family roles, that of providing shelter for the night. Too often, it has seemed that with the federal government also providing a subsistence allowance to cover the purchase of food and clothing, there remained little for husbands to do except watch television (Rodgers, 1991).

In Greenland, the traditional homes of turf and stone have now been replaced. In smaller settlements, there are frame houses with several rooms and coal or oil heating (Helms, 1981), and in the larger towns many of the population live in large apartment blocks.

In Volochanka, accommodation is currently government owned, but it remains quite primitive in concept. Our observations, made in 1991, showed that most people were living in barrack-like buildings containing four to eight apartments per unit (Rode & Shephard, 1994b). Construction materials varied from logs in the older buildings, through stucco on wood to metal sheeting. With the exception of some of the older log buildings, insulation was poor. Occupancy was commonly as high as 10–12 people per unit. Heating was still by the traditional Russian masonry stove, which also served for cooking and the heating of water. None of the buildings was provided with running water or flushing toilets. Water was drawn from an ice hole in the river, some 3 km distant, and was off-loaded from a bulldozer by a bucket, In bad weather, it was common for fresh supplies of water to be unavailable for periods of several days. There also seemed no organized system of sewage disposal. The contents of the toilet bucket were simply thrown onto the 'street' periodically. Combustible rubbish was burned in open heaps as it accumulated.

Medical care

'Modernization' has brought an increased level of formal health-care to many circumpolar communities, although it has been argued that sometimes such services meet the demands of the colonizing bureaucracy rather than the needs of the local community.

In Alaska and Canada, satellite telephone and/or television systems have been used increasingly to extend the capabilities of on-site nurse-practitioners.

In 1970, one nurse-practitioner and a local interpreter served the needs of the Igloolik community from a well-equipped but small (two-trailer) nursing station designed for out-patient use only. Weather permitting, patients with serious illnesses were evacuated to Iqualuit by air. By 1990, Igloolik had a modern, five-bed nursing station, with a two-bed pediatric unit. The professional staff included two nurse-practitioners, a nurse/midwife, and an Inuk health educator. The Station also included a fully equipped dental clinic, staffed by a full-time dentist and assistant, and a small lecture room for health education classes. Family physicians and other medical specialists visited Igloolik at least once a month on a regular rotating schedule that also included other circumpolar settlements.

In Greenland, district doctors were appointed and modern hospitals opened after World War II. As in Canada, air evacuation is currently arranged for patients requiring more specialized treatment in Godthaab, Copenhagen or Reykjavik (Helms, 1981).

We judged that the provision of formal medical services was much more rudimentary in Siberia. In Volochanka, we found a 'hospital' with four rooms and a total of 20 beds, with a staff of 16 people (Rode & Shephard, 1994b). However, the installation had little in the way of medical equipment or modern hospital facilities, and we never saw more than two of the beds in use during our eight-week visit to the community during 1991 (Rode & Shephard, 1994b). Sanitary conditions in the hospital, although better than in the rest of the village, appeared to be quite primitive. Water was often in short supply, and a single sink over a slop bucket served all ambulatory patients, staff and visitors. Even the hospital had no system for disposal of either garbage or sewage. Visits by physicians and other health-care professionals were a rarity. During the course of our survey, a team comprising a general physician, a dentist, a gynecologist, an ophthalmologist and a psychiatrist arrived by helicopter for a total of nine days, but this was the first such visit they had made to the settlement in three years. The local nurse and her assistants seemed fully occupied in dealing with the day-to-day operation of the hospital and out-patient clinic, and the available personnel lacked time for any involvement in health education or preventive care.

Formal education, dominant language and religion

Formal education

Some schools such as that at Wainwright were established early in the twentieth century. However, 'white' dominated education became increasingly prevalent and compulsory in arctic communities subsequent to World War II. Many sad anecdotes are now emerging of Inuit and Amerindian children who were forced by government inspectors to leave their homes and attend distant residential schools where the use of their maternal language was strictly forbidden. But even when schooling was provided within the local community, the new knowledge, language and culture acquired by the students drove a wedge between the children and their parents. Compulsory school attendance also prevented the children from learning the techniques of arctic survival by accompanying their parents on hunting trips.

In 1970, the school in Igloolik offered classes from kindergarten to grade 8, and it adhered strictly to an English language curriculum which had been developed by educational 'experts' living far removed from the North West Territories. The system was centrally administered from Yellowknife, some 2000 km to the west, with little input from either parents or the local community. By 1990, the school taught all grades from kindergarten to grade 12, and the first four years of the curriculum were presented in Inuktittut. Further, a locally elected Education Committee participated in all decisions affecting education within the settlement. School attendance was good in the lower grades, but it dropped to an average of no more than 65% in the final years of schooling. An Adult Education Centre was established in the 1980s. It offered opportunities for academic upgrading and a variety of more practical courses through its affiliation with Arctic College in Iqualuit. On-site training programmes for research technicians were also conducted by the Igloolik Research Centre, which had opened in 1975.

In Volochanka (in 1991), all teaching was in Russian, and at the time of our visit it followed the standard Soviet model (Rode & Shephard, 1994b). We observed a kindergarten and day-care centre for children aged 5–6 years, a primary school for those aged 7–10 years (with residential accommodation for children whose parents were working outside the community), and a secondary school for children aged 11–16 years (the last including an industrial arts workshop and a small gymnasium).

Problems in establishing the relevance of formal education continue to concern many circumpolar communities. For example, only 24% of the Skolt boys (but 58% of the girls) who were studied by Seitamo (1976) liked attending school. Perhaps the most critical factor in acceptance of formal modern education seems a stable family background that supports the child's self-esteem and actively encourages school attendance (Forsius *et al.*, 1985).

Harvey (1976) demonstrated that modification of the curriculum through the inclusion of courses on native land claims, traditional arts and crafts, anthropology, archaeology and circumpolar history greatly reduced the number of expulsions, drop-outs and suicidal gestures in an Alaskan school. Likewise, Seitamo (1981) noted that folk tales, songs, beliefs and themes from traditional Skolt culture were beginning to be collected and used as learning material in an improved school system for Lapp children.

Dominant language

Historically, colonizing authorities, whether English, French, Danish or Russian, have attempted to impose the use of their own language upon the indigenous residents of the circumpolar regions. Nevertheless, the use of traditional languages has persisted in the more isolated circumpolar settlements. Inuktittut was still the dominant language among the adults of Igloolik in 1970. The school and the Hudson's Bay Company store had begun operations around 1965 and 1930 respectively, but most of the older adults had been born in field camps and had received little formal education. Few children under the age of 7–8 years or adults over 45 years understood English. In order to communicate with the native population, outside observers either had to learn the local language, or rely upon informal interpretation by teenagers who had attended residential schools in Chesterfield Inlet or Ottawa. By 1990, the majority of young adults were able to speak a useful amount of English. An Inuit dictionary (Inuktittut/English) had been published, and syllabics were progressively being replaced by anglicized forms of Inuit words, particularly in official government documents. On the other hand, the Igloolik Research Station was busy preserving oral history through taped interviews with older residents.

Berry *et al.* (1981) had similar findings among the James Bay Cree Amerindians. He studied this population in 1967 and 1979. At the time of his more recent survey, older people were persisting with use of the Cree language, but younger, school-educated inhabitants spoke English and used EuroCanadian media.

In Siberia, insistence on use of the Russian language was seen as an important key to political indoctrination during the Soviet era. In many respects, the group that we studied at Volochanka were less acculturated than the Canadian Inuit living in Igloolik, but all spoke fluent Russian (Rode & Shephard, 1994b). All schooling was conducted in Russian, as was all radio and television programming. Government business was transacted only in Russian, and all communications, publications and signs used the Russian language.

Religion

The traditional religion of the Inuit peoples has involved Shamans, worship of the sea-goddess Sedna, and a respect for the sea as the source of all life. Likewise, Amerindian tradition has postulated a profound unity between the earth and human identity. However, in many circumpolar communities, traditional indigenous religions have now given place to the prosyletizing influences of Christianity (in Alaska, Canada and Scandinavia) or to Leninist/Marxist materialism (in Siberia, Chapter 1).

In 1970, Igloolik had a 150 seat Anglican church and a Roman Catholic Mission; the latter comprised a church seating about 100 people, and an adjacent hall for social events. The village was relatively evenly divided, both demographically and geographically, between supporters of the Anglican and the Catholic churches. The Anglicans were served by an Inuit deacon, who conducted services primarily in Inuktittut. The Roman Catholic church was served by a Belgian priest, but he also was fluent in Inuktittut. Indeed, he had translated the Bible and a selection of hymns into the local syllabics, and conducted all church services in Inuktittut. Church attendance was nearly universal in 1970, with several services being held at both churches each Sunday. Over the two subsequent decades, various other religious groups (Bahai, Baptist and Fundamentalist) gained and lost small followings. In the late 1980s, a small group of Inuit established a 'full Gospel' Pentecostal church in Igloolik, converting a small house into a place of worship. However, the religious and social influence of the various churches has waned since 1970, particularly among younger members of the community. Although the population has doubled during the period, there has been no need to increase the size of church buildings. Indeed, the diminished Catholic congregation has moved from the church into the Mission Hall in order to economize on heating costs.

In other circumpolar communities, the imposed religion has varied with the practices of the colonizing society. Across Canada, Anglican, Roman Catholic and Pentecostal churches have been particularly active (Berry *et al.*, 1981), but other smaller denominations have also had 'successful' missions in specific settlements, for example the Moravian mission that has served the Inuit of Northern Labrador since 1760 (Fitzgerald & Ehrenkranz, 1985). In Alaska, a wide range of American denominations have each had their converts, and in Scandinavia, the dominant influence has been that of the Lutheran church. In Siberia, the early influence of the Russian Orthodox church, dating from the seventeenth century, was largely replaced by Marxist–Leninist beliefs; these have now been rejected in their turn, but it is too early to speculate what may take their place.

Transportation, mechanization and outside contacts

Personal transportation

Throughout the circumpolar regions, the trend over the last two decades has been to a mechanization of personal transportation, with a reduction in the ownership of traditional hunting equipment, and an ever increasing range of tools and technology available to serve life within the permanent settlements.

In northern Scandinavia and parts of Siberia, traditional transportation was based on the reindeer. Groups such as the nGanasans, who were living on the open tundra, used the reindeer to pull sledges, but 'edge of the forest' dwellers such as the Dolgans preferred to ride the reindeer (Rode & Shephard, 1994b).

In the Canadian and Alaskan arctic, most of the journeys demanded by traditional winter and spring hunting activities were made by sled and dog team. In 1970, the local hunters of Igloolik kept approximately 500 dogs. There were also some 30–35 snowmobiles in the settlement, but quite a number of these were owned by 'white' residents. There were no private cars, trucks, all-terrain vehicles or bicycles in the village at this time, but the territorial government maintained three large tracked vehicles that were used for grading roads during the summer and for the transport of passengers to the airstrip in the winter months. For summer open-water journeys, approximately 20 of the Igloolik boats had been equipped with outboard engines in the 10–20 HP (7–15 kW) range, and very little use was made of traditional kayaks.

Gessain (1968) expressed the number of inhabitants in the Amassalimuit region of Greenland as a ratio to the number of available kayaks. He found that the ratio had increased from 3.4 people per kayak in 1884 to 5.3 in 1930, and 20 in 1968. He reasoned that in a sea-mammal hunting society, it was impossible to be an effective hunter without owning a kayak, and he thus deduced that there had been a major decline in the hunting activity of that region over the period of his observations.

By 1990, Igloolik had seen an enormous expansion in most mechanical forms of transportation. There were now 160–170 snowmobiles, nine privately owned cars and light trucks, 78 all-terrain vehicles, and 216 bicycles. Some 90 power boats had engines ranging in size from 30–150 kW (40 to 200 HP). Even short trips to the store, church, school or a neighbour's home were commonly made by snowmobile or all-terrain vehicle. A taxi service was a considerable commercial success, although no dwelling was more than 1000 metres from the centre of the hamlet.

Mechanization has been at least equally apparent in Alaskan communities, but the pace of 'modernization' has been much less rapid in Siberia. In 1991, there were still only about 20 snowmobiles in Volochanka (Rode & Shephard, 1994b). Both the machines and the necessary gasoline were very expensive relative to the earning power of the local inhabitants, and only the most successful trappers and hunters could afford such equipment. The nGanasan used snowmobiles mainly to make the one day journey to the best hunting grounds. On arrival, most of the actual hunting was done on foot or on skis. Unlike Igloolik, there were no privately owned cars or trucks in Volochanka (Rode & Shephard, 1994b).

One negative consequence of increased mechanization has been an alarming increase in injuries and accidental deaths, due for example to the collision between powerful boats and submerged ice. A high intake of alcohol or other drugs is particularly hazardous when a driver is operating a powerful vehicle in an inherently treacherous environment.

Although chronic otitis media is commonly blamed for the hearing problems of circumpolar groups, Ling (1976) traced a progressive hearing loss among middle-aged Inuit to a combination of inadequately muffled motors on snowmobiles, plus the frequent use of powerful rifles. There have also been suggestions that prolonged exposure to vibration from the operation of high-speed snowmobiles may be causing damage to the spinal columns of the drivers (Chapter 7).

Scheduled air services

Because of the difficulty in constructing roads over the permafrost, travel to distant locations (originally made by long river and/or sea voyages) has become increasingly dependent on a network of air services. Some staple supplies and larger items of domestic equipment and furnishings still reach isolated coastal settlements by the summer cargo boat, but an ever-increasing volume of merchandise also arrives by air-freight. In the future, large dirigibles and/or ground-reaction vehicles may further facilitate the importation of manufactured goods from 'developed' societies.

In 1970, Igloolik had a recently established Class C scheduled air service to Iqaluit, operating approximately twice per week, but highly dependent on local weather conditions. Night landings were particularly difficult, as the airstrip was illuminated only by oil flares. By 1990, the local airport authority had installed permanent landing lights that could be controlled from both ground and air. There was also a weather advisory service for pilots, and a small terminal building had been constructed for the ticket agent and passengers. A class 2 scheduled service operated to Iqualuit five days per week, and to Yellowknife three days per week. The route to Yellowknife also served Pelly Bay, Gjoa Haven, Spence Bay and Cambridge Bay. Other air routes to Nanisivik, Arctic Bay, Pond Inlet, and Hall Beach all operated at least once per week.

The Alaskan arctic has had a network of commercial air routes for an even longer time, and Fairbanks has now become a major international airport. The major cities of southern Siberia have also been served by substantial fleets of modern jet passenger aircraft for at least two decades, but the north has been reached mainly by military helicopters. Even in 1991, commercial airflights reached Volochanka only once every two to three weeks. Moreover, the planes used for this service were small and antiquated, dating back to the early 1930s. Seemingly, they lacked the power to carry both passengers and goods on the same flight (Rode & Shephard, 1994b).

Outside contacts

Offers of employment have provoked a progressive outward migration of young men from many circumpolar settlements to larger urban centres. In recent years, women also have seen marriage with a member of the dominant colonizing culture as a means of social advancement. For these and other reasons, even very isolated communities have experienced increasing contact with the outside world.

In 1970, hardly any of the Iglulingmuit had relatives living elsewhere in Canada. Moreover, contacts with the few individuals who had migrated to 'modern' communities were limited by poor communications (weather-vulnerable signals transmitted by radio telephone, high frequency radios operated by the Hudson's Bay Company and the Catholic Mission, and occasional airplane visits of indigenous emigrants back to Igloolik).

By 1990, a radio-satellite offered direct dialling telephone service to any city in the western world. Nearly every family in Igloolik was related to, or knew someone living in one of the larger centres of the NWT or southern Canada. Many of the Inuit (particularly those eligible for subsidized air-fares because of governmental employment) travelled south regularly for holidays or to visit friends and relatives. Even the schoolchildren had made a one-week educational visit to Toronto.

Partly for political and partly for economic reasons, communications in Siberia have remained less advanced. In 1991, we counted about 50 telephones within Volochanka, but telecommunication was limited to links with Dudinka and Norilsk. Messages to a more distant destination had to be relayed by a third party (Rode & Shephard, 1994b). We presume that this reflected restrictions on population movement imposed by the former Soviet government and/or problems of censoring conversations that might be conducted in a local dialect.

Land-based and wage economy

Land-based economy

The progressive concentration of the circumpolar population in large urban settlements has had a strongly negative impact upon hunting, fishing and other forms of land-based economy. This reflects in part the resultant environmental pollution and an over-exploitation of natural resources. Moreover, centralization of the population has inevitably forced those living in permanent settlements to travel very long distances in order to reach good hunting and fishing grounds.

The decrease in availability of kayaks in Greenland has been noted above. Helms (1981) estimated that in Eastern Greenland the population had reached the limiting size (700 people) that could be supported from local resources as early as 1920. Previously, the Danish government had resisted the idea of providing 'market' foods from fear of weakening the population. After 1920, the Danish government began to sell such foods to the indigenous population. Three quarters of energy needs still came from 'country' sources in 1945, but by 1978, the figure had decreased to less than 22%, equivalent to full provision for some 600 of the 2600 population.

In 1970, about two thirds of the male Inuit of Igloolik were active in full– or part-time hunting (Godin & Shephard, 1973a). They used dog-teams in winter, and low-powered launches in summer. The game exploited by such expeditions has been discussed in Chapter 1.

By 1990, a third of the Igloolik males over the age of 40 years still claimed to be hunters, but less than 10% of those under 40 years were active in hunting. Mechanization had also greatly curtailed the length and the average energy cost of individual hunting expeditions. In 1970, even a short round trip to the floe edge by dog-sled occupied a full day, but by 1990, high-powered snowmobiles could reach the most distant hunting grounds within a few hours. In addition to the direct influences of mechanization on the energy cost of hunting, there have been other, less direct effects upon activity patterns within the settlement. For instance, the care, feeding and running of a traditional dog-team demanded much physical work of earlier generations of hunters, and at many points in the day there were quite intense bouts of physical exertion. The dog teams consumed large quantities of seal and walrus meat. The summer walrus kill was followed by the butchering of the meat. Sides of meat were sewn into their own skins, forming 'sausages' that weighed 50–60 kg each. The sausages were loaded into boats, unloaded and buried in hard gravel, only to be dug up again as needed later in the season. In the winter months, slabs of frozen meat also had to be carved into slices of appropriate size to provide the dogs with a day's feed.

The decrease of hunting activity over the past two decades has affected daily energy expenditures not only for the men, but also for their wives. From 1964 to 1969, a total of 11 814 seal skins were traded at the two stores in Igloolik. Given some 40 hunting families, this would imply that the female partner in each of these households had prepared an average of about 6 sealskins for sale each week. In addition, she would have prepared the seal and caribou skins she needed for making the family's clothing, and she would have cleaned the skins of other animals (fox, wolf, and polar bear) for commercial trade. By 1990, the seal boycott had become sufficiently effective that neither the Coop nor the Hudson's Bay Company store purchased a single sealskin in that year. In consequence, the female task of preparing skins had been largely abandoned.

In Volochanka (Rode & Shephard, 1994b), the traditional local economy had been based upon reindeer herding and fur farming, supplemented by private hunting, trapping and fishing. Such activities appeared to prosper within the Soviet system through the mid-1960s. However, some five to six years ago the local reindeer herd was virtually destroyed, in part by an outbreak of brucellosis, and in part by deliberate slaughter for food.

Wage economy

As part of the move to an integration of the indigenous population into 'modern' society, parents now commonly encourage their children to learn skills of the 'white' culture which can be combined with land-based economic strategies. Nevertheless, Inuit children and their parents see more value in manual outdoor labour than in clerical functions, and many students in circumpolar schools wish to become carpenters, heavy equipment operators, or tradespeople (O'Neil, 1985). The fact that higher education must still be pursued at a distant residential school or university probably remains a factor in this decision.

By 1990, most of the food in Igloolik was being purchased from commercial sources (Chapter 4), and most of the clothing was also imported rather than made by women in the community. In Volochanka, food and clothing at the government store remained strictly rationed through 1991, although for those with money additional items could be purchased on the black market, particularly from an adjacent army garrison (Rode & Shephard, 1994b). It is hard to predict how the wage-based economy will emerge from the turbulence of glasnost and peristroika. At the time of our visit in 1991, a few foxes were still being kept, but the manager of the farm indicated that he was experiencing difficulty in obtaining sufficient fish or meat to feed the animals. A tannery, a boot factory and a fur clothing shop were also suffering from a shortage of raw materials. A local herd of 12 milk cows had been slaughtered just one month before our arrival in the community. We were unclear whether this was because the herd was diseased, or whether the slaughter was merely a means of obtaining meat. As in much of the Soviet system, the one shop in Volochanka had mainly empty shelves. Basic supplies such as bread, pasta, flour, kasha, salt, tea, sugar and butter could be purchased against the necessary coupons. However, we did not see any supplies of fresh or canned meats, fish, vegetables or fruit throughout our eight week stay (Rode & Shephard, 1994b).

Recreation and entertainment

Traditional recreations of the Inuit have included drum dances, blanket-tossing and wrestling (Glassford, 1970a, b). Children have also practised the skills they will need in later life (for example, ice-fishing and the handling of a dog-team, Shephard, 1974a). However, acculturation has brought indoor facilities, allowing the adoption of 'white' pursuits such as basketball and volleyball. For adolescents lacking the funds necessary to purchase a powerful snowmobile, such pursuits have recently become an important source of self-esteem.

Physical facilities

In 1970, the only formal recreational facilities in Igloolik were a small village hall (with a floor area of about 70 m^2) and the Catholic mission. The former offered twice weekly film shows, and a weekly square dance. The Catholic Mission provided facilities for table tennis and an occasional bingo game.

Over the past two decades, the construction of additional physical facilities has brought recreational opportunities close to the level typical of 'modern' communities in southern Canada. A school gymnasium was opened in 1971. This included a stage and a regulation-sized basketball court. A new recreation hall and an indoor swimming pool were added in the early 1980s, and a full-sized indoor skating and curling rink was completed in 1991. The school staff now offers regularly scheduled sports activities throughout the academic year, including indoor soccer, floor hockey, basketball, volleyball and badminton. The swimming pool is open in the summer months only, for lessons and supervised recreational swimming. A public lending library began operations in the late 1970s, and now has an extensive collection of books (although most are in the English language).

In other countries, the construction of leisure facilities has been less lavish. In Volochanka, there was only a small library and a small gymnasium, both at the village school.

Television

In Canada and the US, leisure entertainment is now dominated by video-casettes and multi-channel television, although in general the quality of available programming has lagged far behind the available technology.

A few television sets and video-casette players appeared in Igloolik in the late 1970s. At first, video-casettes were ordered by mail. Regularly transmitted television programming was first received via radio-satellite in 1983. By 1990, eight Canadian and US channels were available to cable subscribers. At least 134 of 139 households had at least one television set, and 81 also had a video-casette recorder. A CBC station in Iqualuit provided local news broadcasts and some regional television programmes in Inuktittut. The two main stores in Igloolik reported selling over 100 Nintendo machines, and game cartridges and video cassettes could be rented at three video-outlets in the settlement.

In Volochanka, also, most homes had televisions sets by 1991. However, reception was limited to the two standard Russian channels. These offered no regional radio or television programming, and no transmissions in the aboriginal language.

Alienation

Berry (1976) has argued that behaviour is a function of culture, which adapts to the ecology of a given habitat, but also takes account of acculturative influences. Often, 'modernization' modifies the traditional resources of a habitat (for instance, the construction of major hydro-electric reservoirs or oil pipelines may interfere with the normal migration patterns of caribou herds, restricting opportunities for hunting). Because of 'modernization,' historic patterns of adaptation may no longer be appropriate or effective tactics for survival, and this increases the stress occasioned by the inroads of a 'foreign' culture.

Those experiencing acculturative stress commonly show abnormal psychological traits (Forsius, 1980). These may become manifest through alcohol and drug abuse, suicide attempts, other forms of violence (including homicide and the physical and sexual abuse of women and children, Hart-Hansen, 1976), and other forms of criminal behaviour.

Substance abuse

The extent of substance abuse among the circumpolar people has sometimes been over-stated. For example, Klausner *et al.* (1980) made the poorly substantiated suggestion that 72% of the population of Point Barrow were alcoholics. Nevertheless, substance abuse has become widely prevalent in many arctic communities, with alcoholism, and glue and gasoline sniffing being particularly frequent problems.

In 1970, Lynge (1976) used the traditional diagnostic criteria of the psychiatrist to establish an alcoholism or suspected alcoholism rate of 6% for the 2000 Inuit who were living in western Greenland. This figure is certainly higher than desirable, but is comparable with that for many urban societies. Among the Inari Lapps, the age-standardized male mortality ratio for alcohol poisoning is now 270, relative to a value of 100 for Finland as a whole (Nähyä, 1991). Likewise, we observed much drunkenness in Volochanka, and more than 20% of those dying in Volochanka over the period 1965–1991 were said to be in a state of alcohol intoxication (Golcova, pers. commun., 1993). The abuse of marijuana, stimulants, cocaine, depressants, tranquillizers, hallucinogens and heroin has also become more prevalent among indigenous Alaskan populations than in major urban centres in the US (Bowman *et al.*, 1985; Segal, 1985).

In 1970, Igloolik saw only occasional episodes of alcohol abuse, with no other types of substance abuse. The creation of an Alcohol Education Committee in the mid 1970s introduced a measure of control over the import of alcohol. Currently, there are no local bars or liquor outlets in the community, and every order to an external vendor of alcohol must receive a time-limited permit from the Alcohol Education Committee. However, substance abuse has unfortunately become widespread among the young, with resort to a wide range of 'soft' drugs and volatile materials.

Suicide and violence

Suicide was not unknown in traditional circumpolar cultures. A number of settlements had their cliff or Nakaivik, where (with the support of close relatives) a person who had become unproductive because of age or illness jumped to their death in periods of food shortage (Lynge, 1981). However, with the spread of Christianity, such practices became regarded as sinful, and suicide remained uncommon until the 1970s, when the mental stresses associated with acculturation had become severe in many circumpolar communities. The epidemic of suicides over the past two decades is in marked contrast with the traditional Nakaivik ritual. Modern acts of suicide often occur without warning, and they are usually made in loneliness, while the victim is in a state of confusion and/or depression.

Kraus & Buffler (1976) found that the overall suicide rate for the indigenous populations of Alaska had already increased from 13 per 100 000 per year in 1961–1965 to 33 per 100 000 per year by 1970. The attempted suicide rate in one Alaskan town was set at 1450/100 000 per year, ten times greater than the rate for the large and supposedly stressful urban centre of Los Angeles. Other indigenous circumpolar populations in Canada, Greenland and Scandinavia have shown similar disturbing trends in the face of acculturative stress. Over the period 1985 to 1987, Status Indians in northern Saskatchewan had an annual suicide rate of 34 per 100 000, deaths being concentrated in those aged 15 to 34 years of age (Szabo, 1991). Alcoholism, drug or solvent abuse and/or depression were identified as contributing causes in about a half of the cases that were reviewed. In 1971, the annual suicide rate for the North West Territories as a whole was around the Canadian national average, at 10 per 100 000. By 1978, it had risen to 35/100 000 per year, and in 1983 Rodgers (1991) reported an annual rate of 167/100 000 among Canadian Inuit who were living on the Arctic north shore, in a region where oil exploration was being undertaken. Abbey *et al.* (1991) surveyed 13 isolated communities in the Baffin region between 1986 and 1989. They found that 19% of women and 31% of men had seriously contemplated suicide. By 1990, Igloolik had also experienced episodes of murder, suicide, rape and family violence.

In Upernavik, the incidence of homicide and suicide increased more than sixfold from 1950–1954 to 1970–1974 (Green & Kromann, 1981). Over the period 1977–1986, Thorslund (1991) found that suicide rates for those aged 15–30 years had risen to 100–200/100 000 per year in some Greenlandic towns. Village 'epidemics' brought occasional peak suicide rates for a single year to figures of over 400 per 100 000.

Woman and child battering have also become a problem in many northern communities (Pasquale, 1991; Schultz *et al.*, 1991). In one recent survey of the Baffin region, 3.5% of women reported sexual abuse, and 4.6% admitted to being victims of spousal assaults (Abbey *et al.*, 1991).

Other types of crime

The traditional culture of most circumpolar groups was based on sharing, and there was little theft. However, urban living, a wage economy, and an increasing ownership of consumer durables has brought a growing number of break-ins and thefts to communities such as Igloolik. Shoplifting has also become a serious problem for managers of the local stores.

Causes of alienation

Berry (1976) examined three Amerindian groups, the Cree, the Tsimshian and the Carrier Amerindians, all of whom were undergoing quite rapid acculturation to the lifestyle of Southern Canada. He argued that stress was greatest for the Cree, since they had a strong interest in preserving their traditional culture, and their migrant lifestyle differed widely from the sedentary habits of 'modern' Euro-Canadians. The traditional Tsimishian had a sedentary, stratified culture which diverged much less from that of the colonizing society, and he reasoned that they should thus be experiencing much less stress as they adjusted to the patterns of life imposed by the colonists. The traditional culture of the Carrier Amerindians was intermediate between that of the Cree and the Tsimishan, and in Berry's view they should thus be experiencing an intermediate level of stress. However, he apparently did not offer any objective data to support his hypothesis concerning the ranking of stress.

This is an important criticism, since acculturation does not always result in pathological reactions, even if the process is a rapid one (Chance, 1968; Vallee, 1968). Critical determinants of response include the individual's perceptions of events, the effectiveness of the coping mechanisms that are adopted, any sense of guilt the individual may feel at the repudiation of camp life (Sampath, 1976), and any disruption of the integrity of the family (by alcoholism, attendance at a distant residential school or extended hospitalization, Forsius, 1980), together with any incongruity between the individual's present occupational or cultural goals and the likelihood of achieving them.

Problems have been frequent in those who have elected to identify themselves with the dominant culture, but who have lacked the opportunity to interact with it or have failed to become integrated into it for want of employment or social contacts (Vallee, 1968). Wintrob *et al.* (1981) found that many of the older Cree children in the James Bay region had high occupational aspirations, but most of them also lacked confidence in their ability to achieve these goals. In the Inuit community studied by Rodgers (1991), there were no suicides among continuing hunters or among those who had become wealthy by working on oil rigs. Problems were concentrated among those who had anticipated that acculturation would bring them training and well-paid employment, but who now found themselves a poor minority within an alien society. Likewise, Thorslund (1991) commented that in Greenland problems had been greatest in villages where the process of 'modernization' had been slow, and there had been little opportunity for the population to find wage-earning employment. Psychological disturbances were particularly common among Inuit women who wished to retain their traditional culture but who married men who had adopted a modern lifestyle (Forsius, 1980).

Options for the future of circumpolar societies

A study of Cree Amerindians living in the James Bay region suggested to Wintrob *et al.* (1981) that the indigenous peoples of the circumpolar regions had four options: integration, assimilation, rejection, or marginalization. In their view, integration was most consistent with a successful synthesis of identity, elements of both traditional and modern cultures being identified and internalized. Over the course of 12 years, increased cultural control had allowed both integration and substantial cultural revitalization in the James Bay region; this has been a common finding in a number of the more isolated northern communities.

Assimilation is a common fate of minority cultures, but even in southern Canada, such a process remains incomplete and is difficult to achieve for several generations. The choice of this path is likely to lead to a prolonged identity conflict and a growing sense of alienation, as is the option of marginalization.

Resistance to the dominant culture may lead to withdrawal of the individual from society, a loss of self-esteem and identity confusion. But in a few instances, resistance has also been a path to a revitalization of indigenous cultural values. For example, Andrew & Sarsfield (1985) described how one band of Inuit living in the Ntesinan, a region of Labrador, had moved out of government settlements for much of the year. Apparently as a direct consequence of the changed lifestyle, alcohol abuse had suddenly stopped. An improved diet, rigorous physical activity, and the stable emotional environment associated with a functioning Inuit society had seemingly restored community health, socio-cultural values and economic success.

In Volochanka, also, we noted some early signs of resistance and revitalization: an nGanasan dictionary was now being compiled by the village elders, a museum of Dolgan and nGanasan artifacts had been established in the local school, and there was much talk among the villagers of forming an nGanasan autonomous region.

Plainly, the population of many circumpolar communities has become too large to suggest a full return to a traditional lifestyle. However, the honouring of historic values and a preservation of local culture to date seem the methods that have proved the most effective in restoring self-esteem and building the permanent circumpolar settlements into healthy, productive and self-supporting communities.

4 Secular trends in diet, metabolism and body composition

In this chapter, we examine secular trends in diet, metabolism and body composition, including changes in the type of food consumed ('country' versus 'market' sources), the energy costs of 'country' versus urban activities, developing problems from the contamination of local food and water resources, secular trends in body composition, and specific indicators of metabolic health such as glucose tolerance and the blood lipid profile.

One major nutritional change associated with the 'modernization' of indigenous circumpolar populations has been a progressive shift from 'country' to 'market' foods. There has also been a decrease in the total food needs of the individual as daily energy expenditures have declined, and in at least a substantial minority of the population, a substantial intake of refined carbohydrates and/or alcohol has displaced more nutritious food items. Nevertheless, serious malnutrition remains much less common in the arctic than in many developing countries, in part because of financial support from central governments, and in part because many of the indigenous communities continue to supplement store purchases by protein-rich items of 'country' food. However, the continuing isolation of arctic settlements and the high costs of air-freight have precluded substantial purchases of fresh fruit and green vegetables by the average villager, so that deficiencies in the blood levels of certain vitamins are common. In some communities, a high intake of refined carbohydrate and lack of vegetable fibre has also had an adverse effect upon dental health.

'Country' food versus 'market' food

One initial premise of the International Biological Programme Human Adaptability Project was that the indigenous residents of the circumpolar habitat would obtain most of their food from 'country' sources, and that the biological characteristics of the individuals concerned, either inherited or acquired, would give them success in acquiring the quantities of food needed for survival in a very challenging environment.

In some arctic communities, sparse supplies of 'country' food did indeed present a limit to colonization in the period before the coming of 'white' settlers, and there are anecdotal reports of episodic starvation among the Inuit of Eastern Canada and Greenland (Chapter 1, Berkes & Farkas, 1978; Clark, 1974; Hart *et al.*, 1962). However, there is little evidence that any unusual biological characteristics contributed to success in hunting and thus an ability to exploit what remains a very hostile habitat. Our own observations of the Igloolik community in 1969/70 showed that young hunters who were physically very fit might spend many days on the trail, yet they would return to the settlement with little game, whereas older and more experienced hunters would be very successful despite much more limited physiological capacities. Skill in tracking game and an accumulated knowledge of interactions between the local fauna and their habitat were much more important to the success of the hunter than was the development of an outstanding strength or a very large maximal oxygen intake.

In many circumpolar communities, the territory that was exploited varied with the season. In Siberia, for example, caribou were sought and/or herded on the open tundra during the summer and fall, but in winter the local residents retreated to the shelter of the forests. Likewise, in the Canadian arctic, families migrated from their winter settlements to temporary spring camps along the coast-line, in order to fish for char at the river mouths and then to hunt sea mammals at the floe edge. In many regional variants of the circumpolar economy, inter-individual physiological differences that might offer an advantage when exploiting one sector of the habitat during one particular season were less useful at another season, when the person was engaged elsewhere in a different type of hunting or fishing. Thus, the dominant need for cognitive skills and the diversity of the habitat greatly restricted the potential for emergence of 'useful' genetic adaptations.

As discussed in more detail below, a second hope of the International Biological Programme Human Adaptability Project was to develop models of energy flow for various indigenous communities, so that estimates could be made of the populations that could be sustained through optimal use of local resources. In several of the environments that were investigated by the Human Adaptability Project, valuable models of energy flow were established (Harrison, 1982). Investigations planned for the circumpolar habitat (Kemp, 1971) were hampered by the untimely death of the principal investigator. Analysis was further compromised by the great diversity of hunting patterns. It was necessary for an observer to spend several weeks on an arduous field trip in order to obtain even a limited amount of information on a few hunters who were undertaking one type of hunting in one particular season and in one particular year (Chapter 1, Godin & Shephard, 1973a).

Nevertheless, data on the usage of 'country' foods was obtained (Chapter 1). Our studies were sufficient to demonstrate that even at the beginning of the IBP Human Adaptability Project in 1969/70, no more than 30–40% of the Igloolik community's energy needs were being satisfied from the 'country' resources of the Foxe Basin (Godin & Shephard, 1973a; Table 4.1). Doolan *et al.* (1991) obtained rather similar data for Dene Amerindians living at Fort Good Hope; in the summer months, 26% of energy needs were met from 'country' sources, and in winter the figure rose to about 32%. Partly because of lesser acculturation, and partly because of a lower population density, Doolan *et al.* (1991) estimated that in the smaller settlement of Colville Lake, some 63% of food needs still came from traditional sources.

Other reports from the Ainu on Hokkaido (Koishi *et al.*, 1975), Alaskan Inuit settlements (Bell & Heller, 1978; Rennie *et al.*, 1962), Greenlandic Inuit (Bang *et al.*, 1976; Helms, 1981), and Norwegian Lapps (Gassaway, 1969; Ogrim, 1970) all showed that only a small fraction of the energy needs of the indigenous populations were being obtained from 'country' sources even at the beginning of the IBP-HAP investigation. Mackey & Orr (1985) collected detailed records of the 'country' food hunted and harvested by coastal residents of Makkovik, in northern Labrador. At the time of their survey (1980/81) a total harvest of 28 397 kg of country mammals, fish and birds and 832 kg of berries was distributed among a population of 296 individuals. It yielded 96 kg of meat per person-year, not much less than the Canadian national average intake of meat, fish and poultry, and it probably provided each of the local inhabitants with a total of about 5.5 MJ of energy per day. Unfortunately, Mackey & Orr (1985) did not indicate how many of the villagers were children, but if we assume that a half of their total sample were under 15 years of age, then we might infer that as recently as 1980/81, 'country' foods made a somewhat larger contribution to total energy needs in Makkovik than in many of the circumpolar communities evaluated in 1969/70.

Evidence based on total protein intake, blood lipid profiles and detailed dietary records shows that the proportion of food obtained from traditional sources has continued to decrease in recent years. The reasons are complex (Kuhnlein, 1984a). On the one hand, the use of 'country' foods has been compromised by a rapid expansion of local populations, a decrease in the availability of game, the adoption of more expensive methods of harvesting (snowmobile and launch as opposed to dog-team and kayak), and collapse of the market in animal pelts (which had previously met a substantial part

Table 4.1. *Potential income of food energy from hunting and fishing. Data for Igloolik region in 1969*

Game	Number of animals	Weight per animal (kg)	Total weight (kg × 10³)	Edible (%)	Metabolizable value (kJ/kg)	Total (kJ × 10⁶)
Polar bear	16	700	11.2	40	4395	20.9
Ringed seal	3648	50	182.4	25	5358	242.8
Bearded seal	55	250	13.8	27	5358	20.9
Walrus	150	700	105.0	26	4898	134.0
Caribou	800	70	56.0	40	4019	92.1
Fish	5000	5	25.0	50	7367	92.1
Total						602.8

Note: The total potential income of 'country' foods was estimated to satisfy 30–40% of energy needs for the Igloolik Inuit community.
Source: Godin & Shephard (1973a).
See original report for sources of individual pieces of information.

of the costs incurred in hunting). On the other hand, there has been an ever-growing availability of 'market' foods, linked to the dominant 'white' culture through social contacts and the ever-persuasive mass media; 'western' food has been perceived as both sophisticated and convenient.

Helms (1981) argued that by 1920, the population of Eastern Greenland had already grown to a size (700 people) where it was in precarious balance with local food resources. By 1978, less than 22% of the food needed by the Greenlanders was being obtained from 'country' sources, but because the total population had grown further to 2600 people, the local habitat was still in effect feeding 600 of the inhabitants, close to the potential capacity of the region.

Szathamary et al. (1987) studied four Dogrib Amerindian communities who were in the early stages of acculturation; they noted that 'modernization' led the indigenous groups to a progressive supplementation of their traditional food items (caribou, fish and birds) by 'market' foods. At the time of their survey, the oldest members of the population (> 65 years) still obtained only 12% of their energy needs from non-traditional foods, but in adults under the age of 46 years, 41% of energy was derived from non-traditional foods. Either there was a reversion to traditional nutritional practices with aging, or (more probably) the younger generation had undergone greater acculturation of lifestyle and diet. Likewise, Aubrey et al. (1991) found that the protein intake of Amerindians in Northern Ontario averaged 133 g/day in those who were over the age of 45 years, but the average protein intake had dropped to 93 g/day in adults under 45 years. The respective percentages of total daily energy needs met from 'country' foods were 17 and 6%. A cross-sectional study of lipid profiles suggested a similar age-related gradient of dietary practices among the Igloolik Inuit (Rode et al., 1995); given their lower plasma levels of polyunsaturated omega-3 fatty acids, it seemed as though the younger generation had shifted progressively from a diet based upon marine mammals to 'market' foods. Scandinavian investigators, also, have documented a progressive shift from 'country' foods to carbohydrates (particularly sugar) with acculturation. This trend has been observed both in Greenland (Helms, 1981) and in Northern Sweden (Haglin, 1991). In keeping with these several observations, Sabry et al. (1989) concluded that the younger members of an Amerindian band in the Wood Buffalo region had a lesser regard for 'country' foods than did their elders.

Although most circumpolar communities no longer meet or indeed could meet all of their food energy needs from 'country' resources, many observers have argued that traditional hunting and trapping activities should still be encouraged as settlements become 'modernized'. 'Country' food remains important to the indigenous populations, not only as a source of protein and polyunsaturated fats, but also because of its socio-cultural and religious significance (Kuhnlein, 1984b). Furthermore, communal hunting helps to promote values integral to the northern ethos, such as hard work and the sharing of resources. Early studies suggested that regular hunting encouraged the adoption of a healthy, physically active lifestyle, although in our most recent survey of the Igloolik Inuit we found that the hunters were no more fit than other members of the community (Rode & Shephard, 1992a, 1993). Finally, hunting activity serves as a valuable buffer against economic dependence, since the replacement of local meats by imported food would be very expensive relative to the cash income of most circumpolar residents.

In regions of the tundra where the local population rely heavily upon wild caribou as their source of protein, the local meat has the dietary advantage of being much leaner than the force-fed beef that has been reared inside the barns of southern Canada or the United States. In consequence, the plasma lipid profile of the circumpolar populations often remains quite favourable, even in groups such as the nGanasan who derive most of their protein from land-based rather than marine mammals (Rode *et al.*, 1995). In coastal settlements such as Igloolik, where marine mammals remain readily available to the hunters, the lipid profile of the older residents is further enhanced by a high intake of omega-3 fatty acids.

Energy expenditures and food needs

Problems of data collection

It is quite difficult to evaluate either energy expenditures or energy intake in traditional circumpolar communities (Chapter 1). A wide variety of types of hunting, trapping and fishing are undertaken on differing terrains, each on the basis of season and weather conditions, and for many of these items there is no conveniently available table of energy costs that can be applied to observations of activity patterns. Detailed data collection is thus required, and this implies that observers will be prepared to share the rigours of hunting expeditions for long periods just to obtain data on a few hunters in one season of a single year (Chapter 1).

Likewise, difficulties of communication between observer and subject, unusual sources and methods of preparing foods, and a wide seasonal variation in dietary patterns all hamper application of the traditional procedures of a nutrition survey. Again, it is necessary for an observer to live with a number of circumpolar families throughout at least one year in order to appreciate the full range of food that they consume.

A third possibility is to make 24-hour heart rate recordings, subsequently converting the data to oxygen consumption or energy costs on the basis of a step-test heart rate/oxygen consumption calibration curve determined for the individual in the base laboratory (Leonard *et al.*, 1994a; Spurr *et al.*, 1988; Chapter 1). Much of the data collected using this approach is of doubtful value, since the daily activities pursued during circumpolar life often have a high isometric component, which inflates heart rate in relation to oxygen consumption.

Basal metabolism

Much of the total energy consumption on any given day is attributable to basal metabolism. It is thus critical to know whether the high metabolic rate once commonly reported for circumpolar populations (Table 1.4) was based upon fact, and if so, whether it has been modified by subsequent acculturation. Critical evaluation of the literature suggests that uncertainty pervaded many of the earlier estimates of basal metabolism. As early as 1969/70, Godin & Shephard (1973a) noted that when the metabolic cost of simple daily activities was measured on Inuit subjects who had become accustomed to wearing a Kofranyi–Michaelis respirometer, energy expenditures were not appreciably greater than in 'white' city-dwellers. Although a crude indicator, this suggested that the Inuit no longer had any substantial elevation of basal metabolic rate. Nevertheless, our data collected in the spring of 1982 showed that both men and women had a basal metabolic rate that was some 15% higher than 'white' immigrants living in the same environment (Table 1.5). There was also a suggestion that the discrepancy between Inuit and 'white' subjects was greater in the older age groups, where less acculturation of diet had occurred.

The Food and Agricultural Organization (FAO) apparently were sceptical about the extent of any increase of basal metabolism among circumpolar populations, and in a joint report with the World Health Organization in 1973 they reversed their earlier decision (FAO, 1957) to allow a 3% increase in the recommended supply of food energy for every 10 °C decrease in mean annual temperature below the reference temperature of 10 °C.

If indeed recent generations of Inuit have shown a decrease of what had previously been a high basal metabolic rate, then the change could be explained by a lesser cold exposure (Shephard, 1993a) and a greatly diminished consumption of 'country' foods. In particular, a replacement of dietary protein and fat by carbohydrate would lead to a decreased specific dynamic action.

Katzmarzyk *et al.* (1994) found that when their sample of Siberian Evenki and Keto were lying at rest, the summer metabolic rate still averaged 7.0 kJ/min. in the men and 5.2 kJ/min. in the women, considerably higher than the values reported for men and women living in Scotland. However, other studies of indigenous groups in the Upper Volta (Bleiberg *et al.*, 1980; Brun *et al.*, 1981) and New Guinea (Norgan *et al.*, 1974) have reported similarly increased rates of metabolism, suggesting that sometimes the explanation of high readings is test anxiety rather than any intrinsic elevation of basal metabolic rate. The high population averages of Katzmarzyk *et al.* (1994) were due entirely to members of the 'brigades' who herded reindeer, and resting values were not elevated in permanent residents of the village. This probably reflects a temporary metabolic response to greater cold exposure and/or a higher protein intake during the herding trips, but it could conceivably be explained by a greater habituation of the permanent village residents to the activities and equipment of the investigating scientists.

Hunting and herding

Casual observation has sometimes suggested that traditional hunters engaged only in moderate intensities of physical activity. Nevertheless, when hunting conditions were good, this moderate physical activity was sustained over much of a 24-hour period so that the total daily energy cost of most types of hunt was quite high (Table 1.2). In 1969/70, Godin & Shephard (1973a) calculated that the average energy expenditure for eight of the traditional types of hunting activity undertaken in the Igloolik area was 15.4 MJ/day (Table 1.2). Nevertheless, they also emphasized that hunting only occupied an estimated 161 days per year, so that when values were calculated on an annual basis, the average energy consumption of the hunter dropped to 12.6 MJ/day.

Even the averaged figures for the Igloolik hunters were somewhat larger than the dietary estimates of Haglin (1991) for the current generation of reindeer-herding Saami Lapps (Table 4.2). Haglin accumulated data from earlier reports, noting figures of 12.3, 12.1 and 9.1–15.9 MJ/day for the men of this population who had been tested in 1962, 1967 and 1975, respectively. The Saami findings provide some evidence of a trend to a reduction of energy expenditures in recent years, and we may link this to a mechanization of reindeer herding and other village activities over the period in question.

Helms (1981) applied dietary methods to Eastern Greenlandic Inuit,

Table 4.2. *Estimated daily energy consumption of various circumpolar populations*

Population	Energy consumption (MJ)	Author
Alaska natives, males (1987/88)	11.5	Nobmann (1991)
Amerindians, Canada		Aubrey *et al.* (1991)
Caribou		
<45 years, male & female, 1986	7.9	
>45 years, male & female, 1986	8.4	
Inuit		Godin & Shephard (1973a)
Males, hunting	15.4	
Hunters, averaged over year	12.6	
Males, labouring	14.0	
Labourers, averaged over year	12.5	
Males, sedentary	10.5	
Males, elderly	9.6	
Females, single	9.6	
Females, married	10.0	
Females, elderly	8.3	
Inuit females, 1987/88	9.1	Kuhnlein (1991)
Mountain Lapps		Haglin (1991)
Males, 1987	10.3	
Females, 1987	7.6	
Males, winter 1975	9.1	
summer 1975	9.5	
1975	15.9	
Females, winter 1975	7.2	
summer 1975	6.9	
1975	11.0	
Males, 1967	10.5–13.6	
Females, 1967	7.6–8.6	
? Males, 1962	11.1–13.8	
Evenki		Leonard *et al.* (1994a)
Males	13.4	
Male herders	16.2	
Male labourers	11.0	
Females	8.5	
Females, brigade	11.8	
Females, village	7.7	
Chukchi males		Nobmann (1991)
coastal, 1982/84	13.2	
tundra, 1982/84	11.3	

estimating the total energy intake at 8.7 MJ/day in 1945 and 10.0 MJ/day in 1978. Unfortunately, these values seem to be averages, based on a mixture of records from men, women and children. The figures are thus hard to interpret, although if a half of the population were children, such energy expenditures would be regarded as very high. Likewise, the apparent increase in average energy consumption from 1945 to 1978 could reflect no more than a change in shape of the population pyramid, with a decrease in the proportion of young children.

In Siberia, acculturation to the lifestyle of permanent settlements began with enforced collectivization in the early 1930s, but the mechanization of village life now lags far behind that of other circumpolar communities (Rode & Shephard, 1994b). Leonard *et al.* (1994a) used both dietary records and the heart rate break-point method to estimate the daily energy intake of male Evenki and Keto. The dietary records suggested an average energy expenditure of 13.4 MJ/day in men and 8.5 MJ/day in women. The authors noted that whereas the men of the community were often engaged in energy expensive activities such as fencing, herding and fishing, the women were occupied mainly in traditional domestic chores. Among the female Evenki, heart-rate-based estimates (8.8 MJ/day) were of the same general order as the diet-based values, but in the male Evenki the heart rate data (at 11.9 MJ/day) indicated substantially lower energy expenditures than the dietary survey. Somewhat surprisingly, the heart-rate-based estimates were also lower for reindeer herders than for those employed within the village (respective averages 11.7 and 12.1 MJ/day). It seems probable that the heart rate data were in error, since the dietary figures led to the more reasonable conclusion that energy intake was higher in both the herders (16.2 versus 11.0 MJ/day) and their wives (11.8 versus 7.7 MJ/day) than in permanent residents of the villages. Possibly, cold conditions led to a slowing of the speed of the heart rate recorders, and thus an under-estimation of heart rates.

Nutritional status

Although most of the traditional 'country' foods harvested in the circumpolar regions had a good intrinsic nutritional value, some items (including salmon heads, eggs, fish, seal flippers, beaver tails, caribou, whale and seal meat) were often consumed in fermented format, particularly in the fall. Occasionally, this practice led to outbreaks of botulism (Nobmann, 1991). In recent years, some 'country' foods have also become nutritionally less desirable because they have become heavily contaminated by environmental pollutants (see below).

The current patterns of food intake among the circumpolar indigenous circumpolar populations retain a number of valuable features relative to 'developed' societies, but in most settlements the intake of fresh fruits, vegetables and grains is less than desirable. This reflects in part the high cost of air-freight; given a larger disposable income, most arctic residents would have no problem in buying greater quantities of such items. A second factor is that many indigenous families have had only brief experience in the buying of 'market' foods, and an inappropriate choice of purchases has contributed to a worsening of nutrition with acculturation. Finally, alcoholism has become a problem in a number of circumpolar communities (Chapter 3); in the affected individuals a substantial fraction of daily energy needs is now taken in the form of alcohol, and the intake of more important nutrients is correspondingly limited.

In 1970–1972, Nutrition Canada (1975) studied a sample of some 1800 Amerindians from 29 different bands. The survey was marred by a response rate of only 30%, but the data were adequate to demonstrate a deficient intake of iron, calcium and vitamin D. Desai *et al.* (1974a, b) also found poor iron, vitamin C and vitamin E levels in a survey of Amerindians living in the Yukon Territory.

Doolan *et al.* (1991) concluded from their survey of Hare Dene/Metis living in the subarctic Sahtu region of central Canada that traditional 'country' foods still made a substantial contribution to satisfying several daily dietary needs, including protein (whitefish, caribou and moose flesh), iron (caribou, rabbit, moose flesh, and moose blood) and zinc (moose, caribou and rabbit flesh and whitefish). However, a shift from 'country' to 'market' foods was increasing the total fat content of the diet, and in particular the ratio of saturated to polyunsaturated fat. Many other studies of arctic and subarctic communities have demonstrated similar changes in nutritional patterns with acculturation.

Nobmann (1991) surveyed Alaskan Indians, Aleut and Inuit in 1988. Their average protein intake remained high, more than 200% of the recommended daily allowance (RDA). Cholesterol intake had also reached high levels in the men (513 mg/day, as compared with the recommended intake of 300 mg), but often this was still offset by a high intake of fish and poultry. As in earlier studies of North American circumpolar groups, the calcium intake was poor (86% of RDA in men, 65% in women) and the iron intake was also low in the women (81% of RDA). In comparison with the Chukotka natives (Nobmann *et al.*, 1994), the Alaskan populations had shown much greater acculturation to the dietary practices of 'developed' society.

Haglin (1991) attempted to compare the current Saami Lapp diet with their previous eating practices. In the 1600s, the Lapps had lived almost entirely on reindeer meat and fish; this diet had provided a high phosphate intake (2000–3000 mg/day) and a very low calcium intake (200–400 mg/day). Health consequences of the low calcium intake had included a high prevalence of rickets in children, and a short stature in adults. These findings are in keeping with the low bone calcium readings observed by Mazess & Mather (1974) in photon absorptiometric studies of North Alaskan Inuit. Currently, the Saami derive only 14% of their daily energy needs from protein, 61% from carbohydrate, and 25% from fat. Nevertheless, the intakes of zinc, phosphate, vitamins B6 and B12 and selenium are still higher than in the diet of an average city-dweller. Calcium intake has also risen to 780 mg (an adequate level for all except pregnant and post-menopausal women), but the present diet contains less than recommended amounts of Vitamins C and E, folacin and fibre.

Current challenges are to encourage the circumpolar peoples to maintain the high intake of polyunsaturated fatty acids (which until now has offered protection against atherosclerosis) and to avoid an excessive consumption of refined carbohydrates in processed food and drinks. A high sugar intake has already led to a growing prevalence of obesity, diabetes mellitus and dental caries among Amerindian settlements in both subarctic and temperate habitats (Kriska *et al.*, 1993), although perhaps because of a high phosphate/carbohydrate ratio, the Inuit, circumpolar Amerindians and Lapps until recently have had a low prevalence of diabetes (Haglin, 1991; Kirjarinta & Eriksson, 1976; Mouratoff *et al.*, 1969). Unfortunately, this situation is changing as dietary acculturation continues (see below).

Contamination of food and water resources

In recent years, fears regarding a progressive contamination of local foods with polychlorinated biphenyls (PCBs), toxaphenes, cadmium, mercury and lead have imposed growing constraints upon advocacy of a traditional 'country' food diet.

Polychlorinated biphenyls and toxaphenes

Polychlorbiphenyls and toxaphenes persist in the environment for long periods, and indeed become concentrated as they pass through the food chain because they are stored in animal fats, including the fat of sea mammals. Kuhnlein (1991) found substantial quantities of both contaminants in the 'country' food available to Baffin Island Inuit, but except on occasional days the intake of these substances did not surpass currently recommended limits for the general population.

Doolan *et al.* (1991) completed a benefit/risk analysis of the consumption of 'country' foods for the Sahtu Dene and Metis of Fort Good Hope in the North West Territories, including in their analysis the risks that had arisen from the contamination of water resources. Polychlorinated biphenyls are the major source of concern in this particular community. The Canadian government has set a maximum PCB residue limit of 2000 ng/g wet weight for fish, and 200 ng/g for meat. Baked caribou is the only Dene food source with too high a PCB content based upon current standards. Doolan *et al.* (1991) found that the great majority of Dene residents consumed less than 5% of the maximum accepted daily intake of PCBs, as calculated from the product of observed residues and reported eating patterns. The one individual who may have exceeded the current ceiling was a hunter who reported eating an improbable quantity (6 kg) of fish and meat over a 24-hour recall period. It seems unlikely that this recollection was accurate, and it would certainly be extremely difficult to sustain such a level of protein intake throughout the year.

Nevertheless, the published ceilings are based upon the assumption that people consume an average of only 20 g of fish and 48 g of meat per day. Given that circumpolar populations such as the traditional Inuit and Dene Amerindian have substantially larger intakes of fish and meat, contamination limits which are probably acceptable for 'white' city-dwellers may be too broad to protect the health of indigenous populations. The risk of ingesting an excessive dose of PCBs may be even greater for coastal populations such as the Inuit, because PCBs undergo bioconcentration in such traditional food items as fish oil, caribou fat and bone marrow, and maktuq (raw skin) from the narwhal (unicorn whale).

Cadmium

Food derived from marine mammals has also been thought to contain excessive amounts of cadmium (Tarp & Hansen, 1991). In fact, the main source of cadmium in Greenlandic Inuit appears to be cigarette smoking rather than the consumption of 'country' foods, although the blood cadmium level of non-smokers is increased measurably by a preference for foods derived from marine mammals.

Mercury

Dumont & Wilkins (1985) and Kosatsky & Dumont (1991) expressed a concern that the large hydro-electric developments in James Bay had exposed northern Cree Amerindians to increasing and potentially dangerous levels of methyl mercury over the past two decades. Small quantities of inorganic mercury, present in the local rock, dissolve into the lake water, and when plants, bacteria and the submerged debris associated with reservoir construction act upon this solution, it is converted to methyl mercury. The organic mercury enters the food chain, and accumulates in the fish which have formed a major part of the Amerindian diet.

In recent years, mercury concentrations have continued to rise in fish taken from the James Bay region, but human blood mercury levels have declined. It is unclear how far the risk to local residents has been reduced as a result of health education programmes designed to control fish consumption, and how far there has been an unplanned decline in fishing due to acculturation of the local residents to a 'market' diet (Kosatsky & Dumont, 1991; Spitzer *et al.*, 1988; Wheatley & Wheatley, 1981). Certainly, blood mercury levels are now much higher in those who have persisted with inland trapping than in those who have adopted other lifestyles. Levels also show a seasonal variation, peaking in the autumn.

Hansen *et al.* (1991) noted high mercury levels in pregnant Inuit women who were living in remote areas of Greenland where there had been no direct exposure to factories, cars or other immediate human sources of mercury. They attributed their finding to the high concentrations of methyl mercury found in marine mammals. As in the James Bay region, the blood mercury levels of the Greenlandic Inuit have declined progressively over the period of observation, from 1984 to 1988, and again this has been attributed to a progressive decrease in the consumption of 'country' food with acculturation. The authors commented that the young Greenlandic Inuit reported a preference for 'western' food, and that only 23% of the energy needs of the population were now obtained from 'country' sources.

High blood mercury levels have also been reported among the indigenous populations of Alaska (Galster, 1976).

Lead

Hansen *et al.* (1991) observed high lead levels in their sample of pregnant Greenlandic Inuit. No evidence was obtained that the lead had been derived from 'country' foods. It was suggested that the most likely source was the long-distance transport of aerosols from factories in 'developed' countries.

Body composition

Acculturation to a sedentary, urban lifestyle has been associated with adverse changes in the body composition of many circumpolar residents, particularly an increase in the percentage of body fat. Given that most indigenous populations have also decreased their intake of food energy in recent years, the main explanation of the fat accumulation seems to be a substantial decrease of habitual physical activity. This verdict is borne out by parallel decreases in both muscular strength and aerobic fitness of circumpolar groups (Rode & Shephard, 1992a; Chapter 5). Nevertheless, various methodological problems have hampered the precise determination of body composition in circumpolar populations.

Weight for height

In a 'developed', urban society, a fair idea of the prevalence of obesity can be formed by examining the population distribution of weight for height, or of body mass relative to the actuarial ideal (Metropolitan Life, 1983; Society of Actuaries, 1959). A few individuals will be described as overweight because they have a good muscular development rather than an excessive accumulation of body fat, but these misclassifications will be offset by a similar number of individuals with less than average muscularity who are wrongly classified in the opposite sense. Such indices are thus acceptable epidemiological tools; the most popular is the body mass index (body mass in kg/[height in m]^2).

Weight for height analyses have been less helpful when assessing traditional circumpolar populations. A large proportion of the indigenous arctic residents have had a well-developed musculature. Standing height has also been an unreliable reference point for some of these people because of a short limb length (although currently, the ratio of sitting height to standing height is not very different from that for city-dwellers). Application of urban body mass index standards (BMI) to the populations of Eskimo Point, Pelly Bay, Frobisher Bay and Coppermine (all in the NWT) led Nutrition Canada (1975) to conclude that a high proportion of the Inuit tested were obese and at a high risk of developing ischemic heart disease. Other data on the body mass index of circumpolar residents are summarized in Tables 4.3 and 4.4.

Shephard & Rode (1973) noted that in the Igloolik Inuit, a substantial excess of body mass (8–10 kg over actuarial norms in adult males) was associated with extremely low skinfold readings. They thus concluded that

Table 4.3. *Body mass index of selected circumpolar populations*

Locale and year	Body mass index (kg/m²)	Sample size	Author
Alaska			
Point Barrow, 1885	25	51M	Newman (1960)
Aleutian Islands, 1951	25	28M	Newman (1960)
St. Lawrence Island, 1950	23	27M	Newman (1960)
Alaska, 1952	23	17M	Newman (1960)
Northern Alaska, 1958	23	82M	Mann *et al.* (1962)
Canada			
Coppermine, 1901, 1923	28	20M	Newman (1960)
Chesterfield Inlet, 1939	25	30M	Newman (1960)
NWT, various sites, 1964–70	25	844M	Schaefer (1977)
	23–27	539F	Schaefer (1977)
Eskimo Point, Pelly Bay, Iqualuit,	26	102M	Nutrition Canada (1975)
Coppermine	23–27	96F	Nutrition Canada (1975)
Keewatin Inuit, 8 sites, 1990/91			
Males <35 years	24.7	199M	Young (1994)
Males 35–49 years	27.6		
Males >50 years	26.7		
Females <35 years	25.1	235F	
Females 35–49 years	28.5		
Females >50 years	28.6		
Cree-Ojibway Amerindians		704M	Young & Sevenhuysen (1989)
Males <35 years	25		
Males 35–50 years	28		
Males >50 years	28		
Females <35 years	27		
Females 35–50 years	30		
Females >50 years	30		
Chukchi, 4 sites, 1991		170M	Young (1994)
Males <35 years	24	192F	
Males 35–50 years	25		
Males >50 years	24		
Females <35 years	28		
Females 35–50 years	26		
Females >50 years	25		
Greenland			
SW Greenland, 1949	24	121M	Newman (1960)
E. Greenland, 1954	24	162M	Newman (1960)
E. Greenland, 1958	24	133M	Newman (1960)

M = male, F = female.

Table 4.4. *Body mass index (BMI, kg/m^2) and triceps skinfold thicknesses (T, mm) or average of three summed skinfolds (S, mm)*

Age (years)	NW Alaska		Foxe Basin		Lapps		Igloolik 1970		Igloolik 1980		Igloolik 1990		Volochanka	
	BMI	T	BMI	T	BMI	T	BMI	S	BMI	S	BMI	S	BMI	S
Males														
20–29	25.5	8	24.4	5	22.3	7	24.4	5.5	23.8	7.1	24.1	10.4	22.8	7.0
30–39	25.6	8	25.1	5	23.9	9	24.9	6.3	25.8	8.4	24.3	10.9	23.6	7.6
40–49	27.5	9	25.6	5	23.4	8	25.3	5.4	26.9	10.1	28.1	15.6	23.4	7.3
50–59	27.2	8	23.6	6	24.2	10	25.8	7.9	26.4	8.6	27.1	15.7		
Females														
20–29	24.8	16	23.6	8	23.1	15	23.2	8.5	23.1	12.0	23.3	15.1	26.3	15.5
30–39	26.4	20	24.6	11	24.9	17	23.9	9.2	25.4	13.5	25.0	21.4	28.3	24.8
40–49	29.2	22	24.5	10	26.4	19	23.7	7.0	27.9	16.4	29.0	28.5	29.8	25.0
50–59	28.6	20	25.6	15	27.0	20	27.5	19.0	24.0	11.2	31.7	34.8		

Comparison of data of Auger *et al.* (1980) for NW Alaskan Inuit, Foxe Basin Inuit and Lapps, collected in 1970, and of Rode & Shephard (1992a, 1994a), collected on Inuit at Igloolik in 1969/70, 1979/80 and 1989/90, and on nGanasan at Volochanka in 1991.

muscularity and/or a short stature rather than obesity was responsible for the excess body weights seen among traditional Inuit.

The data from the three Igloolik surveys have shown a tendency for BMI to increase with acculturation among older adults of both sexes. But in younger individuals, the BMI has either remained constant or even decreased, despite substantial increases in skinfold readings (Table 4.4). This suggests that the younger segment of the Igloolik population have less lean tissue and more body fat than their older peers. The BMI of male subjects from Igloolik is now substantially higher than values for age-matched nGanasan men of Volochanka, with associated higher skinfold readings in the Inuit group. However, in the women all three age-comparisons show similar BMIs and skinfold readings in Siberia and the Canadian arctic.

In 1989, Leonard *et al.* (1994a) and Katzmarzyk *et al.* (1994) studied 47 Siberians with an average age of around 33 years. They were drawn mainly from two Evenki villages, but a few representatives from one Keto village were added in 1989. The BMI of this sample was similar in men and women. At a value of 23 kg/m^2, it did not differ appreciably between the Evenki and the Keto, although the Keto women had somewhat less subcutaneous fat than the Evenki women (respective averages of four skinfolds, 9.5 versus 17.3 mm). The skinfold data for the Evenki (an average of 7.3 mm in men and of 17.3 mm in women) are in good agreement with the observations of Rode & Shephard (1995) for the nGanasan of Volochanka. The BMI values for the male Evenki are also very similar to our figures for the nGanasan, but averages for the Evenki women are considerably lower than our results for the nGanasan. BMIs did not show any great difference between the Evenki who were pastoralists (22.5 ± 2.5 kg/m^2) and those who were living permanently in the settlements (21.9 ± 2.9 kg/m^2), although skinfolds in the brigade-living males (6.8 ± 2.5 mm) were marginally less than in the male villagers (8.0 ± 3.8 mm).

In the Inuit of Igloolik, the 1989/90 survey showed that hunting no longer conferred any advantage in terms of either body mass index or skinfold readings (Table 4.5). However, those who had elected to participate in village sport programmes had substantially lower skinfold readings than their peers who were not so involved (Table 4.5).

Young (1994) and Young & Sevenhuysen (1989) have provided other data on the body mass indices of arctic residents (Table 4.3). A survey of the Keewatin Inuit covered 406 men and 468 women from eight small coastal settlements, mostly on the west side of Hudson's Bay. About two thirds of those sampled were of pure Inuit heritage. A second survey covered arctic and subarctic groups of Cree-Ojibway Amerindians living in six land-based settlements to the south-west of Hudson's Bay. Finally, data on the

Table 4.5. *Influence of hunting activity and of participation in voluntary sports programmes upon body mass index (BMI) and skinfold readings (average of three folds, S, mm) in Igloolik Inuit examined in 1989/90*

| | Men | | | | Women | | | |
| | 17–19 yr | | 20–29 yr | | 17–19 yr | | 20–29 yr | |
	BMI	S	BMI	S	BMI	S	BMI	S
Hunters	24.9	10.7	25.6	13.5				
Non-hunters	23.5	9.5	23.9	10.1				
Sport participants	23.2	8.0	23.5	8.8	23.8	15.0	23.4	13.5
Non-participants	23.7	10.1	24.3	10.8	22.9	17.2	23.3	15.2

Source: Rode & Shephard (1992a).

Chukchi were drawn from four coastal settlements along the Bering Strait. The body mass index increased with acculturation, being lowest in the Chukchi, larger in the Keewatin Inuit, and yet larger in the Cree-Ojibway Amerindians. The body mass index of the male Inuit tended to be lower in those who had only a limited education, little cash income, and spent time on the land (in other words, the least acculturated segment of the community). In contrast, the main determinants of BMI in the women were current smoking ($29.4 \, kg/m^2$ in non-smokers, $25.9 \, kg/m^2$ in current smokers) and knowledge of Inuktittut (fluent $27.3 \, kg/m^2$, not fluent $25.4 \, kg/m^2$).

Ekoé *et al.* (1990) compared various measures of obesity between Inuit and Cree Amerindians in Nouveau Québec, a territory extending from $50°$ to $61°$ N. The Inuit occupied the more northern part of the region, along the shores of Hudson's Bay and Ungava Bay, and the Cree Amerindians lived further south. Data were collected in 1982–1984. Although no specific comment was made regarding the relative acculturation of these two populations, greater contact with 'white' civilization and thus greater obesity might be anticipated in those who were living closer to the large cities of southern Canada. In keeping with these hypotheses, the body mass index was on average significantly lower in the Inuit (males, $25.7 \pm 4.9 \, kg/m^2$; females, $25.9 \pm 5.7 \, kg/m^2$) than in the Cree (males, $26.8 \pm 5.4 \, kg/m^2$, females, $29.8 \pm 6.8 \, kg/m^2$).

Beall & Goldstein (1992) examined a sample of 750 Mongolian pastoral nomads. As in the other studies from the former Soviet Union, the body mass index of the men was relatively low (at the 25th percentile of US norms), but that for the women was close to the US median value.

Skinfold thicknesses

Early data from Alaska suggested that the triceps skinfold thicknesses of young adult Inuit men increased from values of 6.0 mm (observed in the isolated interior settlement of Anaktuvuk Pass in 1963) to 11.0 mm (found in the more acculturated settlement of Wainwright, seven years later, Rennie *et al.*, 1970). In an early study of Lapps (Andersen *et al.*, 1962), skinfold readings averaged 7.0 mm, but with these two exceptions the values observed in young men from circumpolar populations averaged only 5.5–6.5 mm at the time of the IBP survey.

Cross-sectional data obtained in Igloolik in 1969/70 showed a small but statistically significant gradient of skinfold readings with acculturation, whether the latter was measured in terms of the MacArthur index, or in terms of wage employment (Rode & Shephard, 1973b; Shephard, 1980). Repetition of our observations showed substantial and statistically significant increments of skinfold readings in both men and women from 1969/70 to 1979/80, with further increments in 1989/90 (Table 4.4). Probably as a consequence of the mechanization of hunting, the hunters of 1989/90 no longer had thinner skinfolds than permanent residents of the village, but participation in village sport programmes was associated with a lesser accumulation of subcutaneous fat. A comparison between Igloolik and the less acculturated nGanasan settlement of Volochanka showed that in 1991, the nGanasan men were significantly thinner than the Canadian Inuit. However, perhaps because of greater exposure to television images of thin women, the Inuit females aged 20–29 and 30–39 years were actually thinner than their Siberian peers.

Young (1994) had similar findings. Triceps and subscapular skinfolds were much thinner in the Chukchi (average values 8.5 mm in men, 14 mm in women) than in the more acculturated Keewatin Inuit and Cree-Ojibway Amerindians. In both of the latter populations, values averaged about 16 mm in males, and in women respective values were around 28 and 24 mm). In the male Keewatin Inuit, a limited education, a low cash income and time spent on the land were each associated with 3–4 mm lower skinfold readings, but in female Inuit the only significant finding was that non-smokers had a 6 mm thicker average skinfold than the smokers.

Szathmary & Holt (1983) tested some 30% of the Dogrib Amerindians in 1979. The male members of this population still engaged in caribou hunting in the fall, and set trap lines in the winter. The average thickness of 5 skinfolds was 12.2 mm in the men and 23.5 mm in the women. More recently, Ekoé *et al.* (1990) noted higher triceps skinfolds in Cree Amerindians (11.5 ± 6.3 mm in males, 22.7 ± 10.6 mm in females) than in Inuit who were living further north at isolated coastal settlements in Nouveau Québec (8.5 ± 5.6 mm in males, 15.2 ± 7.4 mm in females).

Murphy *et al.* (1992) also noted a substantial increase in obesity among Alaskan Yu'pik Inuit relative to figures that had been obtained 25 years earlier.

Body fat distribution

There have been suggestions that in terms of an augmentation of cardiovascular risk, the critical issue is not the total accumulation of body fat, but rather the pattern of its distribution. A 'male' patterning, with central fat accumulation, is associated with a high risk of cardiovascular disease, whereas a female, peripheral distribution (hips and thighs) is linked with a much lower risk.

Rode & Shephard (1994b) found rather similar chest/waist circumference ratios for Inuit (measured in 1989/90) and nGanasan (measured in 1991) (Table 4.6). In both populations, ratios were quite high relative to norms for urban society, but this did not seem to have had an adverse effect upon health, perhaps because the absolute amounts of fat found over both the chest and the abdomen were still relatively small. A comparison of skinfold ratios for Igloolik (Table 4.7) indicated that both sexes had a more 'masculine' distribution of fat in 1989/90 than in 1969/70, the trend to a more central fat patterning being shown particularly by the ratio of subscapular to triceps skinfolds.

Young (1994) found somewhat higher (and more 'masculine') waist/hip ratios in the Keewatin Inuit (0.86–0.93, male; 0.82–0.89, female) than in the Chukchi (0.82–0.88, male; 0.82–0.87, female). In both sexes, waist/hip ratios were particularly high in the Cree-Ojibway Amerindians (0.94–0.99, male; 0.90–0.93, female). Since this last group had undergone the greatest 'modernization', Young's data again suggested that acculturation had an influence upon adoption of male fat patterning. Young (1994) fitted multiple regression equations to data for the Keewatin Inuit. In male subjects, the waist/hip ratio was related only to age, but in females it was also negatively related to the level of education.

Beall & Goldstein (1992) found a central (abdominal) pattern of fat distribution in Mongolian nomads, particularly in the female members of this population. They argued that abdominal obesity could be considered a useful adaptation to a cold environment, since it gave a more spherical body shape (with a lower surface/mass ratio). Moreover, abdominal fat was more thermogenic than peripheral fat (Chapter 1.).

Table 4.6. *Chest/waist ratio in Canadian Inuit (measured at Igloolik in 1989/90) and Siberian nGanasan (measured at Volochanka in 1991) (mean ± SD)*

Age group (years)	Males		Females	
	Inuit	nGanasan	Inuit	nGanasan
18–29	1.19 ± 0.05	1.21 ± 0.05	1.15 ± 0.05	1.14 ± 0.05
30–39	1.18 ± 0.07	1.17 ± 0.05	1.12 ± 0.06	1.06 ± 0.07
40–49	1.15 ± 0.06	1.16 ± 0.06	1.09 ± 0.12	1.03 ± 0.07
50–59	1.07 ± 0.06	1.07 ± 0.04	1.05 ± 0.21	1.00

Source: Rode & Shephard (1994b,g).

Table 4.7. *Influence of acculturation to the lifestyle of 'developed' society upon subscapular/suprailiac (SS/SI) and subscapular/triceps (SS/T) skinfold ratios. Data for Igloolik Inuit, obtained in 1969/70 and 1989/90*

Age group (years)	Males				Females			
	SS/SI		SS/T		SS/SI		SS/T	
	1969/70	1989/90	1969/70	1989/90	1969/70	1989/90	1969/70	1989/90
20–29	0.99	0.91	1.39	1.74	0.95	1.25	0.81	1.06
30–39	1.04	0.97	1.48	1.60	1.01	1.09	0.78	1.15
40–49	1.15	0.83	1.36	1.68	1.10	1.15	0.70	1.06
50–59	1.23	0.91	1.49	1.78	0.87	1.06	0.70	1.12

Source: Rode & Shephard (1994a).

Obesity

Young (1994) characterized his subjects as normal (BMI $< 26 \, \text{kg/m}^2$), overweight (BMI $26–30 \, \text{kg/m}^2$) and obese (BMI $> 30 \, \text{kg/m}^2$). Particularly in the men, the proportion with an excessive body mass index was greater in the Inuit and the Amerindians than in the Chukchi (Table 4.8). However, it must be emphasized that the high body mass index could have arisen from muscularity rather than an accumulation of body fat, and that (at least in an earlier generation of Inuit) this was the explanation of the phenomenon (Shephard & Rode, 1973).

Lean body mass

Rode & Shephard (1994b) compared lean tissue mass (as estimated from skinfold thicknesses) between Igloolik Inuit and the nGanasan residents of Volochanka, who seemed to be less acculturated. No differences of lean mass were seen.

Table 4.8. *Cumulative percentages of Chukchi, Keewatin Inuit and Cree-Ojibway Amerindians who are obese* (OB, *body mass index* >30) *or overweight* (OW, *body mass index* 26–30)

Age group (years) and physical characteristics	Male			Female		
	Chukchi	Inuit	Cree	Chukchi	Inuit	Cree
<35 OB	16	24	39	19	27	32
OB + OW	18	32	48	48	38	53
35–49 OB	35	43	44	34	29	34
OB + OW	35	68	68	52	64	85
>50 OB	18	39	46	31	33	37
OB + OW	27	62	73	42	72	84

Values approximated from graphs of Young (1994).

Body fat

There have been few attempts to measure body fat by more direct means than the measurement of skinfold readings. Shephard *et al.* (1973) used deuterated water to estimate the body fat content of Igloolik Inuit (Table 4.9). Their methodology required assumptions about tissue hydration (which may differ between Inuit and 'white' people); however, the data suggested that the total body fat content was greater than would be predicted by applying urban prediction equations to the skinfold data.

More recently, Rode & Shephard (1994d) have applied the underwater weighing technique to a substantial sample of Igloolik Inuit. Again, there are some problems of interpretation, since the density of the lean tissue may differ from that of 'white' people; however, predictions obtained from the skinfold equations of Durnin & Womersley (1974) agreed with the hydrostatic data to within 1–3% of body fat.

Metabolic health

Physiologists are increasingly recognizing the importance of looking at biochemical markers of metabolic as well as cardiovascular health. Items such as the glucose tolerance curve and the blood lipid profile must be considered, in addition to traditional measures of cardiovascular function such as the maximal oxygen intake (Shephard & Bouchard, 1994). Often, the 'modernization' of indigenous populations in other habitats has brought about a much larger deterioration in metabolic health than in cardiovascular function. For example, diabetes mellitus now has a very

Table 4.9. *Body composition of the Igloolik Inuit. Average values for 1969/70, based on the deuterated water dilution technique*

Age (years)	Sex	Number	Body fat (kg)	Body water (kg)	Lean mass (kg)
15–19	M	11	7.9	39.3	53.7
	F	9	10.7	31.8	43.4
20–29	M	12	10.6	43.0	58.7
	F	12	12.0	33.5	45.8
30–39	M	10	8.5	46.0	62.8
	F	10	14.3	31.1	42.1
40–59	F	10	15.3	31.8	43.4

Source: Shephard *et al.* (1973).

high prevalence among the Pima Amerindians (Bogardus *et al.*, 1990; Knowler *et al.*, 1991; Kriska *et al.*, 1993), the Zuni Amerindians (Long, 1978) and the indigenous populations of Hawai (Aluli, 1991), Samoa (Greksa & Baker, 1982) and Micronesia (Zimmet, 1979).

Glucose tolerance

A variety of indices has been used in examining the glucose tolerance of circumpolar populations, and this has hampered comparisons between different ethnic groups. In some instances, hospital or nursing station records have been searched for known cases of diabetes mellitus, although it has been recognised that diagnostic criteria may differ from one region of the world to another, and that such an approach may miss as many as a half of those with a poor glucose tolerance curve. Other studies have looked at fasting blood glucose levels or the concentration of glycosylated haemoglobin, and yet others have noted the prevalence of blood glucose readings in excess of 8.8, 10, or 11.2 mM/l two hours following a standard glucose tolerance test. Young & Krahn (1988) commented that the fasting blood glucose was a simple yet valuable epidemiological tool, and in their study of Northern Amerindians, no additional cases were diagnosed by carrying out a two-hour glucose tolerance curve. Glycosylated haemoglobin was also quite reliable, with a sensitivity of 93.3% and a specificity of 98.1% relative to standard glucose tolerance curves.

Irrespective of habitat (hot, temperate or cold), the Amerindians currently have a high prevalence of non-insulin dependent, ketosis-resistant diabetes. In many of the arctic and subarctic Amerindian settlements, this disorder first became commonplace in the 1960s (Table 4.10). It is speculated that because of the low carbohydrate content of traditional diets, circumpolar populations have lacked well-developed mechanisms for metabolizing simple carbohydrates such as glucose, sucrose and lactose. The change of diet and/or a decrease of habitual physical activity associated with 'modernization' have increased the availability of glucose, and this has revealed what had remained for many centuries a latent inherited metabolic problem. In contrast to the Amerindians, the Chukchi and most groups of Inuit until recently have had a relatively low prevalence of diabetes mellitus. An early study of 10 Ainu men (Kuroshima *et al.*, 1972) also found low resting plasma glucose concentrations; moreover, the peak readings reached during a standard glucose tolerance test were lower than in 40 Japanese control subjects (6.39 ± 0.05 mM/l versus 7.17 ± 0.03 mM/l). Nevertheless, statistics from Alaska show that the prevalence of diabetes mellitus among male Inuit increased from 0.5 per 1000 in 1957 (Scott & Griffith, 1957) to 5.8 per 1000 in 1987 (Murphy *et al.*, 1991).

In addition to between-population comparisons, several within-group analyses have pointed to a strong adverse influence of acculturation in recent years. Young *et al.* (1990) found a steep inverse gradient in the prevalence of diabetes mellitus with latitude (and thus persistence of a traditional lifestyle). In Canadian Amerindians, the prevalence decreased from 1.3% at 55–60° N to 0.5% at 60–65° N, 0.4% at 65–70° N, and 0.2% above 70° N. Among the Canadian Inuit, there was a west to east gradient of diabetes prevalence, rates being highest for the 'modern' western towns of Inuvik and Aklavik, and lowest in the small and remote settlements of the eastern arctic ($r^2 = 0.213$). A classification of Amerindians based on the 'modernity' of their habitat also found diabetes rates of 2.7%, 2.3% and 1.3% for urban, rural and remote locations, respectively. Likewise, Schraer *et al.* (1988) found that in Alaska the prevalence of diabetes mellitus among the Inupiaq Inuit was higher in Point Barrow (1.40%) and Kotebue (1.66%), on the acculturated north shore, than in the more remote west coast settlements around North Sound (0.67%). Among the Yu'pik Inuit, the prevalence of diabetes again increased as a function of contact with 'white' settlers, being greater in the south-west of Alaska (Bristol Bay, 1.04%) than in the remote west coast region of Yukon/Koskokwim (0.58%).

Blood lipid profile

Measurements of blood lipid concentrations are subject to considerable inter-laboratory differences unless the methodology is carefully standardized. Appropriate control procedures were developed for use in the multicentre

Table 4.10. *Prevalence of impaired glucose tolerance and/or diabetes mellitus in circumpolar populations*

Population and year	Age (years)	Sample	Prevalence	Author
Ainu (?1970)	17–44	10M	Less than Japanese controls, rest, 30 min peak glucose	Kuroshima et al. (1972)
Aleuts (?1972)	10–39	96M, 111F	1%(M), 1.9%(F)[1]	Dippe et al. (1976)
	>40	79M, 49F	10%(M), 27%(F)	
Aleuts (1983)	25–44	9800M&F	1.2%[2]	Schraer et al. (1988)
	45–64		5.1%	
	65+		12.4%	
Aleuts (1987)	25–44	9868 M&F	0.8%[2]	Young et al. (1992)
	45–64		6.1%	
	65+		12.0%	
Amerindian				
Alaskan (1987)	25–44	26,671M&F	0.6%[2]	Young et al. (1992)
	45–64		4.7%	
	65+		10.8%	
Athabascan (1969)	>20	152M, 154F	1.3%(M), 2.6%(F)[3]	Mouratoff et al. (1969)
Cree-Ojibway (1983)	25–44	1650M&F	5.3%[2]	Young et al. (1985)
	45–64	820M&F	12.6%	
	65+	360	9.6%	
	>20	309M. 395F	8%	Young (1994)
Dogrib (1979)	47±16(M)	60M, 97F	13.3%(M)[1]	Szathmary & Holt (1983)
	44±16(F)	7.2%(F)		
James Bay (1989)	>20	8840	5.2%[2]	Brassard et al. (1993)
NWT (1987)	25–44	11194	1.1%[2]	Young et al. (1992)
	45–64		1.2%	
	65+		1.6%	
Yukon	25–44	4249	0.5%[2]	Young et al. (1992)

Chukchi				
(1983–85)	>35	222M	4.9%(M)	Stepanova & Shubnikov (1991)
(1987)	25–44	6090	0.1%	Young et al. (1992)
	45–64		0.6%	
	65+		0%	
Inuit				
Alaska (1972)	>20	378M, 327F	0.5%(M), 7.0%(F)[3]	Mouratoff & Scott (1973)
(1972)	>40	153M, 133F	0.7%(M), 6.1%(F)	
(1972)	>20	140M, 180F	5.0%(M), 9.4%(F)[3]	Mouratoff & Scott (1973)
	>40	96M, 96F	5.2%(M), 10.3%(F)	
(1987)	25–44	41640M&F	0.3%[2]	Young et al. (1992)
	45–64		1.9%	
	65+		4.5%	
Greenland (?1964)			<0.1%[2]	Sagild et al. (1966)
NWT (1987)	25–44	18733M&F	0.3%[2]	Young et al. (1992)
	45–64		0.7%	
	65+		1.0%	
Lapps				
Norway (1974)	35–49		0.5%(M), 0.3%(F)[4]	Westlund (1981)

Notes: [1]Blood glucose >11.1 mM/L, 200 mg/dl, 2 h after standard plasma glucose. [2]Clinically diagnosed cases of diabetes mellitus. [3]Blood glucose >8.9 mM/L, 160 mg/dl, 2 h post-glucose loading. [4]Diagnostic criterion not specified.
M = male, F = female.

Table 4.11. *Serum cholesterol and triglyceride levels of circumpolar populations*

Population	Age (years)	Sample	Cholesterol (mM/L)	Triglycerides (mM/L)	Author
Ainu	?	24 (? gender)	3.94		Namiki et al. (1964)
Amerindian	>20	68M, 155F	4.9 ± 1.0(M)	2.3 ± 1.9	Brassard et al. (1993)
James Bay			5.0 ± 1.2(F)	1.8 ± 1.0	
Asahikawa	17–45	17M	4.17	0.85	Itoh (1974)
Niikappu (half-breeds)	22–58	10M	4.35	0.97	
Evenki	33	75M	3.61 ± 0.67	0.98 ± 0.40	Leonard et al. (1994a)
	33	33F	3.83 ± 0.78	0.98 ± 0.45	
Nenets	?	147M	5.01 ± 1.33	1.18 ± 0.61	Tkachev et al. (1991)
nGanasan	18–29	16M	3.26 ± 1.01		Rode & Shephard (1994b)
	30–39	7M	4.12 ± 0.91		
	40–49	4M	4.12 ± 0.54		
	18–29	22F	4.04 ± 0.91		
	30–39	16F	4.27 ± 1.22		
	40–49	8F	4.04 ± 0.78		
	50–59	13F	4.20 ± 1.19		
Inuit					
Baffin Island	14M	3.73			Corcoran & Rabinowitch (1937)
Devon Island	6M	3.42 (non-fasting)			Bang & Dyerberg (1972); Bang et al. (1976)
Greenland	>30	61M, 69F	5.58 ± 0.24	0.43 ± 0.03	
Hudson's Bay	?	7M	3.66		Corcoran & Rabinowitch (1937)
Igloolik	18–29	38M	3.63 ± 0.98		Rode & Shephard (1994b)
	30–39	18M	4.20 ± 0.85		
	40–49	17M	3.86 ± 0.93		
	50–59	17M	3.86 ± 1.50		
	18–29	22F	4.04 ± 1.01		
	30–39	16F	4.27 ± 1.22		

Group	Age	Sample	Value		Reference
Kasigluk (inland SW Alaska)	40–49	8F	4.04 ± 0.78		
	50–59	13F	4.20 ± 1.19		
	18–74	79M&F	6.40 ± 0.93(M)		Draper (1976)
		6.73 ± 1.24(F)			
Keewatin	25–64	274M	4.54 (M, BMI <26)	0.87	Young (1994)
			5.05 (M, BMI 26–30)	0.93	
			5.44 (M, BMI >30)	1.30	
	25–64	279F	5.13 (F, BMI <26)	0.93	
			4.96 (F, BMI 26–30)	1.17	
			5.22 (F, BMI >30)	1.38	
Nunapitchuk	18–74	79M&F	6.40 ± 0.93	0.69	Draper (1976)
Point Hope	6–12	78M, 90F	5.18	0.73	Feldman et al. (1972)
	13–17		5.64	0.97	
	18–24		5.67	0.80	
	25–35		5.80	0.91	
	36–60		5.98		
	61–82		6.48	0.99	
Point Hope	18–74	82M&F	5.49 ± 1.01(M)		Draper (1976)
			5.44 ± 0.98(F)		
Québec	23–38		4.71–6.58		Carrier et al. (1972)
Wainwright	18–74	66M&F	4.84 ± 1.06(M)		Draper (1976)
			5.31 ± 1.25(F)		
Lapps					
Inari, Finland		828M	6.65	1.08	Björkstén et al. (1975)
		F	6.78	1.01	
Finland	35–49		7.67(M)		Westlund (1981)
			7.35(F)		

M = male, F = female.

Canadian/US Lipid Clinics trial of cholesterol-sequestrating drugs, but these controls have only been applied to the more recent circumpolar studies, such as those of Rode *et al.* (1995). A high alcohol intake is a further factor that has complicated the interpretation of blood lipid profiles in many circumpolar communities.

Early data for Inuit living in the Canadian arctic (Corcoran & Rabinowitz, 1937), the US (Draper, 1976, 1980) and Siberia (Leonard *et al.*, 1994a) all showed quite low population averages for serum cholesterol (Table 4.11) with low prevalence rates of hyperlipidemia. The data of Corcoran & Rabinowitch (1937) were apparently collected in the non-fasting state, but they nevertheless documented quite clearly that the Canadian Inuit had extremely low plasma cholesterol concentrations prior to World War II. Even at this period, there was some latitudinal gradient, the lowest readings being observed on Devon Island (76° N), with somewhat higher values being seen among those living on Baffin Island and around Hudson's Bay, to the south.

At the time of the IBP-HAP survey, Maynard (1976) and Draper (1976) both commented on an inverse latitudinal gradient of serum cholesterol among the indigenous populations of Alaska. They found that the prevalence of hypercholesterolemia was 39% in the men and 43% in the women of Kasigluk/Nunapitchuk, but in Wainwright prevalence was only 6.7% in the men and 13.8% in the women, and in Point Hope the figures were 10.3 and 14.0%.

Low plasma cholesterol readings have persisted in many circumpolar communities despite a diet that in some cases contains a very high quantity of meat and fat (Table 4.12). However, it does not seem either possible or necessary to explain this paradox in terms of some dramatic genetic adaptation of cholesterol metabolism. Voevoda *et al.* (1991) studied the genealogy of Chukotka natives, estimating heredity coefficients of only 24% for total serum cholesterol, 37% for HDL cholesterol, and near zero for triglycerides. Feldman *et al.* (1972) compared details of cholesterol metabolism between Point Hope Inuit and 'white' subjects. They found similar cholesterol turnover rates (5.4 versus 6.5 days for the fast pool and 65.1 versus 72.7 days for the slow pool, with a total turnover of 1315 versus 1412 mg per day). Moreover, the absorption of cholesterol was actually more complete in the Inuit (50%) than in 'white' (37%) subjects.

Dietary influences have probably made the most important contribution to optimization of the lipid profile in the indigenous circumpolar groups. The flesh of free-ranging animals such as the caribou has a very low fat content relative to 'market' meats. Leonard *et al.* (1994a) estimated that only 18–19% of energy intake was fat-derived among Evenki reindeer-herders

Table 4.12. *Sources of food energy in traditional diet of Eastern Greenlandic Inuit (based on data of Helms, 1981, for year 1945: average for entire population of Angmagssalik)*

Food source	Intake per day (g)	Energy yield (kJ)
Seal meat	380	2200
Polar bear meat	10	106
Seal blubber	100	3632
Fish	145	412
Birds	10	121
Bread and cereals	104	1425
Sugar, sweets	42	715
Dairy products	1	13
Fats and oils	3	83
Preserved fruit and vegetables	2	24
Total		8730

Source: Helms (1981).

on the Siberian taiga: their average fat intake was no more than 65 g/day in men and 43 g/day in women. Fish and marine mammals, which form a large part of the diet of traditional coastal Inuit (Eaton & Konner, 1985; Rode & Shephard, 1994b) are also rich in polyunsaturated as opposed to saturated fatty acids (Bang & Dyerberg, 1980; Bang et al., 1976; Connor & Connor, 1990). Adoption of a fish oil or fish diet is well recognized as one method of reducing plasma triglyceride concentrations, and (if eicosapentanoic acid outweighs docosohexanoic acid) it also increases plasma concentrations of the desired HDL-2 form of cholesterol (Blonk et al., 1990; Bonaa et al., 1992; Childs et al., 1990; Cobiac et al., 1991; DeLany et al., 1990; Kestin et al., 1990). Simopoulos (1991) has argued that humans evolved an adaptation to an omega-6/omega-3 fatty acid ratio of around 1.0. Our data for Igloolik suggest that the older, traditional coastal Inuit remain close to this standard. The younger Inuit and the nGanasan of the Siberian taiga are already far from the ideal ratio (Table 4.13). Older, traditional Inuit have approximately a 2:1 balance of 20:5/22:6 omega-3 fatty acids, again a ratio that nutritionists regard as having a favourable influence upon both the lipid profile (Childs et al., 1990) and platelet aggregability (Dyerberg et al., 1978; Hansen et al., 1994).

Acculturation is associated with a progressive increase in cholesterol and triglyceride levels (Bang & Dyerberg, 1972; Draper 1976, 1980; Feldman et al., 1975), probably through the combined influence of a decreased

Table 4.13. *Quantities of total saturated fatty acids omega-3 and omega-6 fatty acids in plasma: a comparison of data for Inuit of Igloolik (I) with nGanasan of Volochanka (V). Mean values, ±SD, shown as percentage of total fatty acids*

Age (years)	SFA		ω-3		ω-6		ω-6/ω-3	
	I	V	I	V	I	V	I	V
Males								
18–29	29.4 ± 2.6	33.6 ± 3.3	6.0 ± 3.2	8.8 ± 2.2	31.4 ± 5.6	35.9 ± 6.5	6.7 ± 3.2	4.4 ± 1.6
30–39	31.0 ± 2.4	33.3 ± 2.5	9.6 ± 4.5	9.5 ± 1.7	28.2 ± 4.9	36.1 ± 6.3	3.7 ± 2.0	3.9 ± 0.7
40–49	30.5 ± 3.2	34.7 ± 2.9	15.5 ± 6.4	10.1 ± 3.3	22.6 ± 5.2	32.7 ± 2.6	2.4 ± 3.0	3.5 ± 1.0
50–59	30.1 ± 4.6	32.0 ± 3.0	16.4 ± 4.7	11.4 ± 1.9	20.2 ± 5.0	37.9 ± 4.6	1.3 ± 0.5	3.4 ± 0.9
Females								
18–29	29.9 ± 2.4	31.4 ± 2.9	5.5 ± 2.9	9.2 ± 3.5	32.1 ± 4.9	39.7 ± 7.1	7.6 ± 3.8	4.9 ± 1.9
30–39	31.5 ± 2.4	31.4 ± 1.3	9.6 ± 4.4	6.6 ± 1.0	28.3 ± 4.8	33.4 ± 3.9	3.7 ± 2.1	5.1 ± 0.2
40–49	32.3 ± 1.6	33.9 ± 1.2	13.9 ± 4.8	9.6 ± 1.9	24.0 ± 5.8	30.9 ± 4.6	2.0 ± 1.1	3.2 ± 0.1
50–59	31.4 ± 2.3		15.9 ± 4.9		21.5 ± 3.0		1.6 ± 1.0	

Source: Based on data of Rode & Shephard (1994g) collected in 1991.
For further details of data, see original report.
SFA = saturated fatty acids. ω-3 = omega-3 fatty acids; ω-6 = omega-6 fatty acids; ω-6/ω-3 = ratio of omega-6 to omega-3 fatty acids.

Table 4.14. *Coefficients of correlation between omega fatty acid levels and other demographic and fitness-related variables. Data for Igloolik Inuit, tested in 1991*

Variable	ω-3	ω-6	ω-(7 + 9)	ω-6/ω-3
Males				
Age	0.69	−0.64	ns	−0.65
Body mass	0.28	−0.22	ns	ns
Total cholesterol	ns	ns	−0.30	ns
Free cholesterol	ns	ns	−0.25	ns
Triglycerides	−0.34	ns	0.26	0.28
PC/FC	−0.36	0.32	ns	0.31
Females				
Age	0.73	−0.71	ns	−0.69
Body mass	ns	ns	ns	ns
Total cholesterol	ns	ns	ns	ns
Free cholesterol	ns	ns	ns	ns
Triglycerides	ns	ns	ns	ns
PC/FC	−0.60	0.55	ns	0.6

Source: Rode & Shephard (1994g).
PC/FC = phosphocreatine/free cholesterol ratio.

consumption of omega-3 fatty acids and a reduction of habitual activity (Rode *et al.*, 1995). In support of the first hypothesis, Rode *et al.* (1995) noted that in Igloolik, omega-3 fatty acid levels were much higher in old than in younger adults (Table 4.12). The main age gradients were for the 20:5 ω-3 and 22:6 ω-3 components, derived primarily from marine foods (Bang *et al.*, 1976), and (in the opposite sense) for 18:2 ω-6 components, derived from vegetable oils. The consumption of polyunsaturated vegetable oils such as sunflower oil may have an adverse effect on health, since this dietary practice seems to encourage the production of reactive species of oxygen, particularly in active individuals (Huertas *et al.*, 1994).

Correlation matrices suggest that in younger subjects triglyceride levels have risen as omega-3 fatty acid levels have fallen (Table 4.14). It seems most reasonable to attribute these age gradients to the fact that younger individuals have become more acculturated, and are eating more 'market' food. It may also be that they eat less raw fish than their older peers, preferring to fry foods in cooking oils. Such cooking greatly reduces the omega-3 fatty acid content of marine foods. Finally, since the data is cross-sectional in type, it remains conceivable (although not very probable) that younger Inuit will revert to a traditional 'country' diet rich in omega-3 fatty acids as they become older. In marked contrast with the Inuit, the

Table 4.15. *Plasma concentrations (mg/dl, ±SD) of 20:5 and 22:6 omega-3 fatty acids in plasma in Igloolik Inuit (I) and Volochanka nGanasan (V)*

Age (years)	20:5 n-3 fatty acids		22:6 n-3 fatty acids	
	I	V	I	V
Males				
18–29	2.0 ± 2.1	2.4 ± 1.2	2.4 ± 1.0	4.0 ± 0.9
30–39	4.2 ± 3.5	3.0 ± 1.0	3.4 ± 1.0	4.0 ± 0.9
40–49	8.7 ± 4.9	3.5 ± 1.7	4.7 ± 1.3	4.1 ± 1.1
50–59	9.5 ± 3.5	3.8 ± 0.4	4.7 ± 1.7	5.1 ± 0.8
Females				
19–29	1.8 ± 1.8	2.7 ± 2.2	2.3 ± 0.9	4.4 ± 1.3
30–39	4.1 ± 3.0	1.2 ± 0.3	3.6 ± 1.1	3.1 ± 0.4
40–49	6.6 ± 2.8	3.0 ± 0.5	5.1 ± 1.6	4.3 ± 0.6
50–59	9.1 ± 3.9		4.5 ± 1.0	

Source: Rode & Shephard (1994g).
Data collected in 1991. See original paper for details of other plasma fatty acids.

nGanasan show no age-related gradients of omega-3 fatty acids, and even the oldest nGanasan have lower plasma levels of omega-3 fatty acids than the traditional Inuit. The nGanasan also have a less favourable ratio of 20:5/22:6 omega-3 fatty acids than their Inuit counterparts (Table 4.15). Nevertheless, the omega-6/omega-3 ratio of the tundra residents (3–5) remains far below the values of 10–25 that are commonly seen in urban North America (Simopoulos, 1991).

A substantial fraction of the body's cholesterol pool is synthesized in the liver. The rate of synthesis is increased when food intake exceeds energy expenditures. Another factor that may have contributed to a favourable lipid profile among arctic residents is thus the achievement of a balance between food intake and energy expenditure. It is less easy to over-eat a high fat than a high sugar diet, and the high level of energy expenditure demanded by traditional hunting activities probably contributed to their energy balance (Godin & Shephard, 1973a).

In their study of the Evenki, Leonard *et al.* (1994b) found that the more acculturated villagers tended to be fatter and to have higher serum lipid levels than the active reindeer herders. In contrast with the general experience of 'developed' societies, Leonard *et al.* (1994a) also found that the HDL cholesterol level of the Evenki women was no higher than that of their male partners. This finding, again, could be related to the very large differences in daily energy expenditures between Evenki men and women.

Other metabolic characteristics

Other metabolic characteristics of the circumpolar peoples reflect a lack of milk and sugar in their traditional diet: a lactose intolerance (seen in Lapps, Sahi *et al.*, 1976; Greenlandic Inuit, McNair *et al.*, 1972, and Alaskan Inuit, Bell *et al.*, 1973, Duncan & Scott, 1972), and a sucrose intolerance (seen in Greenlandic Inuit, McNair *et al.*, 1972, and northern Alaskan Inuit, Bell *et al.*, 1973, but not in south western Alaska, where substantial harvests of berries have always been available). It is unclear how far these metabolic defects have been changed by acculturation to a 'market' diet.

A low bone calcium has been found among Inuit populations in both Alaska (Mazess & Mather, 1974) and Canada (Mazess & Mather, 1975). Examination of archaeological specimens suggests that the deficiency in bone calcium has been characteristic of arctic residents for many centuries. Comparisons of Yu'pik with Inupiaq Inuit suggest that abnormalities are less prevalent in the Yu'pik of Kodiak Island and St. Lawrence Island than in the Inupiaq Inuit of more northerly settlements, including Point Hope, Baffin Island, Southampton Island and Greenland, where the traditional 'country' foods have provided only limited amounts of calcium. Recent nutritional data (Kuhnlein, 1991) suggest that an increased intake of calcium may be one of the few tangible benefit of the transition from 'country' to 'market' foods.

Conclusions

A combination of a high level of physical activity, animal flesh with a low fat content, and (particularly in the case of coastal residents) a high intake of polyunsaturated omega-3 fatty acids has protected the hunter–gatherers of the circumpolar regions against ischemic heart disease. The limited body fat stores of the indigenous populations has also been associated with a low prevalence of diabetes mellitus, although a lack of milk and sugar in the traditional diet has allowed the development of a latent, inherited lactose and sucrose intolerance in a number of circumpolar groups.

Acculturation has been associated with a progressive shift from 'country' to 'market' food sources, and because of a limited understanding of nutritional principles, indigenous populations have sometimes spent their limited food budgets unwisely. Particularly at higher latitudes, the quantity of 'country' food currently obtained is still sufficient to meet body protein needs. However, a decrease of habitual physical activity and a greatly increased intake of refined carbohydrates have led to a growing prevalence of obesity, hyperglycemia and hypercholesterolemia. The resulting risk of ischemic heart disease has been further exacerbated by a decreased consumption of marine foods, and thus a reduced intake of omega-3 fatty acids. As with the 'white' residents of large cities, the current generation of indigenous circumpolar people needs to take more voluntary exercise, and to reduce their intake of sugar and saturated fats if they are to avoid the chronic 'diseases of civilization'.

5 Secular trends in physical fitness and cold tolerance

This chapter begins by describing certain features of fitness patterns that were observed among indigenous circumpolar residents at the time of the IBP-HAP surveys, particularly seasonal effects and inter- and intra-community differences related to the early phases of acculturation to a sedentary, urban lifestyle. Information on secular trends in aerobic power, anaerobic power and capacity, lean tissue mass and muscle strength is then drawn from communities where cross-sectional surveys have been repeated or longitudinal analyses undertaken. Changes in cold tolerance are considered in similar fashion, and finally the possibility of developing community programmes to conserve physical fitness is explored in the context of indigenous societies.

Lifestyle, season and fitness: the IBP-HAP findings

Fitness levels

Early information on the physical fitness of circumpolar populations, obtained mainly by non-standard methodology, was summarized in Chapter 1. Despite limitations in the available data, the conclusion was drawn that in the 1950s and 1960s, the small samples of people that were tested had a higher fitness level than would have been expected in sedentary city dwellers of similar age. However, it was unclear whether those examined were representative of their communities, and it was also uncertain how much the individual fitness levels may have been biased downward by such factors as malnutrition and the ravages of chronic disease, particularly tuberculosis (Andersen & Hart, 1963).

Aerobic power

The IBP-HAP data on aerobic power is summarized in Table 5.1 and Fig. 5.1. At most of the circumpolar test sites, direct measurements of maximal

Table 5.1. *Direct measurements of aerobic power on circumpolar populations, as observed by IBP-HAP surveys and related studies from the late 1960s*

Population	Age (years)	Sex and number tested	Heart rate (per min.)	Gas exchange ratio	Ventilation (l/min.BTPS)	Maximal oxygen intake (l/min. STPD)	Maximal oxygen intake (ml/kg·min. STPD)
Hokkaido: Ainu							
Cycle ergometer	16–19	M(5)	196 ± 5	1.18 ± 0.10	95.8 ± 6.6	2.72 ± 0.39	46.5 ± 5.3
	20–29	M(3)	189 ± 8	1.23 ± 0.10	85.6 ± 23.1	2.32 ± 0.40	44.4 ± 2.3
	30–39	M(6)	189 ± 11	1.20 ± 0.13	81.3 ± 10.2	2.34 ± 0.34	39.5 ± 7.4
Igloolik Inuit							
Step Good maximum	23 ± 8	M(12)	185 ± 13	1.13 ± 0.03	—	3.52 ± 0.56	52.3 ± 3.6
Fair maximum	27 ± 13	M(14)	178 ± 15	1.04 ± 0.03	—	3.03 ± 0.83	48.9 ± 5.3
Poor maximum	24 ± 8	M(7)	176 ± 3	0.97 ± 0.03	—	3.27 ± 0.72	54.1 ± 3.0
Fair maximum	14 ± 2	F(5)	182 ± 16	1.05 ± 0.03	—	1.95 ± 0.30	39.2 ± 7.0
Nellim Lapps							
Cycle ergometer	15–19	M	—	—	—	~2.8	52
	20–29	M	—	—	—	2.88	~49
	10–14	F	—	—	—	~1.4	35
	20–29	F	—	—	—	1.76	~33
Kautokeino Lapps							
Cycle ergometer	10	M ⎱(49)	197	—	—	~1.4	~53
	18–30	M ⎰	197	—	102.8	~3.5	~53
	50	M	181	—	—	~2.7	~45
skiing	15–38	M(5)	194	—	101.3	2.68	57.4

Point Hope Inuit[a]

Treadmill						
11–18	M(14)	180	—	94.8	2.41	44.9
24	M(2)	186	—	99.8	2.75	42.3
35	M(2)	181	—	102.1	3.07	41.6
13–18	F(10)	189	—	69.3	1.86	33.5
24.7	F(3)	182	—	73.5	1.95	32.0

Upernavik Inuit

Cycle ergometer						
14–19	M(4)	186	0.96	—	2.36	40.1
20–29	M(3)	191	1.04	—	2.75	40.7
30–45	M(4)	174	1.04	—	2.35	37.2

[a] Data selected from larger sample on basis of terminal heart rate.
Source: Data accumulated by Shephard (1980).

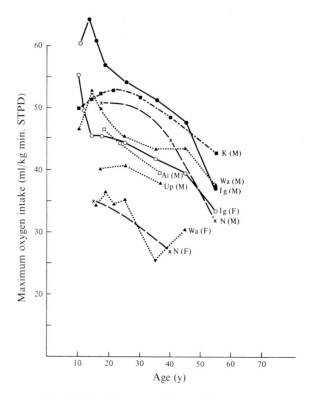

Fig. 5.1. Aerobic power of circumpolar populations, as observed by IBP–HAP surveys and related studies during the late 1960s. Direct measurements using cycle ergometer on Ainu (Ai), Upernavik Inuit (Up) and Kautokeino (K) and Nellim (N) Lapps. Linear extrapolation of step test data on Wainwright Inuit (Wa). Åstrand nomogram applied to step test data and corrected downward by 8% on Igloolik Inuit (Ig). M = male; F = female. Data accumulated by Shephard (1980).

oxygen intake were only obtained on small samples of subjects, with younger adults preponderating.

Several investigating teams chose cycle ergometry as the exercise test mode, although bicycles were not commonly available in circumpolar households of the late 1960s. In consequence, the subjects' quadriceps muscles were not well-prepared for all-out cycling, and tests were commonly halted by local leg exhaustion rather than a central limitation of cardio-respiratory function. For example, in supposed maximum testing of the Nellim Lapps, the peak heart rates averaged only 180 beats/min. and the peak blood lactate readings averaged only 7 mM/l (Karlsson, 1970). In contrast, when more acculturated Ainu aged 16–19 years performed a maximal test, they reached an average heart rate of 196 beats/min., with a respiratory gas exchange ratio of 1.18 (Ikai *et al.*, 1971). Likewise, Kautokeino Lapps aged 18–30 years had average heart rate readings of 197 beats/min. during maximal cycle ergometer exercise (Andersen, 1969; O. Wilson, unpublished data). Comparable peak figures for young 'white' adults living in an urban society would be a heart rate of 195 beats/min. and a plasma lactate concentration of 10–12 mM/l (Shephard, 1982). Andersen & Hart (1963) admitted that the peak oxygen intakes that they were reporting for the Arctic Indians of Old Crow, in the Yukon Territory, could have under-estimated the true maxima for this population by as much as 20%.

In some early studies, the peak oxygen transport of circumpolar populations was further limited by clinical anemia (Brown, 1954; Sellers *et al.*, 1959), but more recent studies of Inuit at Bathurst Inlet (Davies & Hanson, 1965), Wainwright (Sauberlich *et al.*, 1970) and Igloolik (Rode & Shephard, 1971, 1973d; Sayed *et al.*, 1976a) have all shown very normal haemoglobin readings.

Observations on the exercise tolerance of Inuit at Point Hope, Alaska, were made on a treadmill rather than a cycle ergometer (Rennie *et al.*, 1970), but again it was necessary to estimate the true maximal oxygen intake of the population by arbitrarily selecting subjects with adequate terminal heart rates from among a larger sample that had made poor maximal efforts. Even after taking this precaution, the maximal heart rate of young adults averaged only 186 beats/min. Shephard (1974b) carried out direct maximal testing of a sub-sample of the Igloolik Inuit, using a 45.7 cm (18 inch) stepping bench that subjects ascended at a progressively increasing rate. Again, heart rates peaked at 185 beats/min. in those who were judged to have made a 'good' maximal effort, and at only 176 beats/min. in those whose effort was rated as 'fair'.

Some of the highest directly measured maximal oxygen intakes were observed by Andersen *et al.* (1962). They tested Kautokeino Lapps by a mode of exercise that was very familiar to that population, and involved use of the arms as well as the legs (Nordic skiing). Since much of the traditional physical work in the arctic involves use of the arm rather than the leg muscles, this may be important to a successful test. The Kautokeino community seems to have been very well-motivated. Young adults reached an average heart rate of 197 beats/min., even when maximal exercise was performed on a cycle ergometer. Rennie *et al.* (1970) also tested responses to arm work on Inuit at Wainwright, Alaska. However, in this community, heart rates were quite high during submaximal exercise, and the predicted peak oxygen intake for arm work was only about 2 l/min.

Having due regard to the various technical problems of measuring maximal oxygen intake in circumpolar groups, there seems to be a fair agreement between the directly measured results obtained on small samples and the findings from much larger populations that have been evaluated using submaximal prediction procedures (Fig. 5.1). In the Igloolik Inuit, both the discrepancy between direct and indirect measurements (typically, an 8–9% over-estimate by the indirect method) and the variance of maximal oxygen intake predictions (an average of 12% within a given age decade) were much as anticipated in a 'white' community.

In some settlements such as Igloolik, where the process of rapid acculturation to an urban lifestyle began relatively recently, the overall conclusions drawn from the IBP-HAP survey were similar to those based upon earlier examinations of circumpolar groups, namely that the aerobic fitness at any given age was substantially greater than would have been expected in a sedentary urban community. Such findings were associated with large lung volumes (10% above prediction in males, 18% above prediction in females, Rennie *et al.*, 1970; Rode & Shephard, 1973c) and large blood volumes (102–115 ml/kg, Brown, 1954; 95 ml/kg, Shephard, 1980), but a normal radiographic heart volume, a peak stroke volume that averaged only 126 ml in young men and 78 ml in young women, a normal diffusing capacity at rest and during exercise, and a normal arterio-venous oxygen difference of 133–151 ml/l (Rode & Shephard, 1973d; Shephard, 1980). In contrast, levels of aerobic power were relatively low in the more acculturated Alaskan settlements such as Point Barrow and Wainwright.

Gender-related gradients of aerobic power were at least as large in the circumpolar peoples as in urban populations, and the overall rate of ageing of aerobic fitness as deduced from cross-sectional curves was similar to the decrease of 5 ml/[kg·min.] per decade that has been described in sedentary city-dwellers (Shephard, 1987). However, a more detailed analysis of findings in the large sample of Inuit tested at Igloolik suggested that (in contrast to southern Canadians) the aerobic fitness of both sexes was well-maintained until around the age of 50 years, with a steep decline of condition in the oldest members of the community. It seemed unlikely that the oldest members of the community were more acculturated than those who were younger. Possibly, they may have faced more disease and more periods of malnutrition than younger cohorts. However, in the case of the men, the main explanation seemed that when a hunter became a grandparent, he was no longer expected to participate in hunting expeditions, but was nevertheless given the first choice of game by his sons (Rode & Shephard, 1971; Shephard, 1974b). The habitual physical activity of the traditional Inuit thus decreased suddenly around the age of 50 years. The oldest group of women had also ceased travelling to distant hunting camps, where household equipment was not available and much hard physical work was required. There was a sharp post-menopausal increase in body fat, so that in the women, the age-related decrease in relative aerobic power (ml/[kg·min.] was greater than the decrease in absolute values (l/min.).

Measurements of oxygen consumption during submaximal stepping allowed a calculation of mechanical efficiency for the Igloolik Inuit. The observed values were less than we had found earlier in young 'white' adults (a net efficiency of 14–15% rather than 16%), and we suggested that this might reflect the limited number of staircases in the Igloolik community (Rode & Shephard, 1971). Bicycles did not reach Igloolik until the late 1970s, and during the tests of 1969/70 the net mechanical efficiency when operating a cycle ergometer (an average of 18.4% in males and 18.1% in females) was also lower than the 23% that would have been anticipated in Southern Canada. Likewise, Aidaraliyev & Maximov (1991) commented that the mechanical efficiency of Europeans at a cycle ergometer loading of 150 Watts was 29% higher than that of Chukchi and Yu'pik Inuit. A similar phenomenon has been noted in other isolated indigenous populations, Andersen *et al.*, 1963; Rode & Shephard, 1973c). In keeping with our hypothesis of an association between mechanical efficiency and acculturation to an urban lifestyle, the Ainu showed net cycle ergometer efficiencies of 19–20% (Shephard, 1980), and values for the Kautokeino Lapps were similar to those seen in a 'white' community (Andersen *et al.*, 1962).

Anaerobic power and capacity

The only data on the anaerobic power of circumpolar people have been collected on male Inuit at Wainwright (Rennie *et al.*, 1970). Subjects performed the Margaria staircase sprint (Table 5.2). Although there were few staircases in Wainwright, scores for the Inuit of this settlement were some 20% greater than would have been expected in Italian adults, this advantage persisting over most of the adult lifespan.

Rennie *et al.* (1970) attempted to measure the anaerobic capacity of Alaskan Inuit using simple field tests of running performance, but scores were compromised by clumsiness in running and a lack of competitive sense in the local culture. Low peak lactate values were also seen during maximal testing of the Nellim Lapps, as noted above. However, Rode & Shephard (1971) observed peak respiratory gas exchange ratios (R) of 1.14 ± 0.04 in Igloolik Inuit who made good maximal efforts, and in 20–29 year old Ainu the peak R values averaged 1.23 (Ikai *et al.*, 1971). During submaximal exercise, Rode & Shephard (1971) noted very normal ventilatory equivalent values, some 25 l/l in the men, and 28 l/l in the women. There was little evidence that disproportionate hyperventilation developed even at high

Table 5.2. *Anaerobic power (staircase sprint) grip strength and leg extension force of circumpolar peoples. Mean ± SD, based on data collected in the late 1960s*

| Age (years) | Wainwright anaerobic power | | Igloolik | | | |
| | m/s M | m/[kg·s] M | Grip strength (N) | | Leg extension (N) | |
			M	F	M	F
20–29	1.67 ± 0.20	24.2 ± 2.9	510 ± 48	292 ± 39	866 ± 171	676 ± 175
30–39	1.49 ± 0.13	21.3 ± 2.6	452 ± 79	279 ± 48	872 ± 236	656 ± 162
40–49	1.22 ± 0.31	18.6 ± 5.4	416 ± 49	284	750 ± 153	668
50–59	1.17	17.0	379	268 ± 20	754	682 ± 26
60–69	0.99	14.8				

Source: Shephard (1980).

work-rates. The only other information on anaerobic capacity is the comment of Ikai *et al.* (1971) that the oxygen consumption of the Ainu, examined over the first five minutes of recovery from an exercise bout at 75% of aerobic power was only three quarters of that seen in an urban comparison group. However, their finding seems due as much to a low body mass (an average of 57 kg in the men) as to any reduction in the initial size of increase in the rate of repayment of the oxygen deficit per unit of muscle tissue.

Muscle strength and endurance

Differences of body mass and leverage limit the possibility of making direct comparisons of strength measurements between circumpolar and urban populations. The IBP-HAP findings for the grip strength of the male Igloolik Inuit (Table 5.2) did not appear to be remarkable (Rode & Shephard, 1971), and values for the women of that community were quite poor. Villagers have had relatively little work for their hands while driving a dog team. They may have gained some training effect from the handling of dog leashes, and the owner of a dog team commonly had to untangle the dog traces several times per day, but the latter task demanded cold tolerance rather than physical strength.

The main muscular tasks of the traditional lifestyle involved large muscle activity: the construction of sleds and harpoons, the repairing of boats and motors, the butchering of meat, the building of caches, and the making of skin ropes. Our data showed that the peak leg extension force of the males in Igloolik was somewhat greater than that of 'white' immigrants who were tested on the same apparatus, and the leg strength observed in the female Inuit would be regarded as unusually high in an urban society. A small part of this advantage may have been due to the effect of a shorter limb length upon leverage and thus cable tensiometer readings, but there does also appear to have been a real difference in strength relative to city-dwellers; this view is supported by a large weight-for-height ratio despite a low body fat content. Measurements based upon the dilution of deuterated water (Shephard *et al.*, 1973) indicated that the Igloolik Inuit of 1969/70 had a relatively large lean body mass per unit of standing height. For example, a figure of 354 g/cm was observed in 20–29-year-old men, compared with the figure of about 300 g/cm that would be expected in urban 'white' men of similar age.

In 'developed' communities, women commonly have a 30–40% deficit of leg strength relative to their male peers, as compared with the 10–20% difference that was seen at Igloolik in the IBP-HAP survey of 1969/70. The age-related loss of muscle strength inferred from the 1969/70 cross-sectional survey was relatively normal in the case of male Inuit (a 4–5% decrease of strength per decade), but in the women there was little decrease of strength between 20–29 and 50–59 years of age.

Influence of season

Biological investigators have shown an understandable reluctance to visit the arctic during the winter months, and most fitness data have been collected during the summer. However, the IBP-HAP survey included seasonal comparisons of summer and winter data for the Ainu (Ikai *et al.*, 1971) and the Igloolik Inuit (Rode & Shephard, 1973c).

Despite substantial seasonal differences in hunting patterns, differences in conventional measures of fitness between the spring/summer (May to August) and winter (January to March) periods were remarkably small at Igloolik. Perhaps because of the increased muscular work involved in walking over snow and ice, the knee extension strength was 5–6% greater in winter than in the summer months. This difference was seen not only in hunters, but also among those who were living more permanently in the settlement. The difference approached statistical significance for those aged 20–29 years, and for the subjects as a whole. However, aerobic power was closely similar for summer and winter seasons (Table 3.1). It is well known that any training-induced increase of aerobic fitness is lost over a few weeks of physical inactivity (Shephard, 1977). The fitness data thus support the activity analyses in suggesting that periods of vigorous aerobic activity continued throughout the year.

Skinfold readings tended to be greater in the winter (Table 3.1). Seasonal differences were statistically significant in those aged 20–29 years ($p < 0.01$), in those aged 30–39 years ($p < 0.01$), and in the population as a whole ($p < 0.001$). The winter increase of skinfolds was greatest and most consistently observed in regular hunters. The average increment in the three-fold total amounted to $+4.0$ mm, or 1.3 mm per fold ($p < 0.001$) for this sub-group of the population. One possible explanation would be that the scope of their habitual activities was greatly curtailed by the extreme weather conditions of winter, but this view is not supported by either observation of activity patterns or determinations of aerobic power. In fact, much of their hunting was undertaken in the winter months, when both the ocean and the tundra were solidly frozen. Moreover, the skinfolds increased without any substantial gain of body mass. This paradox is difficult to explain in terms of an associated decrease of lean body mass, since muscle strength also increased in the winter months. One intriguing possibility is that cold exposure may have increased tissue hydration (a view supported by Baugh *et al.*, 1958, and Brown *et al.*, 1954, but not by Bass & Henschel, 1956). It is well recognised that an increase of tissue hydration can increase the thickness of the skin itself, thus augmenting skinfold readings by 15% or more (Consolazio *et al.*, 1963), and this change might be sufficient to account for the 1.3 mm average increment in skinfold readings.

Influence of lifestyle

Even at the time of the IBP-HAP survey, there were substantial cross-sectional differences of aerobic power between the various circumpolar communities. These differences apparently reflected the extent of acculturation to an urban lifestyle, although other differences of habitat and observer technique may also have been implicated.

The Igloolik sample of Inuit was large enough to be sub-divided on the basis of individual lifestyle. Traditional hunters were distinguished from those with a transitional lifestyle and those who had become relatively acculturated, accepting permanent residence and wage-earning employment within the settlement (Table 3.1). Based upon the ownership of dog teams and other hunting equipment, we estimated that in 1969/70, about two thirds of the 147 men and boys over the age of 15 years engaged in some hunting activity, and 29 were serious hunters. The latter group made many arduous field trips per year, each lasting from three days to several weeks, and in consequence they were absent from the village about 40% of the time. Both in summer and in winter, there was a substantial cross-sectional gradient of aerobic power and body fatness across the three sub-groups of the community, with the acculturated Inuit being less fit than their traditional and transitional peers.

When observations were repeated in 1979/80 (Rode & Shephard, 1984a) about 44 of the 60 who were still identified as 'hunters' engaged mainly in recreational hunting. Nevertheless, inter-group differences of aerobic power and skinfold readings persisted. The advantage of aerobic power seen in the traditional Inuit, some 10–12%, was certainly no greater than would be anticipated from regular vigorous physical activity, although we could not rule out the possibility that an inherited advantage of fitness had led a sub-set of the villagers to select hunting as a career, and to persist with this endeavour in the face of acculturating influences.

A second type of cross-sectional analysis tested the association between selected measures of fitness and an index of acculturation developed by Ross MacArthur (Chapter 3). Skinfold correlations are significant in both men and women in Table 3.3, but correlations with aerobic power and muscle strength were seen only in the women.

In 1989/90, we made further cross-sectional comparisons of fitness, based on occupation and lifestyle (Rode & Shephard, 1992a, 1994a). We first divided the Igloolik population into those who held full-time jobs and those who did not. The two groups were in most respects similar, but a few differences favoured those who had won permanent jobs within the village. Thus, the 17–19-year-old and 30–39-year-old workers were heavier than their peers (by 8.7 kg, $p < 0.027$, and 7.9 kg, $p < 0.058$, respectively), and the 50–59-year-old workers had a higher aerobic power (2.62 l/min.) than their peers (2.14 l/min, $p < 0.041$). Similar comparisons between women with full-time jobs and those who were otherwise occupied (predominantly with child care) did not reveal any significant differences in fitness levels.

The male sample of 1989/90 was once again classed by their involvement in hunting. Three village elders identified 24 men who were still engaged in hunting on a regular basis, and 134 other men of comparable age who were working, studying, or living permanently in the village on welfare support. As noted in Chapter 3, the hunting activities of the Igloolik community had become highly mechanized by 1989/90. Because of the use of high-speed snowmobiles and power boats, journeys to the hunting grounds had been cut from several days to a few hours. Much less meat was being butchered, because there were no longer large teams of dogs to feed, and any carcasses that were brought back to the settlement were carried on powered sleds or all-terrain vehicles rather than handled physically by the hunters.

In the 1989/90 survey, the 17–19-year-old and 20–29-year-old hunters were significantly taller than their peers ($+8.6$ cm, $p < 0.038$, and $+4.9$ cm, $p < 0.032$), but (perhaps because of vertebral compression sustained during the operation of snowmobiles (Chapter 7)), this advantage was not maintained in older age groups. The 17–19-year-old hunters were also heavier than their peers ($+11$ kg, $p < 0.018$), and perhaps because of controlling powerful snowmobiles, the 20–29-year-old hunters had a greater grip strength than other Igloolik men of similar age ($+84$ N, $p < 0.021$). However, in marked contrast with our findings in 1969/70, the aerobic powers of 20–29-and 40–49-year-old hunters were less than those of their settlement-dwelling peers, by margins of 9.9% ($p < 0.024$) and 24% ($p < 0.001$) respectively.

The physiological data for 1989/90 seemingly reflect decreases in the physical demands of hunting, and such findings support the view that the high levels of fitness seen in earlier surveys were causally related to the energy demands of traditional hunting.

Secular changes in aerobic fitness and muscular strength

Early cross-sectional comparisons between and within communities showed an association between high levels of physical fitness and maintenance of a traditional lifestyle. However, such observations are susceptible to two distinct explanations: (1) the traditional view, espoused during the conception of the IBP-HAP, was that a specific genetic adaptation allowed various circumpolar groups to colonize an unpromising habitat; (2) alternatively, the physical demands of the harsh environment may have had a more short-term training effect upon the individuals concerned. The second of the two hypotheses seems more plausible, for at least two reasons. Firstly, the magnitude of the physiological differences observed between active hunter–gatherers and individuals living a more sedentary life within permanent settlements has never been larger than could have been induced by the vigorous training of an urban population. Secondly, the challenges imposed by a traditional lifestyle in the arctic environment have been so varied that it is difficult to conceive of any single genetic adaptation that could have given a major physiological advantage to circumpolar groups in the quest for overall survival.

The relative importance of nature and nurture might best be examined by making repeated observation of one or more circumpolar populations during the course of their acculturation to a sedentary, urban lifestyle. In theory, there are many potential obstacles to such an analysis:

1. Accurate assessment of community activity patterns continues to be a difficult and costly undertaking (Chapter 4).
2. Apparent changes in fitness scores with time may be attributable to modifications of equipment or personnel over the period of observation. A new observer may introduce personal variants of technique, not foreseen in the development of the original standardized protocol (Chapter 2). Moreover, because of differences in personality or understanding of the local language, one observer may achieve a better rapport than another with the test subjects, thus inducing greater effort in tests that demand cooperation and maximal effort on the part of those who are being evaluated.
3. Acculturation may lead to secular differences in standing height or body mass (Chapter 7), raising largely unresolved issues about the most appropriate method of allowing for the influence of such variables upon aerobic power and muscle torque (Shephard *et al.*, 1980).
4. Acculturation generally increases access to and consumption of 'market'

foods (Chapter 4). Sometimes, this reduces the likelihood of gross malnutrition and episodic starvation. But if a limited cash income is spent unwisely, reliance on store purchases may worsen certain aspects of nutrition, for instance reducing the intake of high quality protein and vitamins.

5. The loss of cultural identity that is commonly associated with acculturation may give rise to drug abuse (cigarettes, alcohol, solvents and other drugs), with an adverse impact upon overall health and thus physical fitness.

6. Population levels of fitness in women may be influenced by secular trends in regard to the average number of pregnancies and the accepted duration of lactation (Shephard, 1980).

7. Episodic contact with representatives of larger populations may increase the incidence of both acute and chronic disease (Chapter 8, Allen, 1973; Keenleyside, 1990). On the other hand, increasing public health education and better access to medical facilities may lessen the health impact of diseases that were brought to arctic communities by earlier generations of travellers.

Our continuing surveys of the Inuit community of Igloolik provide the only available long-term follow-up of arctic residents. The physical activity patterns of the Igloolik population have been carefully documented over a 20 year period, and broadly based measurements of physical fitness have also been obtained by the same observers, using the same equipment, in 1969/70, 1979/80 and 1989/90 (Rode & Shephard, 1992a). The average size of the adults has remained relatively static over this period. The community was well-nourished at the time of the first survey, and although there has been a continuing shift from 'country' to 'market' food sources, the overall nutritional status has remained good. Cigarette consumption has increased somewhat over the 20 years of our observations (Chapter 6), but alcohol and drug abuse are not the major problems that they have become in some larger circumpolar settlements. The average number of pregnancies per female has diminished progressively, but the influence of this variable upon the individual's level of physical fitness is well documented for this community (Shephard, 1980), as is the impact of chronic diseases such as tuberculosis (see below).

Many of the same individuals were tested in each of the three surveys. In theory, the data could thus be analyzed both cross-sectionally and longitudinally. However, a substantial proportion of subjects missed at least one of the three tests. In order to minimize the impact of this complication, we elected to make a 3 × 4 analysis of variance for groups of unequal size, testing for the effects of birth cohort (those aged 20–29 years in 1969/70, 1979/80, and 1989/90, respectively), age group (20–30, 30–40, 40–50, and 50–60 years), and cohort/age-group interactions.

The 1989/90 data were also compared with data collected by the same observers on a population of Dolgans and nGanasan in 1991/92. The Siberian groups living in the village of Volochanka began the process of acculturation to a permanent settlement earlier than the Inuit, but they had made much less material 'progress' since their collectivization in the 1930s, essentially conserving a hunting/reindeer herding economy. Analyses of variance tested for inter-population differences, and age × population interactions (Rode & Shephard, 1994b).

Aerobic fitness

Step test predictions of relative aerobic power (ml/[kg·min.]), based upon the direct measurement of oxygen intake during near maximal stepping exercise, showed a substantial (9–17%) decline in the aerobic fitness of adult male Inuit in Igloolik over the period from 1969/70 to 1989/90. Much of the loss in physical condition had already been incurred by 1979/80 (Rode & Shephard, 1984a), but the data suggested that further small decrements of aerobic power occurred over the subsequent decade (Fig. 5.2, Rode & Shephard, 1992a, 1994a).

Possible technical factors that could have simulated a deterioration of aerobic fitness were discussed by Rode and Shephard (1984a). The directly measured peak heart rate in those of our sample who apparently made a good effort had increased from 184.5 beats/min. in 1969/70 to 194.4 in 1979/80, so that at the second assessment the step test predictions under-estimated directly measured maxima by an average of some 3%. Although there is some evidence that a high level of aerobic fitness in itself can be associated with low peak heart rates, the more usual explanation of a less than anticipated maximal heart rate is difficulty in eliciting maximal effort. We concluded that because of nervousness on the part of the Inuit and our own limited knowledge of Inuktittuk in 1969/70, it was likely that (as in other circumpolar studies) our first series of direct measurements under-estimated true, centrally limited maxima by some 5–10%. A deficit of this order would explain why the heart rates recorded during submaximal effort were higher in 1979/80 than in 1969/70, whereas the directly measured peak oxygen intakes were similar for the two surveys.

The secular trend to a loss of relative aerobic power (ml/[kg·min.]) in the more recent cohorts of Inuit could not be attributed simply to an increase in their body mass, since a comparison of absolute maximal oxygen intake

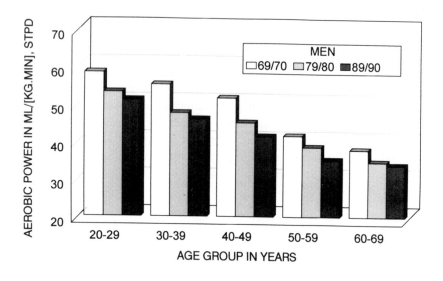

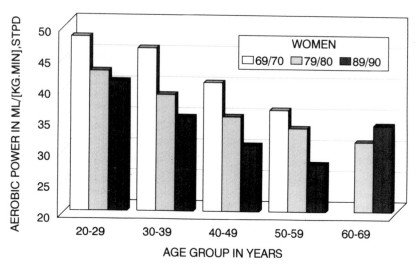

Fig. 5.2. Secular trend to decrease of relative aerobic power (ml/[kg. min.]) in Igloolik Inuit. Step test predictions, classified by cohort (those aged 20–29 years in 1969/70, 1979/80 and 1989/90, respectively), age group and gender. From Rode & Shephard (1992a).

data (l/min.) for 1969/70, 1979/80 and 1989/90 showed decreases of a similar order and statistical significance to the change in relative values (ml/[kg·min.]). In women, the 20-year loss of aerobic fitness at any given age (15–25%) was even greater than in the men (Fig. 5.2). Among the younger women, losses of relative and absolute aerobic power were of similar magnitude, but among the older women, an increase of body mass apparently had contributed to the functional loss over the two decades (20-year decrements of absolute aerobic power 10% at age 40–49 years, and 16% at 50–59 years). The analysis of variance confirmed that there were highly significant effects not only of age decade ($p < 0.001$ in men, < 0.003 in women), but also of cohort ($p < 0.001$ in both sexes).

There is little evidence that the decrease in aerobic power at any given age can be blamed upon altered pattern of nutrition. Haemoglobin levels for male and female Inuit averaged 152 and 140 g/dl in 1969/70, 153 and 142 g/dl in 1979/80, and 151 and 138 g/dl in 1989/90 (Rode & Shephard, 1992a).

Family size decreased over the 20 years of observation; in the younger women, this may have been a significant factor in reducing the habitual physical activity and thus their fitness levels. Cross-sectional data obtained on the Igloolik community in 1969/70 showed a lower aerobic power and leg strength in nulliparous females than in those with multiple successful pregnancies (Table 5.3).

Over the 20 years of observation, the medical facilities in Igloolik were substantially improved, and selective mortality led to the death of some older subjects who in 1969/70 had been seriously disabled by chronic respiratory disease. For both of these reasons, disabling chronic disease was thus more prevalent in 1969/70 than in 1989/90. In our first survey, 28 of 224 villagers had a history of primary tuberculosis and/or hilar calcification. Nevertheless, the cardio-respiratory function of this group was close to 100% of the values observed in the remaining villagers (Rode & Shephard, 1971). A further 17 of the Igloolik population that were tested in 1969/70 had a history of secondary or advanced tuberculosis, and three had emphysema and/or chronic bronchitis; this subgroup had a 14% deficit of cardio-respiratory function relative to their age-matched peers. To the extent that the problems of chronic respiratory disease had been controlled by 1989/90, we would have anticipated a 1.3% improvement in the average aerobic power of the sample, rather than a substantial deterioration in their aerobic performance.

By a process of elimination, it thus seems that the decrease of aerobic power should be attributed to a decrease of habitual physical activity. Certainly, the magnitude of the observed decrease in aerobic function is of the order that might have been anticipated if the Igloolik population had shifted from a very active to a relatively sedentary lifestyle. In the 1979/80 survey, the advantage of relative aerobic power that we had observed in 1969/70 persisted among the continuing hunters (Rode & Shephard,

Table 5.3. *The influence of parity on selected measures of fitness. Mean ± SD of data for the Inuit community of Igloolik, collected in 1969/70*

Number of children	Sample size	Age (years)	Predicted aerobic power		Skinfold (mm)	Knee extension force (N)	Handgrip force (N)
			l/min.	ml/[kg·min.]			
0	11	16.5 ± 1.5	2.49 ± 0.38	47.4 ± 7.3	10.8 ± 3.8	552 ± 202	275 ± 29
1	5	20.4 ± 3.6	2.99 ± 0.29	49.7 ± 4.0	10.9 ± 2.8	641 ± 179	298 ± 28
2	1	25.0	2.56	47.9	6.7	588	235
3 or more	11	25.7 ± 2.3	2.80 ± 0.50	48.0 ± 8.3	8.0 ± 2.5	662 ± 182	288 ± 47

Source: Shephard (1980).

1984a), but by 1989/90, this group was less fit than those who were living permanently in the village.

Although the secular decline in aerobic power was distributed across all age groups, the change was most marked in those aged 20–49 years (Fig. 5.2). The convex-shaped aging curve which we had noted in the cross-sectional data of 1969/70 had been replaced by a curve of more convex form. The reason seems that whereas in 1969/70, the youngest adults were physically the most active members of the community, by 1989/90 acculturation to a 'modern' lifestyle had had a greater impact upon the activity patterns of the young adults than on those who were older.

Comparison of fitness levels between the Inuit and the nGanasan populations (Table 5.4) was facilitated by there being no significant differences of stature between the two populations. The likelihood that apparent inter-population differences in fitness arose from technical factors was also minimized, since the same observer used (where possible) the same equipment to test the two ethnic groups. In the men, values for relative aerobic power were closely comparable between the two communities, but because the Igloolik Inuit were heavier, they also had significantly larger values for absolute aerobic power than the nGanasan, their advantage amounting to 6.5, 8.4 and 17.6% in the three adult age-groups that were compared. Although there had been much less mechanization of daily life at Volochanka than at Igloolik, we must conclude that the added physical demands of daily living in that part of Siberia were not pursued with sufficient vigour to yield an advantage of relative aerobic power.

The young Inuit women were slimmer than the nGanasan, and had a 17% greater relative aerobic power at age 20–29 years ($p < 0.018$), but this difference was reversed in the oldest age group, significantly so for absolute aerobic power ($p < 0.032$). The ANOVA for female subjects thus showed significant effects of both age and age × population.

Muscular strength

Handgrip

Four of five cross-sectional age-group comparisons showed statistically significant decreases of average handgrip force among the adult male Inuit of Igloolik over the period from 1969/70 to 1989/90. The functional loss over the 20-year interval increased from 7% in those aged 20–29 years to 27% in those aged 60–69 years (Fig. 5.3).

In the women, decreases were of a similar order. In 1979/80, values for those aged 50–59 years were already significantly less than in 1969/70, and

Table 5.4. *Comparison of predicted maximal oxygen intake (ml/[kg·min.]) and muscular strength (Newtons) between the Inuit residents of Igloolik (1989/90) and the nGanasan residents of Volochanka (1991/92)*

Age group (years)	Inuit						nGanasan					
	Max. O$_2$ intake		Grip strength		Leg ext.		Max. O$_2$ intake		Grip strength		Leg ext.	
	M	F	M	F	M	F	M	F	M	F	M	F
20–30	47.9	44.7	456	229	556	391	47.8	38.1	411	254	746	387
30–40	43.8	36.6	417	226	539	402	43.1	33.7	370	222	605	281
40–50	40.0	33.0	442	210	600	380	38.3	37.2	398	237	612	328

Source: Rode & Shephard (1994g).

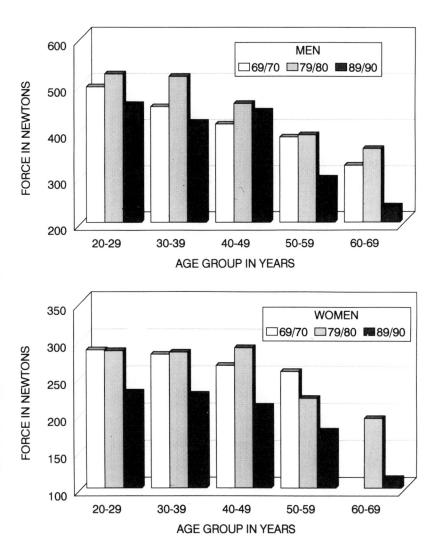

Fig. 5.3. Secular changes in handgrip force among Igloolik Inuit. Stoelting handgrip dynamometer measurements, classified by cohort (those aged 20–29 years in 1969/70, 1979/80 and 1989/90), age group and gender. From Rode & Shephard (1992a).

by 1989/90, the average handgrip force of the female Inuit in all age categories was significantly reduced relative to the initial survey. Again, the impact of 'modernization' was greatest in oldest individuals, losses over the 20-year interval rising from 20% in the third decade of life to 31% in the sixth decade of life.

Analyses of variance analogous to those performed for aerobic power showed significant effects of age-group ($p < 0.001$ in both sexes) and of birth cohort ($p < 0.001$ in both sexes). Secular trends to a decrease in habitual activity patterns probably accounted for most of these changes. The men no longer undertook much of the heavy handling activity associated with traditional hunting, and the women had virtually abandoned the traditional craft of cleaning skins with a metal scraping tool (the ulu). In young adults, the loss of handgrip force was probably somewhat less than in older members of the community because of the force that was needed to control high-powered snowmobiles.

A comparison of handgrip data for the Inuit and the nGanasan tends to support these hypotheses (Table 5.4). The men of Igloolik had significantly greater handgrip readings in all age-categories, the respective margins being 10.7, 12.9 and 10.9%, whereas the women of Volochanka had an insignificant trend to a greater grip strength than their peers in Igloolik.

Leg extension force

Among the Igloolik Inuit, the secular trend to a decrease in leg extension force at any given age was even more striking than was the loss of grip strength (Fig. 5.4). The men aged 20–29 years and 50–59 years showed significant losses of leg extension force from 1969/70 to 1979/80, and by 1989/90, leg extension scores were significantly lower in all age categories except for a small group who were then aged 60–69 years. The average decrease in age-specific leg extension force over the 20-year interval amounted to 37% in young men, but increased to 45% in the oldest men.

Among the women, leg extension scores for all age groups were already significantly lower in 1979/80 than in 1969/70, and values for female subjects aged 20–29, 30–39 and 40–49 years showed further significant reductions in 1989/90.

Analyses of variance showed significant effects of age ($p < 0.001$ in men only), cohort ($p < 0.001$ in both sexes) and age group × cohort ($p < 0.044$ in men only). We suggested in the first part of this chapter that the high leg strengths observed in 1969/70 reflected the need to walk over rough snow and ice for much of the year. It is thus significant that walking around Igloolik was greatly reduced by the introduction of snowmobiles, all-terrain vehicles, and a taxi service. In the women that were studied in 1969/70, leg strength was also related to family size (Table 5.3), probably because young children were carried on the back in the traditional amauti for much of the day. This form of physical activity was reduced progressively over the 20 years of observation, both by a decrease in family size, and by the adoption of a settled urban lifestyle, with a domestic environment where children could be left to play safely on the floor.

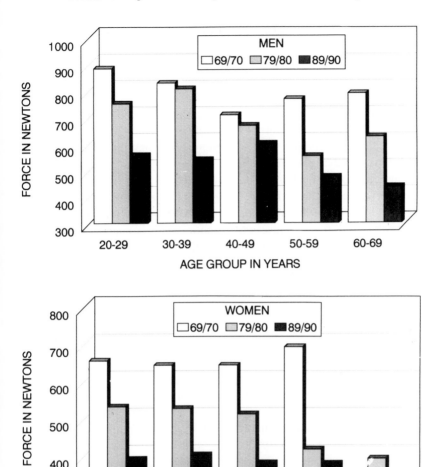

Fig. 5.4. Secular changes in knee extension force. Cable tensiometer measurements at 120 degrees extension. Data classified by cohort (those aged 20–29 years in 1969/70, 1979/80 and 1989/90), age group and gender. From Rode & Shephard (1992a).

A comparison of data for Igloolik and Volochanka (Table 5.4) showed greater leg strength for the nGanasan men (who still walked during much of their daily outdoor activity). The nGanasan women did not carry babies on their backs, and perhaps for this reason they had a lesser leg strength than the women of Igloolik.

Community programming to sustain fitness

Many aspect of physical fitness appear to have been declining in the circumpolar regions. It thus seems desirable to correct this trend, in order to avoid an epidemic of the 'diseases of civilization' of the type which has afflicted many other indigenous communities (Chapter 8).

The local environment cannot yield sufficient 'country' foods to allow all of a rapidly expanding total circumpolar population to return to a traditional hunter–gatherer lifestyle. But even if this were practicable, because 'modern', mechanized methods are now used in the pursuit and transport of game, the current generation of hunters have a poorer level of physical fitness than those who have adopted an urbanized lifestyle. It thus seems desirable to introduce and promote what is a novel concept for indigenous circumpolar populations: the deliberate pursuit of physically active leisure.

Governments have adopted two approaches in fostering greater levels of physical activity in arctic communities. In some instances, there has been a deliberate attempt to sponsor regional games and competitions, involving both the types of competition current in 'white' society (Paraschak, 1982) and traditional Inuit games (Glassford, 1970a, b). Such initiatives have had varying success. Glassford (1970a) commented that traditional Inuit games offered a method of determining dominance largely by allowing the individual to compete against himself or herself. Glassford further demonstrated that the more traditional members of the Inuit community had a preference for such pursuits, relative to the competitive games of 'white' culture (Glassford, 1970b).

In some instances, participation in 'white' sports has become an important vehicle of acculturation (Adams, 1978; Bennington, 1978). Paraschak (1982) described the 'white' model adopted by Canada Sport North and the Arctic Winter Games. A total of 26 sporting associations were funded by a government lottery, each association organizing competitive leagues throughout the arctic. Paraschak argued that this form of organization presupposed a sense of time, a willingness to play according to rigid rules determined by national associations, availability of personal equipment, sufficient urbanization to organise age-matched teams, and easy transportation to competitions held in neighbouring communities. Almost none of these conditions were yet satisfied in the arctic. Moreover, most settlements already had a strong sense of local community which did not need to be fostered by any externally imposed club structure.

One alternative option has been to encourage regular participation in local sports and active pastimes (for instance, indoor soccer, floor hockey, basketball and square dancing), without regard to formal league structures.

In some communities, there are now adequate physical resources to exploit such a tactic, but again a lack of time sense limits the utilization of facilities that adopt fixed hours of operation, and a rapid turnover of 'white' personnel limits continuity of programming. Igloolik illustrates the progress that has been made in this direction. In 1970, the only recreational facilities were a small hall (about $70\,m^2$) that offered twice weekly film shows and weekly square dances, and the Catholic mission (which provided facilities for table tennis and the occasional bingo game). A school gymnasium, complete with a large stage and a regulation-sized basketball court was completed in 1971, a new recreation hall and an indoor swimming pool were opened in the early 1980s, and a full-sized indoor skating rink and curling rink were completed in 1991. At the school gymnasium, regularly scheduled community sports are now organized by school staff throughout the academic year. Potential pursuits include indoor soccer, floor hockey, basketball, volleyball, and badminton.

In order to make a preliminary assessment of the impact of such community recreational programmes upon individual fitness levels, we divided the physical data on teenagers and young adults from the 1989/90 Igloolik survey into two categories: those who had participated regularly in either community sports or other types of vigorous activity such as square dancing or running a dog team, and those who had not. The main inter-group difference was in terms of relative aerobic power (ml/[kg·min.]). The active males in all age-categories showed 21–30% higher values than their sedentary peers, with p values as small as 0.001 in some age groups. Likewise, the active female subjects had a substantial advantage of aerobic power in all except the youngest age category (where only two girls were active), maximal oxygen intake values being 23–32% larger in the active groups. The sedentary groups were also consistently fatter than their active peers, the difference being statistically significant for males in the 17–19 year age-category, for all male subjects and for all female subjects. In contrast, inter-group differences in handgrip and knee extension force were small and inconsistent (Table 5.5).

The advantage of aerobic power currently shown by members of the Igloolik sports programmes is of the order that would be anticipated with participation in a vigorous aerobic training programme (a 20–30% gain of maximal oxygen intake, Shephard, 1977, 1993b). These data thus strongly suggest that involvement in the community sports programmes is helping the participants to conserve the high levels of fitness that were noted in earlier generations of Canadian Inuit. Nevertheless, longitudinal experiments are needed to exclude the alternative possibility, that differences of fitness have arisen because taller, heavier and fitter individuals became selectively involved in active community pursuits (Shephard et al., 1978a). A difference

Table 5.5. *The effects of physically active leisure upon the fitness of Igloolik Inuit. A cross-sectional comparison of active (A) and sedentary (S) individuals, completed in 1989/90*

					Age group (years)					
	13–14		15–16		17–19		20–30		30–39	
Variable	A	S	A	S	A	S	A	S	A	S
Male subjects (31 A;132 S)										
Height (cm)	159.3	153.9	166.0	162.1	164.3	163.2	164.9	164.2	162.2	163.9
Body mass (kg)	54.2	49.0	59.1	57.1	62.6	63.1	63.8	65.4	66.4	64.9
Skinfolds (mm)	21.0	27.4	22.5	23.9	24.1	30.4*	26.5	32.5	28.0	33.5
Grip force (N)	319	216	402	363	413	398	460	454	415	418
Knee extension (N)	392	366	461	442	486	517	579	540	628	527
Aerobic power (ml/[kg·min.])	69.1	53.7*	68.3	56.1*	66.3	52.1*	62.0	47.8*	54.3	44.8*
Female subjects (11 A; 84S)										
Height (cm)	147.5	151.0	151.3	153.2	151.7	154.2	152.0	153.3		
Body mass (kg)	44.6	46.5	52.1	54.7	54.8	54.4	54.0	54.7		
Skinfolds (mm)	41.0	34.0	49.0	55.4	45.0	51.5	40.5	45.6		
Grip force (N)	142	155	211	196	170	199	258	227		
Knee extension (N)	304	341	388	391	380	406	449	386		
Aerobic power (ml/[kg·min.])	44.3	45.6	50.0	40.6*	52.8	40.0*	50.0	40.1*		

Source: Rode & Shephard (1992a, 1993). For details of statistics, see original reports.
*Significant inter-group difference ($p < 0.05$).

of height between active and sedentary boys suggests that there may also have been some selective recruitment of large and early-maturing adolescents to community sport teams, much as has been observed in urban societies (Shephard *et al.*, 1978a).

As in most 'white' populations, the big challenge is now to increase the percentage of the community who are involved in active leisure pursuits. Currently, participants total only 11.6% of females and 19.0% of males. Optimal techniques for the promotion of a physically active lifestyle have yet to be determined for indigenous circumpolar peoples. Given the increasing rejection of externally imposed 'white' norms of behaviour, the answer is unlikely to lie in required school programmes of physical education or government-sponsored propaganda. The solution will probably be found through the development of appropriate indigenous role models. Possibly, local leaders will be willing to accept the message of the physiologists studies, that involvement in community sports programmes can conserve the fitness that allowed the famous hunters of the settlement to exploit the arctic habitat.

Changes in cold tolerance

Tkachev *et al.* (1991) found a large seasonal fluctuation of T4 levels in a sample of Nenets living in the Archangelsk region. Winter values exceeded summer values by about 16%, and there was an associated decrease of free fatty acid levels. Tkachev argued that as a result of many generations of life in this region, the Nenets had developed an unusual capacity to metabolize fat during the winter months, thus providing themselves with added protection against cold exposure. Itoh (1974) also found that following norepinephrine injection, the Ainu showed a larger rise of free fatty acids than was seen in Japanese controls. Other mechanisms increasing resistance to a cold environment have been discussed in Chapter 1. However, it is still not clearly established whether such adaptations are greater than would be anticipated in 'white' subjects who had been exposed to similar extremes of climate (Chapter 1). Are these examples of immediate acclimatization, or genetic adaptation?

Acculturation to a 'modern' lifestyle has affected cold exposure in a number of divergent ways. As hunting activity has decreased, there has (of necessity) been an increasing reliance upon store clothing. In general, such clothing provides much less effective insulation than traditional winter garments, thus tending to increase the intensity of cold exposure. On the other hand, a reduction in the number of hunting expeditions and the development of indoor recreations has reduced the number of hours of exposure to cold, so that some loss of acclimatization might be anticipated. Nevertheless, even the most sedentary of circumpolar residents cannot totally avoid exposure to severe cold during the winter months, and the minimum exposure needed to sustain acclimatization remains unclear.

In those operating high-speed snowmobiles, local exposure to cold air may have increased in recent years, since the high air-speeds of the snowmobiler tend to displace the film of still and insulating air this is in close contact with the skin of the face under the hood of a parka. Virokannas *et al.* (1984) and Ervasti *et al.* (1991) found that up to 68% of reindeer herders who used snowmobiles regularly had experienced frostbite, most commonly on the face. There is some limited evidence that the inhalation of colder air is having adverse consequences for the lungs and the pulmonary circulation (Chapter 6).

One data set which has shown a clear secular trend to a decrease in cold tolerance with technical innovation is that of Hong *et al.* (1986). These authors demonstrated that when the Korean diving women (Ama) began wearing 'wet suits' as a protection against cold water, they lost much of the ability to generate heat by non-shivering thermogenesis that had previously characterized their diving sessions. Likewise, Dessypris *et al.* (1981) commented that a previously observed difference in plasma concentrations of thyroid-stimulating hormone between the Mountain, Fisher and Skolt Lapps and people living in southern Finland had disappeared in recent years. They suggested improvements of home heating in Lapland as a probable explanation.

Conclusions

Acculturation to a settled lifestyle has reduced the habitual physical activity demanded of indigenous circumpolar populations, and it has also reduced the duration of exposure to extreme cold. One immediate practical consequence has been a decrease in the ability of the indigenous groups to meet the climatic challenges of the circumpolar habitat without technical support. Aerobic power and muscular strength, initially high, have declined to around urban levels, body fat content has increased, and there is some limited evidence of a loss of cold tolerance. Population growth precludes a return to traditional patterns of 'country' living, but much of the traditional physical fitness of the circumpolar community can apparently be conserved by participation in community sports programmes. The challenge now is to make such initiatives relevant to a larger fraction of the indigenous residents of the arctic regions.

6 Secular trends in respiratory hazards, lung function and respiratory disease

Acculturation of the indigenous circumpolar populations to an urban lifestyle has changed many factors which could potentially influence lung function. In addition to possible secular changes in body dimensions (Chapter 7), cigarette smoking has become an almost universal habit among the older children and adults of many indigenous circumpolar groups. Furthermore, the per capita consumption of cigarettes has increased as the cash income of the arctic residents has risen. The overall duration of cold exposure has decreased for most circumpolar people as a consequence of the move from overnight encampments to life in permanent settlements, but the operation of high-speed snowmobiles may have increased the periodic inhalation of very frigid and dry arctic air for at least some of these individuals. Exposure to the smoke from oil lamps has greatly diminished (Beaudry, 1968), but this source of air pollution has been replaced by exposure to quite high concentrations of second-hand cigarette smoke from an early age. Physical fitness has diminished substantially in some circumpolar populations (Chapter 5). Finally, minor respiratory infections remain endemic among young children, although tuberculosis (which was a major consequence of early contacts with 'white' immigrants) has now been largely controlled. This chapter discusses the course of these various secular changes in relation to alterations in the lung function of circumpolar populations, which are seen both in children and in adults.

Secular changes of body size

According to proportionality theory, static lung volumes such as vital capacity should be a cubic function of stature (Von Döbeln, 1966). Thus, a small secular change in the average standing height of a population might make a considerably larger difference to its average lung volumes.

Traditionally, the adult population of the circumpolar regions has been some 10 cm shorter than their 'white' peers (Chapter 7), and their 6% disadvantage of height might be expected to translate into a 16–17% deficit

151

in lung volumes (assuming also that their sitting height was a normal fraction of standing height). Currently, the ratio of sitting to standing height does seem to be normal (see Chapter 7). Rennie *et al.* (1970) attributed the large height-standardized lung volumes of Alaskan Inuit to their large ratio of sitting to standing height, but in fact the IBP-HAP data for the Alaskan north shore showed ratios of 53.2% in males and 54.2% in females, only marginally more than the figures of 52.8% and 53.3% found for the Canadian population as a whole, by Nutrition Canada in 1970–72 (Jetté, 1983).

Most authors have expressed pulmonary function data as a percentage of norms based upon age, gender and a linear height coefficient. Such a method of adjusting for differences of stature is adequate to allow for the small inter-individual discrepancies encountered within 'white' populations, but the difference between a linear and a cubic height coefficient might become quite important if the norms developed on urban populations were to be extrapolated to predict values for much shorter individuals such as the Inuit and Lapp populations.

Until the 1970s, it seemed that the circumpolar populations were becoming progressively taller, and that they were indeed destined to lose their traditional short stature. However, in more recent years, longitudinal studies of the Inuit population have shown a substantial loss of stature, beginning early in adult life (Chapter 7). It is unclear whether this is attributable to increased kyphosis, compression of the intervertebral discs, or vertebral collapse, but irrespective of cause, it seems likely that it could exaggerate the normal age-related decline in lung function.

Secular changes in cigarette consumption

The exact date when circumpolar populations began smoking is uncertain. Probably, the habit was initiated as contact was made with 'white' sailors and trappers, and it has become endemic with the establishment of a wage economy. Blondin (1990) suggested that in the Dene nation, the habit began in the 1930s. Native medicine people had warned against the smoke that would come from the south, and initially people were cautious in their approach to tobacco, but with time consumption rose and inhalation of the smoke became more frequent.

Rode & Shephard (1973a) found that by 1969/70, the majority of the Igloolik teenagers and adults were already smoking on a regular basis. The per capita consumption of the adult smokers (age 20–59 years) increased from an average of 8.7 cigarettes/day in women and 14.2 cigarettes/day in

Table 6.1. *Prevalence of cigarette smoking in three successive surveys of the Igloolik Inuit (1969/70, 1979/80 and 1989/90). Data classified by age and percentage of smokers in each age category*

Age (years)	Males			Females		
	1969/70	1979/80	1989/90	1969/70	1979/80	1989/90
9–10	0	25	0	0	20	–
11–12	0	74	11	0	39	38
13–14	13	88	52	17	93	67
15–16	50	100	71	80	95	87
17–19	81	96	87	75	77	95
20–29	97	95	93	100	92	96
30–39	83	71	79	92	96	81
40–49	88	78	63	100	94	68
50–59	86	80	80	50	100	58
60–69	50	67	29	0	25	50

Source: Rode & Shephard (1992a).

men in 1969/70 to 12.0 and 15.3 cigarettes/day in 1979/80. By 1989/90, men aged 40–49 years were consuming an average of 23.3 cigarettes/day, although the average consumption of smokers in the Igloolik population had dropped marginally, to 11.1 and 13.1 cigarettes/day for women and men respectively. The habit of cigarette smoking remained almost universal from an early age, but among male subjects one hopeful sign was that the percentage of smokers within any given adult age decade had shown a small decrease from 1969/70 to 1979/80, with a further small decrease by 1989/90 (Table 6.1). Millar (1990) provided further data on the smoking behaviour of school students in the Canadian arctic over the period 1985–1987. In the age group 10–14 years, 34% of the Dene, Métis and Inuit children were regular or occasional smokers, and in the 15–19 year age group the corresponding figures were 63% for Dene and Métis, and 71% for Inuit.

In Greenland, Misfeldt (1990) calculated that for every person over the age of 15 years, cigarette consumption had risen from an average of 6.5 cigarettes per day in 1960 to a plateau of 10.5 cigarettes per day that was maintained from 1979 through 1985. Westlund (1981) commented that in Finnmark, the prevalence of smoking males in any given age group was 15% higher than in Oslo, although there had been some trend to a reduction in the years preceding his most recent survey. In some other circumpolar populations, the prevalence of smoking among the male subjects has been similar to that which we observed in arctic Canada, but the proportion of females smokers has been lower. Rode & Shephard

(1994b) noted that in 1991, 78% of male nGanasan aged 18–59 years were regular smokers (compared with 80% of Igloolik Inuit in the same age range). However, the prevalence of cigarettes smoking among female nGanasan (56%), although undesirably high, was less than in the Igloolik Inuit. The average daily consumption of the smokers (6.2 cigarettes/day in the women, 11.3 cigarettes/day in the men) was also a little lower in Volochanka than in Igloolik.

Among the adult Cree Amerindians of the eastern James Bay region, Lavallée & Robinson (1991) found that 85% recognized that not smoking was very important to their well-being, and the age-adjusted prevalence of smoking in this population (at 39%) was close to the average for urban Canada. In contrast, Pickering *et al.* (1989) found that 69% of boys and 50% of girls from the same population were self-reported smokers.

Alaskan natives seem to have adopted a more healthy lifestyle than the Canadian Inuit in recent years, at least with respect to smoking. In the mid 1980s, the prevalence of cigarette consumption among the adult indigenous populations of five North-western Alaskan settlements averaged 46% (Lanier *et al.*, 1990; Lee, 1985). Nevertheless, two of the five settlements had a very high prevalence of smokers (79% and 69% respectively) among school students in grades 7 through 12 years, and the great majority of students had tried tobacco by the time that they reached the 6th or the 7th grade. Moreover, the prevalence of lung cancer, historically very low in the circumpolar regions, was by the 1980s exceeding that for urban communities (Lanier *et al.*, 1990).

Bowman *et al.* (1985) undertook a general survey of substance abuse among Alaskan high school students (a mixture of Inuit and 'white' students), and in the highly acculturated Barrow/Kotzebue/Nome area they found that only 16% of students were using cigarettes on a daily basis. However, even figures for the most acculturated of the circumpolar communities have generally shown a higher prevalence of cigarette smoking than in current urban society. It may be that acculturation to an urban lifestyle will ultimately reduce smoking prevalence throughout the arctic regions, as nursing stations are established and preventive health programmes are introduced. However, the proportion of northern nurses and teachers who themselves smoke is still too high. Plainly, there remains a need to extend and adapt programmes to discourage smoking, particularly in the schools of the arctic regions, if the circumpolar communities are not to experience the major epidemics of lung cancer, chronic bronchitis and emphysema that have affected older smokers in 'developed' societies.

In addition to the direct effects of cigarette smoke on the respiratory health of the smoker, Dilley (1985) noted that the birth weights of children born to Alaskan natives were on average 232 g lighter if their mothers had smoked during their pregnancy. Because circumpolar homes are small, heavily occupied by smokers, and tightly secured against the cold, most children are also exposed almost continuously to high concentrations of environmental tobacco smoke, although effects from passive smoke exposure are quickly over-shadowed by the consequences of personal cigarette smoking, which begins at an early age. Because of the virtual absence of both automobile exhaust and other forms of ambient air pollution, Rode & Shephard (1973a) found carboxyhemoglobin levels of only 0.77% in boys and 0.55% in girls at Igloolik, compared with values of 1.12% and 1.09% in Toronto. However, Etzel *et al.* (1991) found that in rural Alaska, there were detectable levels of cotinine in the saliva of almost all children aged three to five years. In a study from the Canadian arctic, Postl (1985) reported that infants were taken to the nursing station more frequently if their mothers had smoked during the first year of their children's lives. The long-term impact of such exposure upon respiratory function remains controversial (Shephard, 1992). Given the high exposure levels in circumpolar homes, environmental tobacco smoke could well play an important role in the development of chronic respiratory infections, which are endemic among circumpolar children.

Fire is a final hazard of the cigarette habit that has increased with acculturation. A combination of wood construction, very dry air and improvized heating systems makes northern homes very vulnerable to fire. One survey estimated that smoking was responsible for 25% of fires and deaths from fires among Amerindians living in the north (Friesen, 1985).

Acculturation and exposure of the respiratory tract to cold air

Early authors, influenced in part by animal experiments and the experience of mountaineers, suggested that a combination of vigorous physical exertion and cold exposure led to pulmonary hypertension (Bligh & Chauca, 1981). Evidence of the disorder among circumpolar residents was deduced from poor lung function, increased hilar markings on chest radiographs and right bundle branch block (Marachev & Matveev, 1978; Schaefer *et al.*, 1980b).

Studies in the community of Arctic Bay apparently showed a substantially faster aging of lung function than what we had inferred from cross-sectional and longitudinal studies in Igloolik (Rode & Shephard, 1985). Although Arctic Bay is located some 400 km north of Igloolik, the climate of the two settlements is very comparable. However, in that era, the caribou hunters of

Arctic Bay travelled further than those of Igloolik in their search for game. It is unclear whether the resulting increase in the duration of exposure to cold air causes a faster deterioration of lung volume, or whether the discrepancy is merely an artifact, due to a selective sampling of diseased individuals in Arctic Bay. According to Schaefer *et al.* (1980b), there is a syndrome of 'freezing of the lungs' in the Canadian arctic, but manifestations are unique to those Inuit who have been life-time trappers.

Acute manifestations of physical activity in extreme cold develop only when the intensity of effort is sufficiently great that the mouth is opened and inspired air by-passes the normal warming mechanism of the naso-pharynx. Typically, the mouth is first opened at a respiratory minute volume of 30–35 l/min. (Niinimaa *et al.*, 1980). Larsson *et al.* (1993) noted that respiratory problems also developed in elite cross-country skiers because substantial respiratory minute volumes were developed under very cold conditions. More than 50% of this class of athletes complained of exercise-induced bronchospasm, although there was no indication that more permanent lung damage had occurred in these individuals. A study of reindeer herders (Reijula *et al.*, 1990) found about 14% complaining of cold-induced dyspnea, this symptom being most prevalent in the oldest ex-smokers.

A similar type of respiratory cold injury had been described in the Siberian arctic (Schaefer, 1980b). Here, it affected both the indigenous peoples and immigrant newcomers who undertook very heavy physical work outdoors during the winter months. A parallel pathology has also been described in sled-dogs: 'frosting of the lungs', sometimes progressing to a death from acute edema (Schaefer *et al.*, 1980b). On the basis of animal studies, it was suggested that the mechanisms of cold-induced damage included a heavy stimulation of peripheral cold sensors, the central nervous system and the efferent innervation of the pulmonary vessels. Possibly, these neural pathways acted synergistically with other stimuli such as exposure to high altitudes or chronic pulmonary infection (Bligh & Chauca, 1981). When working extremely hard, an increase of heat loss and a diversion of blood flow to the active muscles may also reduce the individual's ability to maintain body temperatures (Hirvonen, 1982).

Lloyd (1985) drew a parallel between the direct exposure of circumpolar residents to extremely cold air and the bronchorrhea which is observed in divers who have suffered a major respiratory heat loss when breathing helium gas mixtures. Nevertheless, it remains difficult to dissociate the immediate effects of exposure to cold air from the longer-term consequences of a high prevalence of chronic respiratory disease (which could in itself cause pulmonary hypertension) and exposure to a smoke-filled atmosphere when living in tents and igloos.

In the traditional circumpolar resident, the main protection of the respiratory tract against the inhalation of very cold air was the small zone of warm expired air conserved by the fur-trimmed hood of the parka. In recent years, there has been an increasing use of store-bought clothing. The store-purchased garments have a less effective hood, often with synthetic fur trim. Moreover, the still air that normally surrounds and protects the mouth and external nares can be displaced if a snowmobile is operated at very high speeds. Although 'modernization' has reduced the total duration of cold exposure, it thus seems likely that it has also caused at least brief periods when the coldness of inhaled air is increased.

It is difficult to obtain good data on the prevalence of pulmonary cold injury in arctic residents. Schaefer *et al.* (1980b) based their assessments on the sign of a resting right bundle branch block (RBBB). They reported that in 1969/70, 30% of the men and 13% of the women living in Igloolik showed RBBB. In fact, the condition was diagnosed in 25/115 subjects in the partial sample of the Igloolik population that they examined. Their investigation centred on the sicker segment of the Igloolik community, and only 13 of the 25 subjects with RBBB volunteered for fitness testing by the physiology team. We thus concluded that the true prevalence of resting RBBB was much lower than 25/115 subjects, or 21.7%; probably, it was as low as 25 in all 500 residents of the settlement (5%). Rode & Shephard (1992b) noted that the RBBB was particularly apparent in electrocardio-graphic leads AVR, V4R, V1 and V2. The exercise records in their study were obtained from the CM5 lead, and partly for this reason they found a definite RBBB in only two subjects, although in a further three of the 13 individuals who had been identified as having RBBB by Schaefer *et al.* (1980b) there was some suspicion of QRS broadening or notching of the QRS complex of the electrocardiogram.

When observations were repeated in 1979/80, five individuals showed RBBB on their exercise records (Rode & Shephard, 1992b), but by 1989/90, the number had risen to 29 (19 men and 10 women, 8.3% of those undergoing exercise testing).

In the survey of 1979/80, four of the five subjects with RBBB were men over the age of 50 years, suggesting a possible relationship between the electrocardiographic abnormality and chronic respiratory disease in the older cohort. This view was supported by the fact that forced expiratory volume and forced vital capacity were below norms for age and standing height in the affected individuals. Aerobic power was measured in three of the five men, and (at an average age of 67 years) it averaged 30 ml/[kg·min.] – a normal value for a city-dweller, but below the high values seen in many of the Igloolik population. However, by 1989/90, the characteristics of the condition seemed to have changed. Seventeen of the 29 individuals with RBBB were now under the age of 40 years. The average lung volumes for this group were normal; their aerobic power averaged 47.1 ml/[kg·min.] in the men and 39.6 ml/[kg·min.] in the women at respective average ages of 33 and 37 years, and the majority of the group were free of chronic respiratory disease.

All the five cases of RBBB who were identified in 1969/70 had at that time a QRS complex that exceeded the accepted normal limit of 80 ms, their average QRS duration being 106 ms. In 1979/80, the average QRS duration for the five individuals with RBBB was 119 ms. In 1989/90, three men showed substantial broadening of the QRS complex (average duration 124 ms). Five other men showed substantial QRS notching, but only a slight prolongation of the total QRS duration to an average of 99 ms. The remaining 10 men and 10 women who had been identified all showed only minor QRS notching, with average QRS durations that were well within normal limits (average values of 65 and 57 ms for men and women respectively).

Although the prevalence of RBBB among the Inuit may have been over-estimated by Schaefer *et al.* (1980b), it seems clear that mild examples of the condition occur more frequently than would be expected in an urban community. Moreover, in Igloolik, the prevalence of RBBB has apparently increased with acculturation to a 'modern' lifestyle.

It does not seem that the increased prevalence of RBBB can be attributed to a leftward rotation of the electrical axis of the heart to a more horizontal position as the Inuit have become fatter. In particular, we noted that two of the men who had shown abnormal ECG records in 1979/80 had become much fatter by 1989/90, but this had not led to any broadening of their QRS complexes. It also seems difficult to attribute the recent increase in prevalence of RBBB to such influences as cigarette smoking, inhalation of environmental tobacco smoke from an early age, or frequent respiratory viral infections, since in the most recent survey of the Igloolik population, the individuals concerned all had very normal lung volumes in relation to their age, gender and height. A QRS breadth of more than 120 ms has definite pathological significance, but minor broadening of the QRS complex or notching of the R wave of the electrocardiogram is a frequent accompaniment of normal right ventricular hypertrophy (Beckner & Winsor, 1954; Butschenko, 1967; Mitrevski, 1969). Thus, the phenomenon should probably be interpreted as a normal variant, associated with development of an above-average level of aerobic fitness (Chapter 5).

Lung function in children

Rode & Shephard (1973a, 1984c, 1994e, f) made three cross-sectional surveys of lung function in Inuit children of the Igloolik community. The body build of the children remained relatively consistent over the period covered by the three surveys, allowing data to be compared in terms of the absolute lung volumes attained at any given age, without the complication of introducing height exponents.

Although the children showed substantially lower levels of aerobic fitness as the community became acculturated to a settled lifestyle (Chapters 5 and 7), both their forced vital capacity and their one-second forced expired volume remained very consistent across the three surveys (Fig. 6.1). This is perhaps not surprising. Although various investigators have demonstrated moderate correlations between lung volumes and cardiorespiratory fitness (Dawber *et al.*, 1966; Dreyer, 1920; Ishiko, 1967) or lean body mass (Mercier *et al.*, 1991), programmes of deliberate endurance training have had only a minor influence upon lung volumes, either in adults (Kollias *et al.*, 1972) or in children (Shephard & Lavallée, 1995), unless the training programme has been quite intensive (Engström *et al.*, 1971) or has been localized to the respiratory muscles (Leith & Bradley, 1976). Moreover, with the possible exception of top-level endurance athletes (Dempsey & Manohar, 1992), the respiratory system does not seem to be the major determinant of peak oxygen transport or aerobic fitness (Shephard, 1993b). Any correlation between aerobic power and lung volumes probably reflects mainly a combination of athletic selection by lung capacity (in certain sports such as swimming) and a mutual dependence of lung volumes and aerobic power upon overall body dimensions.

Rode & Shephard (1994b) also reported cross-sectional data on the lung volumes of nGanasan children living in Volochanka. ANOVA comparing the results for the Inuit and the nGanasan showed no effect of ethnic group in the boys, but the Igloolik girls had a larger forced vital capacity than their peers in Volochanka ($p = 0.016$). *Post hoc* analyses showed significant differences at 11–12, 13–14 and 17–19 years. Inspection of the data suggested a slightly later pubertal growth spurt in Volochanka than in Igloolik, probably related to differences in acculturation, and this seemed responsible for the deficit of lung volumes in the Siberian population during the pubertal and post-pubertal years.

The observed lung volumes from the three surveys of Igloolik children may be compared with height-based predictions for city-dwellers, using equations developed by Cotes (1965) and Shephard & Lavallée (1995). The classic normative equations of Cotes (1965) are somewhat suspect, as they

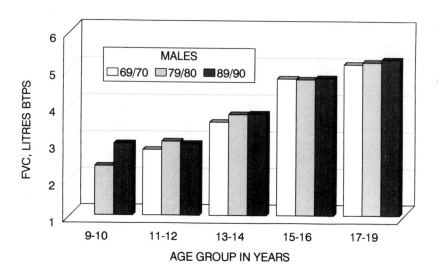

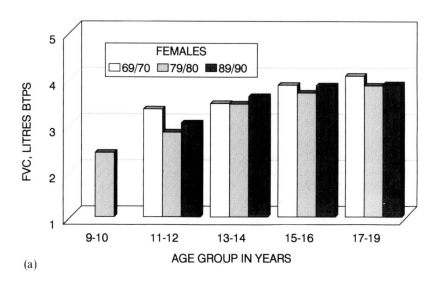

(a)

Fig. 6.1 (a). Forced vital capacity (FVC).
Based upon data of Rode & Shephard (1994e).

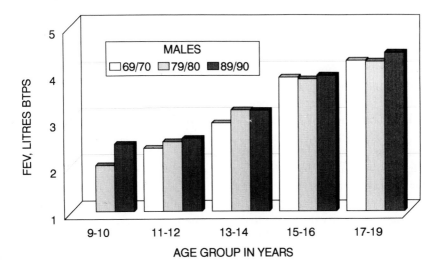

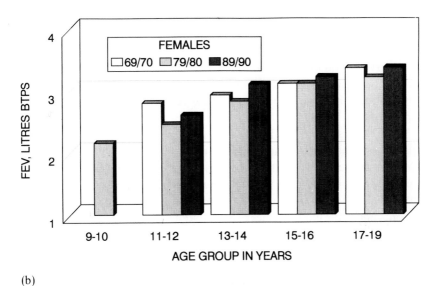

(b)

Fig. 6.1 (b). One-second forced expiratory volume (FEV) of Igloolik Inuit children, as seen in cross-sectional surveys conducted in 1969/70, 1979/80, and 1989/90. Based upon data of Rode & Shephard (1994e).

under-estimate the lung volumes of urban children who are physically active and living in regions with low levels of ambient air pollution (Shephard & Lavallée, 1995). However, even the more modern and gender-specific prediction equations of Shephard & Lavallée (1995) substantially under-predict the forced vital capacity of the Igloolik Inuit children in relation to their height (Table 6.2).

It remains unclear why the lung volumes of the Inuit are larger than those of their peers in developed societies. It was once suggested that the interpretation of lung volumes might be influenced by a short leg to trunk length ratio, but recent anthropometric data do not give much support to the concept that the Inuit have short legs (Chapter 7). In most arctic settlements, there is less exposure to general air pollutants than would be encountered by city-dwellers. On the other hand, exposure to environmental cigarette smoke and other indoor contaminants is likely to be greater in the arctic than in an urban environment, where rooms are larger, occupancy is lower, and windows are often open. Another possible explanation of the larger lung volumes in Igloolik is that participation in a number of the traditional community activities has strengthened the chest muscles. If this were true, we might have expected a diminution of lung volumes with acculturation to a sedentary, urban lifestyle. In fact, there was no significant inter-survey trend in forced vital capacity, and there was a slight tendency for the one-second forced expiratory volume to increase over the 20 years of observation. The large lung volumes of the young Inuit thus have yet to be explained.

When values were standardized to a common height of 150 cm, the Igloolik boys had a 1–4% advantage over the girls, and in Volochanka the male advantage was 6%. Although statistically significant, these differences are less than those we have previously seen in urban samples (Shephard & Lavallée, 1995). As in the study of Anderson *et al.* (1968) on 'white' Torontonians, the lung volumes of both Igloolik Inuit and the nGanasan of Volochanka continued to increase until subjects reached their mid-twenties (Rode & Shephard, 1994b). Fitting logarithmic equations to the cross-sectional data for students aged 11–19 years, the exponents describing the growth of forced vital capacity and one-second forced expiratory volume (Table 6.3) conformed closely to the expected cubic function of standing height both in Igloolik and Volochanka. However, the exponents were somewhat larger than we had observed previously in urban communities (Shephard *et al.*, 1980; Shephard & Lavallée, 1995). Again, a comparison of data for 1969/70, 1979/80 and 1989/90 showed no clear trend to a change of exponent with acculturation.

Data for the percentage of the vital capacity expired in one second and

Table 6.2. *A comparison between the actual lung volumes of the Igloolik Inuit children (data averaged across three cross-sectional surveys conducted in 1969/70, 1979/80 and 1989/90) and height predictions as suggested by Cotes (1965) and Shephard & Lavallée (1995)*

Age (years)	Height (m)	Observed		Cotes prediction		Shephard & Lavallée	
		FVC (l)	FEV1.0 (l)	FVC (l)	FEV1.0 (l)	FVC (l)	FEV1.0 (l)
Boys							
9–10	1.33	2.65	2.22	1.99	1.78	2.20	1.87
11–12	1.43	2.91	2.48	2.41	2.17	2.70	2.37
13–14	1.52	3.68	3.08	2.85	2.58	3.24	2.75
15–16	1.61	4.72	3.89	3.39	3.09	3.91	3.32
17–19	1.64	5.17	4.30	3.55	3.25	4.12	3.50
Girls							
9–10	1.32	2.39	2.16	1.96	1.75	1.97	1.76
11–12	1.43	3.06	2.62	2.43	2.18	2.49	2.23
13–14	1.50	3.50	2.96	2.77	2.50	2.87	2.57
15–16	1.53	3.79	3.16	2.92	2.64	3.04	2.72
17–19	1.55	3.92	3.32	3.01	2.73	3.14	2.81

Note: FVC = forced vital capacity; FEV1.0 = one-second forced respiratory volume.
Formula of Cotes (1965). FVC = $0.907 (H)^{2.75}$; FEV1.0 = $0.708 (H)^3 + 0.11$.
Gender-specific formulae of Shephard & Lavallée (1995) for active children living in a pollution-free environment:
Boys FVC = $0.93 (H)^3$; FEV1.0 = $0.79 (H)^3$
Girls FVC = $0.85 (H)^3$; FEV1.0 = $0.76 (H)^3$
Source: Rode & Shephard (1992a).

Table 6.3. *Exponents describing the relationship between standing height and lung volumes, as seen in three cross-sectional surveys of Igloolik Inuit (1969/70, 1979/80 and 1989/90) and nGanasan tested in 1991*

	Boys		Girls	
Survey	FVC	FEV1.0	FVC	FEV1.0
1969/70	3.28	3.31	3.00	2.48
1979/80	3.47	3.29	3.40	3.21
1989/90	3.74	3.52	2.56	2.34
nGasasan				
1991	3.23	3.27	3.86	3.90

FVC = forced vital capacity; FEV1.0 = one-second forced respiratory volume.
Based on data of Rode & Shephard (1992a, 1994e).

the maximal mid-expiratory flow rates also showed no evidence of a secular trend across the three Igloolik surveys, although (perhaps as a consequence of smoking) the percentage of the vital capacity expired in one second was lower in older than in younger children in each of the three surveys.

Lung function in adults

Surveys in both Wainwright (Rennie *et al.*, 1970) and Igloolik (Rode & Shephard, 1973a) showed that at the time of the IBP-HAP survey the forced vital capacity, maximal breathing capacity and one-second forced expiratory volume of the Inuit all exceeded norms for urban populations. In men, the margin was around 10%, and in women the advantage was over 20%. The three cross-sectional surveys (1969/70, 1979/80, and 1989/90) that were completed on the Igloolik population allowed an examination of the impact of acculturation to a sedentary lifestyle upon these high readings (Rode & Shephard, 1973a, 1984c, 1994e, f). Since many of the same subjects participated in each of the three surveys, it was also possible to examine (age-category × cohort) interactions, and to make longitudinal estimates of the rate of aging of lung function. Additional data were collected from the less acculturated nGanasan population of Volochanka during 1991.

In the case of the Inuit, a simple history of respiratory and general health was also checked against medical and radiographic records held at the Igloolik Health Centre.

Forced vital capacity

Comparisons of the three cross-sectional surveys of the Igloolik Inuit showed a secular trend to an increase of forced vital capacity in the youngest adult age decade (20–30 years, Fig. 6.2), little change in those aged 30–39 or 40–49 years, and a large decrease in those aged 50–59 years. The decrease of lung volumes apparently occurred mainly between 1969/70 and 1979/80, although subject numbers were rather small to be certain of this.

The primary method of analysis of the cross-sectional data was the fitting of multiple regression equations that included height and age terms over the range 20–60 years of age. Static and dynamic lung volumes were consistently affected by height and age, as anticipated, but there was no significant age squared term, except for a weak effect among the women surveyed in 1989/90 ($p < 0.033$). Accordingly, equations fitted simply for age (years) and standing height (cm) were used to compare the rate of aging of forced vital capacity between the three surveys.

At the time of the IBP-HAP survey, the aging coefficient was much smaller than that for urban populations, although it was comparable with data for a number of other indigenous groups who were living in remote locations (Shephard, 1978). Age coefficients for the entire Igloolik sample, including those with a past history of respiratory disease (Table 6.4), increased from 1969/70 to 1989/90 (men: 16.5 versus 52.3 ml/yr; women: 27.3 versus 36.5 ml/yr). Data for women with no history of respiratory disease (26.5 ml/yr in 1969/70, 28.7 ml/yr in 1989/90) showed little secular trend, suggesting that much of the apparent increase in the rate of aging for the entire population could reflect the aging of a diseased cohort with less than average lung volumes. However, men with no prior history of respiratory disease also showed an increase in their age coefficient, from 15.7 ml/yr in 1969/70 to 43.6 ml/yr in 1989/90. In their case, some other factor such as increased smoking or increased inhalation of cold air must have speeded the rate of aging of lung function in the intervening period.

In a typical 20-year-old Igloolik man with a height of 165 cm, the forced vital capacity as predicted from our population-specific regression equation increased significantly from 5.09 l in 1969/70 to 5.48 l in 1979/80, and 5.62 l in 1989/90 (Table 6.4). The corresponding prediction of Anderson *et al.* (1968) for the urban population would be only 4.77 l. In contrast, the predicted forced vital capacity at an age of 60 yr decreased significantly from 4.43 l in 1969/70 to 3.62 l in 1979/80, and 3.53 l in 1989/90 (Table 6.4). The corresponding urban norm would be 3.87 l. Predictions for a 155 cm woman from the Inuit population showed a small increase at age 20 yr, from 3.92 l in 1969/70 to 4.26 l in 1979/80 and 4.14 l in 1989/90, all of these

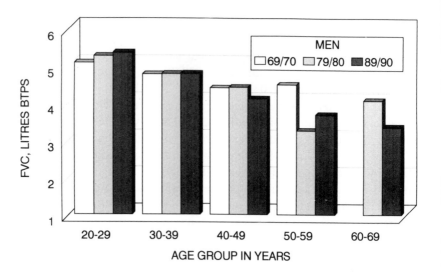

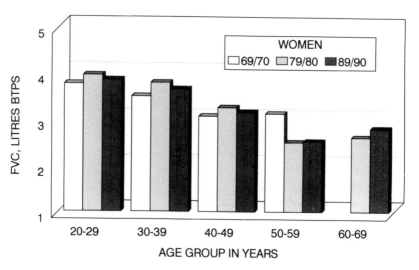

Fig. 6.2. Forced vital capacity of adult Igloolik Inuit as observed in 1969/70, 1979/80 and 1989/90. From Rode & Shephard (1992a).

values being much in excess of the urban norm of 3.12 l (Anderson *et al.*, 1968). In contrast, predicted values at an age of 60 yr showed little change (2.83 l in 1969/70, 2.88 l in 1979/80 and 2.64 l in 1989/90), compared with the urban norm of 2.70 l. In both sexes, the secular trend has been to an increase in the advantage of the young adult Inuit over their urban peers, but particularly in the men an increase in the aging coefficient from 1969/70 to 1989/90 has dissipated much of this advantage among the older adults.

A two-way ANOVA for the Igloolik population (age-group × cohort) showed significant effects of age-group ($p < 0.001$) and of age-group × cohort ($p < 0.011$) in the men, but only of age-group ($p < 0.001$) in the women.

The longitudinal data suggested an acceleration of lung function loss as the Inuit became older (Table 6.5). However, in contrast to the cross-sectional results, there was little difference in the rate of loss of forced vital capacity between decades commencing in 1969/70 and 1979/80. In those initially aged 20–30 years, respective figures for the two cohorts were 17.1 and 6.0 ml/year in the men, and 15.1 and 18.9 ml/year in the women. In those initially aged 30–50 years, values were 73.6 and 45.7 ml/year for the men, and 25.5 and 42.3 ml/year for the women. Corresponding cross-sectional estimates for the age-groups (20–30 years) and (30–50 years) beginning in 1969/70, and in 1979/80 respectively, were 24.0 and 37.9 ml/year in the men, and 30.9 and 35.5 ml/year in the women. Further study is needed to determine whether the discrepancy between the cross-sectional and longitudinal approaches is due to the small number of subjects in the longitudinal comparisons, or whether it is due to the fitting of linear regressions to an accelerating cross-sectional curve.

Average values for the forced vital capacity of nGanasan men and women aged 20–49 years were similar to those of the Igloolik Inuit (and were much larger than would be predicted from urban norms). Lung function was better preserved in the older nGanasan than in the older Inuit, so that the aging coefficients in the multiple regression equation were smaller for Volochanka than for Igloolik residents (Table 6.4). Indeed, the elderly nGanasan retained an advantage over their white counterparts. ANOVA comparing the nGanasan with the Igloolik population of 1989/90 showed a significant age-group × ethnic group interaction in the men, but not in the women.

One second forced expiratory volume

Cross-sectional data patterns for the one-second forced expiratory volume were very similar to those for the forced vital capacity (Table 6.4). At the time of the IBP-HAP survey, values were high in both Igloolik (Rode & Shephard, 1973a) and Wainwright (Rennie *et al.*, 1970), although the ratio of one-second forced expiratory volume to forced vital capacity was below the urban norm (92–100% of the expected percentage in the men, and 92–94% of the expected percentage in the women).

Smoking habits contributed little to the variance of one-second forced

Table 6.4. *Multiple regression equations of the type forced vital capacity or one-second forced expiratory volume (FEV1.0) = a (age, years) + b (height, cm) + c for the prediction of forced vital capacity (l, BTPS) and FEV1.0 in Inuit and nGanasan aged 20–60 years, with typical values for males of height 165 cm and females of height 155 cm at ages shown*

Subject group and n	Constant (mean ± SE)			Typical volume	
	$a \times 100$	$b \times 100$	c	Age 20	Age 60
Forced vital capacity					
All male Inuit					
1969/70 ($n = 69$)	-1.65 ± 0.71	$+5.91 \pm 1.21$	-4.32 ± 2.06	5.09	4.43
1979/80 ($n = 97$)	-4.63 ± 0.64	$+7.50 \pm 1.29$	-5.97 ± 2.18	5.48	3.62
1989/90 ($n = 119$)	-5.23 ± 0.53	$+5.58 \pm 1.04$	-2.54 ± 1.74	5.62	3.53
All male nGanasan					
1991 ($n = 26$)	-2.41	8.50	-8.36	5.18	4.22
Healthy male Inuit					
1969/70 ($n = 49$)	-1.57 ± 0.82	$+5.33 \pm 1.28$	-3.34 ± 2.21	5.15	4.52
1979/80 ($n = 50$)	-3.22 ± 0.79	$+5.72 \pm 1.54$	-3.37 ± 2.51	5.35	4.14
1989/90 ($n = 70$)	-4.36 ± 0.91	$+4.10 \pm 1.24$	-0.34 ± 2.07	5.56	3.82
All female Inuit					
1969/70 ($n = 49$)	-2.73 ± 0.69	$+0.57 \pm 0.29$	$+3.57 \pm 0.48$	3.92	2.83
1979/80 ($n = 70$)	-3.45 ± 0.57	$+6.34 \pm 1.07$	-4.88 ± 1.70	4.26	2.88
1989/90 ($n = 92$)	-3.65 ± 0.38	$+5.23 \pm 0.81$	-3.23 ± 1.27	4.14	2.64
All female nGanasan					
1991 ($n = 25$)	-2.23	$+7.06$	-6.53		
Healthy female Inuit					
1969/70 ($n = 32$)	-2.65 ± 0.90	$+3.45 \pm 1.30$	-0.88 ± 2.07	3.94	2.88
1979/80 ($n = 24$)	-3.15 ± 0.80	$+8.61 \pm 1.61$	-8.41 ± 2.55	4.30	3.04
1989/90 ($n = 55$)	-2.87 ± 0.46	$+6.18 \pm 0.87$	-4.87 ± 1.35	4.14	2.99

One second forced expiratory volume

All male Inuit				
1969/70 ($n = 69$)	-2.22 ± 0.74	$+3.69 \pm 1.27$	-1.61 ± 2.16	4.04	3.16
1979/80 ($n = 97$)	-5.30 ± 0.60	$+5.63 \pm 1.20$	-3.69 ± 2.03	4.55	2.43
1989/90 ($n = 119$)	-5.11 ± 0.49	$+4.62 \pm 0.95$	-1.90 ± 1.60	4.71	2.66
All male nGanasan				
1991 ($n = 26$)	-4.61	$+8.20$	-7.81	4.80	2.96
Healthy male Inuit				
1969/70 ($n = 49$)	-1.85 ± 0.85	$+3.01 \pm 1.34$	-0.56 ± 2.31	4.02	2.87
1979/80 ($n = 50$)	-3.65 ± 0.63	$+5.15 \pm 1.22$	-3.31 ± 2.00	4.46	3.00
1989/90 ($n = 74$)	-3.97 ± 0.82	$+3.27 \pm 1.12$	$+0.53 \pm 1.88$	4.66	3.07
All female Inuit				
1969/70 ($n = 49$)	-2.71 ± 0.65	$+7.90 \pm 2.75$	$+2.48 \pm 0.46$	3.16	2.07
1979/80 ($n = 70$)	-3.62 ± 0.52	$+5.25 \pm 0.97$	-3.90 ± 1.54	3.52	2.07
1989/90 ($n = 92$)	-3.79 ± 0.42	$+4.26 \pm 0.88$	-2.31 ± 1.38	3.54	2.02
All female nGanasan				
1991 ($n = 25$)	-3.11	$+6.26$	-5.43	3.65	2.40
Healthy female Inuit				
1969/70 ($n = 32$)	-2.69 ± 0.91	$+2.17 \pm 1.33$	$+0.36 \pm 2.11$	3.19	2.11
1979/80 ($n = 24$)	-3.42 ± 0.79	$+7.92 \pm 1.59$	-8.02 ± 2.51	3.57	2.20
1989/90 ($n = 54$)	-3.15 ± 0.48	$+5.56 \pm 0.92$	-4.43 ± 1.43	3.55	2.30

Based on data of **Rode & Shephard** (1994b, 1994f), collected in 1969/70, 1979/80 and 1989/90 at Igloolik, and 1991 in Volochanka.

Table 6.5. *Ageing of forced vital capacity and one-second forced expiratory volume (l BTPS/yr × 100), as estimated from longitudinal data. Based on data for two ten-year intervals, commencing in 1969/70 and 1979/80*

Initial age (years)	Males		Females	
	1970–80	1980–90	1970–80	1989–90
Forced vital capacity				
20–30	1.71 ± 4.56	0.60 ± 2.52	1.51 ± 1.82	1.89 ± 1.25***
30–40	6.13 ± 3.58***	4.49 ± 3.55***	1.85 ± 2.44	4.32 ± 1.55***
40–50	8.58 ± 3.40**	4.64 ± 1.93***	3.25 ± 2.04*	4.14 ± 3.10**
One second forced expiratory volume				
20–30	0.42 ± 2.14	2.23 ± 3.46*	1.31 ± 1.45*	1.80 ± 1.74*
30–40	2.63 ± 3.28**	4.87 ± 2.53***	2.73 ± 2.43**	3.35 ± 2.75
40–50	3.85 ± 1.55***	5.96 ± 1.94**	3.03 ± 2.21**	2.30 ± 1.36*

$*p < 0.05; **p < 0.01; ***p < 0.001$.
Source: Rode & Shephard (1994f).

expiratory volume, with the exception of a marginal effect ($p < 0.052$) in the women in 1989/90. Age squared terms were not statistically significant in 1969/70 or 1979/80, but in 1989/90 the introduction of quadratic terms into the prediction equation increased the multiple r^2 from 0.662 to 0.681 for the men ($p < 0.021$), and from 0.766 to 0.786 for the women ($p < 0.010$). Because this effect was small, cohort comparisons were again based on age and height alone in subjects 20 through 60 years of age.

Age coefficients for all Igloolik Inuit aged 20–60 years, as calculated from the cross-sectional data, increased progressively from 1969/70 to 1989/90, (22.2 ml/yr versus 51.1 ml/year in the men, 27.1 versus 37.9 ml/yr in the women, Table 6.4). In the men, the increase in the age coefficient from 1969/70 to 1989/90 was somewhat smaller (18.5 versus 39.7 ml/yr) if subjects with a history of respiratory disease were excluded, but in the women values for the restricted sample (26.9 versus 31.5 ml/yr) were similar to those for all women.

As for the forced vital capacity, the predicted one-second forced expiratory volume for a typical 20-year-old man, initially greater than the urban norm of 3.80 l, increased further (to 4.04 l in 1969/70, 4.55 l in 1979/80 and 4.71 l in 1989/90). In contrast, predictions at 60 years of age decreased (from 3.16 l in 1969/70 to 2.66 l in 1989/90), thus approaching the urban norm of 2.48 l. In women, predicted values at age 20 years, again initially in excess of the urban norm of 2.73 l, increased further over the two decades (from 3.16 l in 1969/70 to 3.54 l in 1989/90). In contrast, values at 60 years of age, initially close to the urban norm of 2.04 l, showed little change (2.07 l in 1969/70, 2.02 l in 1989/90).

The percentage of the forced vital capacity that was expired in one second increased with acculturation, values for 1989/90 averaging 81–83% in the men and 79–85% in the women, compared with 76–79% and 75–80% in 1969/70.

Longitudinal estimates of the aging of one-second forced expiratory volume for the Igloolik community showed substantial curvilinearity (Table 6.5). They also confirmed that there had been a secular trend for faster aging in the decade beginning in 1979/80 when this was compared with the decade beginning in 1969/70. In men, respective averages for the two decades were 4.2 versus 22.3 ml/yr in subjects initially aged 20–30 years and 32.4 versus 54.2 ml/yr in subjects initially aged 30–50 years. In the women, corresponding values were 13.1 versus 18.0 ml/yr in subjects who were initially aged 20–30 years and 28.8 versus 28.3 ml/yr in subjects initially aged 30–50 years. The corresponding cross-sectional estimates for the decades beginning in 1969/70 and 1979/80 were 37.6 and 52.1 ml/yr in the men, and 31.8 and 37.2 ml/yr in the women.

A two-way ANOVA (age-group × cohort) showed significant effects of age-group ($p < 0.001$) and age-group × cohort ($p < 0.005$) in the men, but in the women only the age-group term was statistically significant ($p < 0.001$).

The young adult nGanasan tested in 1991 had a similar one-second forced expiratory volume to the Igloolik Inuit examined in 1989/90. However, as with the forced vital capacity, the one-second forced expiratory volume was better preserved in the older nGanasan than in the older Inuit. A two-way ANOVA again showed an age-group × ethnic group interaction (significant in the men), with the nGanasan developing a substantial advantage in the older age categories.

Maximal mid-expiratory flow rate

The maximum mid-expiratory flow rate was measured on the Igloolik Inuit in 1979/80 and in 1989/90. Smoking habits did not contribute to the variance of maximal mid-expiratory flow rate, and age squared terms were significant only in 1989/90. Respective values of r^2 for the 1989/90 linear and quadratic equations were 0.550 and 0.600 in the men, and 0.605 and 0.653 in the women. To allow comparison with FVC and FEV1.0, cross-sectional analyses were based on age and height alone (Table 6.6).

Overall cross-sectional data for 1979/80 showed age constants of 77 and 58 ml/s per year in the men and women respectively. Corresponding slopes for healthy subjects were 60 and 48 ml/s per year. By 1989/90, overall age

Table 6.6. Equations of the type maximal mid-expiratory flow (MMEF) = a (age, yr) + b (height, cm) + c for the prediction of maximal mid-expiratory flow rate (l/s, BTPS) of Igloolik Inuit aged 20–60, with typical values for males of 165 cm and females of 155 cm height. Data from surveys conducted in 1979/80 and 1989/90

Subject group and n	Constants (mean ± SE)			Typical values	
	a × 100	b × 100	c	Age 20	Age 60
All men					
1979/80 (n = 97)	−7.72 ± 1.06	+5.01 ± 2.14	−1.89 ± 3.61	4.82	1.75
1989/90 (n = 119)	−6.26 ± 1.07	+5.13 ± 2.08	−2.10 ± 3.49	5.11	3.28
Healthy men					
1979/80 (n = 50)	−5.95 ± 1.40	+6.96 ± 2.74	−5.51 ± 4.47	4.78	2.41
1989/90 (n = 70)	−3.71 ± 1.79	+2.07 ± 2.43	+2.38 ± 4.06	5.06	3.57
All women					
1979/80 (n = 70)	−5.82 ± 1.12	+4.92 ± 2.08	−2.54 ± 3.30	3.93	1.60
1989/90 (n = 92)	−5.51 ± 0.87	+3.36 ± 1.86	−0.15 ± 2.90	4.11	1.90
Healthy women					
1979/80 (n = 24)	−4.78 ± 2.02	+11.67 ± 4.07	−13.17 ± 6.44	3.96	2.05
1989/90 (n = 54)	−5.04 ± 1.11	+6.49 ± 2.10	−4.84 ± 3.25	4.21	2.20

Source: Rode & Shephard (1992a).

Table 6.7. *Change of maximal mid-expiratory flow rate (l/s × 100 per year) in Igloolik Inuit, as estimated from longitudinal data collected over a 10-year interval (1980 to 1990)*

Initial age (years)	Males	Females
20–30	+1.62 ± 4.16	−0.75 ± 5.12
30–40	+0.39 ± 7.11	−0.24 ± 5.56
40–50	−5.92 ± 6.08*	−2.43 ± 2.91*

*$p < 0.05$.
Source: Rode & Shephard (1992a).

coefficients had decreased to 63 and 37 ml/s per year in men and women, respectively. Longitudinal data (Table 6.7) showed less decrease of flow rate with age, except in men who were initially aged 40–50 years. Averaging across the three available age categories, longitudinal estimates of losses were only 13 ml/s per year in the men and 11 ml/s per year in the women. A part of the discrepancy between the cross-sectional and the longitudinal data may reflect a learning of technique with a repetition of the measurements.

A two-way ANOVA showed significant effects of age-group ($p < 0.001$ in both sexes), and of cohort ($p < 0.007$ in men, $p < 0.025$ in women).

Other measurements of lung function

Other measurements of lung function were obtained only in 1969/70. At this time, the Igloolik Inuit had a total lung capacity some 20% higher than would have been anticipated in a 'white' person of similar height. Some 10–15% of this difference was due to the large forced vital capacity, and the remaining 5% to an increase of residual volume. However, when subjects with a history of previous respiratory disease were eliminated from the sample, the residual volume/total lung volume ratio corresponded to findings in urban norms (Rode & Shephard, 1973a).

Both the resting pulmonary diffusing capacity and the relationship of diffusing capacity to oxygen intake were much as observed in urban norms (Rode & Shephard, 1973a). Comparing the resting pulmonary diffusing capacity for the third and the sixth decades of life, cross-sectional data showed a decrease of 48% in the men and 39% in the women, possibly a little greater than the 33% loss of resting diffusing capacity that is seen in 'white' norms.

Possible explanations of secular changes in adult data

Observed change

Cross-sectional data for the Igloolik community suggested that there had been a secular trend to a roughly equal decrease of both forced vital capacity and one-second forced expiratory volume in older men and women with acculturation to a 'modern' lifestyle, but the longitudinal data indicated that the only substantial change had been a decrease of one-second expiratory volume in the older men.

Analysis of the cross-sectional data suggested that the forced vital capacity of a typical 20-year-old male had increased by a substantial 0.4 to 0.5 l from 1969/70 to 1989/90. This could conceivably reflect an improvement of respiratory health during childhood, although the absence of inter-trial differences of lung volumes in the children (above) is somewhat against this hypothesis. In contrast to this beneficial change, the cross-sectional data also suggested that the average forced vital capacity of a 60-year-old man had decreased by 0.6 to 0.7 l over the two decades, with parallel trends in the one-second forced expiratory volume. Moreover, the secular trend to a faster aging of male one-second forced expiratory volume was confirmed in longitudinal data, so that it cannot be dismissed simply as an artifact of altered population demographics or curvilinearity of the aging process.

Technical considerations

Artifacts can distort both cross-sectional comparisons of successive cohorts and longitudinal data (Ware *et al.*, 1990), particularly when testing a small community. One consequence of these artifacts is that estimates of the rate of aging of lung function differ between cross-sectional and longitudinal surveys (Glindmeyer *et al.*, 1982; Johnson & Dempsey, 1991; Shephard, 1993c; Vollmer *et al.*, 1988). In urban populations, longitudinal estimates of lung function aging have given both larger (Ware *et al.*, 1990) and smaller numbers than cross-sectional values (Burrows *et al.*, 1986; Glindmeyer *et al.*, 1982; Vollmer *et al.*, 1988), depending on the relative strengths of the various factors biasing the two estimates.

Cross-sectional studies

Successive cross-sectional studies of lung function may be influenced by environmental changes (differential exposure of successive cohorts of subjects to respiratory diseases and air pollutants), secular trends in biological characteristics (particularly adult stature, aerobic fitness and muscular strength) and changes in the lifestyle of a population (particularly the prevalence of cigarette smoking, Shephard, 1993c). Further, if comparisons are based on linear regression techniques, problems may arise from a change in the age distribution of the sample that is included in the analysis.

Longitudinal studies

Longitudinal lung volume data may be biased by the recruitment of a health-conscious sample for a prolonged trial, with a selective retention or loss of subjects with respiratory disease as the observations are repeated. The characteristics of the recording spirometer may also change between successive measurements, and study participants may progressively learn details of measurement techniques. Finally, as in cross-sectional studies, there are effects from any secular changes in lifestyle or environment.

Sampling bias

The reported averages for a given population can be substantially distorted by quite small differences in sample composition (Shephard, 1978, 1980). Such distortion is particularly likely to occur if successive cross-sectional surveys have differing objectives (for example, disease detection versus fitness assessment).

To minimize the inter-trial bias that is an almost inevitable consequence of incomplete sampling (Shephard, 1978, 1980), recruitment of subjects for the three Igloolik surveys was undertaken by the same observer, using the same methodology throughout. Recruitment reached about 50% of adults in each of the three surveys, and a similar figure was achieved subsequently in Volochanka. Although these percentages are less than optimal, they provide larger and better defined samples than in many previous surveys of indigenous peoples.

Linkage of the lung volume measurements to a fitness survey probably biased all three of the Igloolik samples toward fit members of the community, but this should not have vitiated inter-survey comparisons of average lung volumes at any given age, nor should it have distorted comparisons of the rate of aging of lung function from one survey to the next. Higher age coefficients might have been observed if we had tested a medically biased partial sample, but the cross-sectional estimates of aging summarized in Table 6.4 match those that have been reported in other surveys of indigenous populations (Shephard, 1978), so that the data have prima facie reasonableness.

Selective migration has biased some cross-sectional estimates of population aging (Miall *et al.*, 1967). We were fortunate that in 1969/70, the last Inuit of the Igloolik region had just moved from the field camps of the Foxe Basin into permanent settlement housing, and there was as yet little outward migration in search of employment.

Because hunting is weather-dependent, we could not obtain repeat measurements on the hunters according to any rigid schedule. Each of the three Igloolik surveys simply continued from the early fall to the late spring of the following year, when all willing volunteers had been tested. Difficulty in controlling the timing of individual observations was accepted as an unavoidable constraint in the examination of this population.

Methods of data analysis

In keeping with many currently available surveys of urban communities, data on the lung volumes of indigenous populations have generally been reported as regressions on standing height and age. An alternative tactic that one group of investigators has proposed for longitudinal comparisons (the expression of lung volumes as ratios of height-based norms, Ware *et al.*, 1990) was judged inappropriate for the Igloolik community because of substantial secular trends to changes of adult stature (Chapter 7).

Changing stature also precluded the use of one proposed statistical technique for the combining of cross-sectional and longitudinal information on lung volumes (Ware *et al.*, 1990). As an alternative, subjects were categorized by age-group and age-group and cohort effects were analyzed by two-way ANOVAs.

The longitudinal data for the Igloolik population suggested some curvilinearity in the deterioration of lung function with age, and this could be one reason why the longitudinal estimates of losses in forced vital capacity and one-second forced expired volume were smaller than the cross-sectional estimates for young adults, but greater than the cross-sectional estimates for older individuals.

Curvilinearity of the aging process (Anderson *et al.*, 1968; Dockery *et al.*, 1985; Shephard, 1956) is particularly likely to influence the interpretation of cross-sectional results if estimates of aging rates are based on the fitting of linear regressions to data covering a wide range of ages. It is possible to allow for curvilinearity statistically by introducing a breakpoint (24 years in the equations of Anderson *et al.*, 1968) or by fitting a quadratic equation (Ware *et al.*, 1990). The initial approach adopted in our circumpolar studies was to fit both linear and quadratic age terms. Relatively few of the quadratic coefficients were statistically significant. Nevertheless, curvilinearity could have been masked by limitations of sample size, a suggestion that is indeed supported by the apparent trend to curvilinearity in the longitudinal data.

A combination of curvilinearity and an increasing number of elderly survivors in more recent cohorts could generate an artifactual trend to faster aging. However, we minimized the risk of this problem in each of the three surveys of Igloolik and in the subsequent survey of Volochanka by restricting cross-sectional analyses to the age range of 20–60 years.

Test learning

Test learning can affect the scores attained in a longitudinal study, particularly if measurements are repeated frequently. However, the learning of spirometric technique seems at most a minor factor, even with a three- rather than a 10-year inter-test interval (Ware *et al.*, 1990). Since the same observer used the same equipment throughout, differences in motivation and test mechanics were also kept to a minimum in our circumpolar studies.

Learning is probably most marked in terms of the subject's ability to make very rapid respiratory movements, and the learning of this specific technique could have contributed to the discrepancy between cross-sectional and longitudinal estimates of the aging of maximal mid-expiratory flow rate (above).

Cigarette smoking

Cigarette smoking induces an acute, short-term bronchospasm, and more permanent changes in lung structure and function (bronchial congestion, increased mucus secretion and emphysema). The chronic effects of smoking are proportional to the pack-years of cigarettes that have been consumed. They lead to a selective loss of dynamic lung volumes in older smokers (Bossé *et al.*, 1980; Ferris *et al.*, 1976; Fletcher *et al.*, 1976). Thus, in cross-sectional surveys, steeper age coefficients are found for smokers than for non-smokers. Smoking also increases the likelihood of developing chronic chest disease, and it is this association that explains much of the faster aging of lung function in the smoker. In our experience, the effects of cigarette smoking upon forced vital capacity become statistically insignificant if allowance is first made for the chronic chest disease this habit may have induced (Anderson *et al.*, 1968).

The smoking history of most of the indigenous circumpolar populations is unfortunately not known precisely. It seems likely that the oldest cohorts of the population in settlements such as Igloolik have had less pack-years of cigarette exposure than would be predicted from their ages, because there was more limited contact with 'developed' societies and a smaller cash income when they were young adults.

Almost all of the adult population in both Igloolik and Volochanka now smoke, but probably because of limited inter-individual variance in the reported consumption of cigarettes, smoking did not emerge as a significant variable influencing either the mean lung volumes or their rate of aging. The recent adoption of a substantial cigarette habit could be one reason why the overall cross-sectional age coefficients for forced vital capacity and one second forced expiratory volume, as seen in the Igloolik survey of 1969/70, were shallower than those encountered in many urban communities. The aging of a cohort with a high annual cigarette consumption could also contribute to the more rapid aging of respiratory function that is seen in later cohorts from the same population.

Chronic respiratory disease

About a third of the Igloolik sample gave a history of physician-diagnosed or radiographic respiratory disease in each of our three surveys (31% in 1969/70, 56% in 1979/80, and 41% in 1989/90). The diagnosis was usually a prior tuberculous infection, currently controlled.

Separate statistical analyses were performed for all available adults of either gender, and for those adults who were free of chest disease. As discussed below, with a few individual exceptions, radiographic chest disease seemed to have little effect upon pulmonary function. However, it was not possible to test for the possible impact of non-specific viral infections, which are now endemic from early childhood.

Cold air

The secular trend to a faster aging of lung function in the men cannot be explained by any changes in the test environment. Ambient conditions in the test laboratory were identical for men and women, and did not change from one survey to the next. However, acculturation could have changed the pattern of outdoor cold exposure, particularly for male villagers.

Schaefer *et al.* (1980b) argued that in previous decades, the exposure of hunters to arctic air had been sufficient to cause both an accelerated aging of lung function and the development of pulmonary hypertension. Cross-sectional coefficients for the aging of lung function were much smaller in Igloolik than those that Schaefer *et al.* (1980b) had observed in Arctic Bay. Nevertheless, the Igloolik population also showed one possible indicator of pulmonary hypertension, an above-average prevalence of right branch bundle block (above).

Recent reductions of hunting activity have undoubtedly decreased the total *duration* of cold air exposure for the average resident of Igloolik, but the *intensity* of exposure could have increased for some of the hunters, since operation of high-speed snowmobiles now tends to displace the 'cushion' of warm air that previous generations inspired from within the hoods of their parkas.

Snowmobile trauma

There has been a 2–3 cm decrease of stature of Igloolik Inuit over the past two decades (Chapter 7). One probable etiological factor is the vibration incurred during the regular operation of high-speed snowmobiles over rough snow and ice.

However, any such change would have affected lung volumes only to the extent that standing height became unrepresentative of thoracic dimensions. A 1% error in predicting thoracic length would give a discrepancy of < 3% in lung volume predictions, insufficient to account for all of the observed acceleration in the aging of lung volumes. Kyphosis and/or compression of the thoracic spine could also impede thoracic mechanics, although the change of stature seen in Igloolik was rather small to have had any major influence on chest movements.

Indoor air pollution

Traditional Inuit families were exposed to high concentrations of soot from oil lamps (Beaudry, 1968; Schaefer *et al.*, 1980b). However, both the Igloolik and the Volochanka settlements have had electric lighting during the period under review, and the Igloolik residents have also had oil furnace heating for the past 20 years or more.

The homes are small and firmly sealed against the arctic cold, and since most of the family now smoke, the main ambient hazard is exposure to environmental cigarette smoke, which begins at birth. However, any impact of this source of indoor air pollution upon adult respiratory function (Shephard, 1992) is likely to have been over-shadowed by the effects of personal cigarette smoking, which also begins at an early age.

Loss of physical fitness

Over the 20 years of observation, there have been large decreases in the aerobic power and muscle strength of the average adult Inuit in Igloolik, with an associated increase of body fat (Chapter 5). In theory, these changes, also, could have made some contribution to the loss of lung function, although in practice it is hard to explain why most of the decrease in lung volumes would be seen in older members of the population, since the loss of aerobic fitness tended to be more marked in young adults.

Conclusion

Cross-sectional and longitudinal data agree in showing a secular trend to an accelerated aging of lung volumes in the Igloolik Inuit, more marked in the men than in the women. This secular trend does not seem to be explained by any increase in domestic air pollution, nor does responsibility seem to lie with an increased prevalence or an altered age distribution of chronic respiratory disease.

Possible causal factors include an artifact due to increased lung volumes in young adults, the aging of a cohort with a high annual consumption of cigarettes, the adverse effects of endemic non-specific respiratory disease and a loss of physical fitness, altered chest mechanics due to the operation of high-speed snowmobiles and resultant shortening of the spine, or pathological responses to an increased inspiration of cold air. Cigarette smoking, cold exposure and loss of fitness have all affected men more than women over the 20 years of observation, making these some of the more likely explanations.

Respiratory disease

Respiratory diseases may have acute effects upon respiratory function, causing an increased secretion of mucus, inflammation of the airway and bronchospasm. Such changes affect dynamic lung volumes (one-second forced expiratory volume, maximal mid-expiratory flow rate and maximal breathing capacity) to a greater extent than static volumes such as forced vital capacity. They may also worsen the distribution of inspired gas, decreasing pulmonary diffusing capacity and increasing the ventilatory cost of effort.

In contrast, chronic respiratory disorders lead to tissue destruction, fibrosis and calcification of the lungs, affecting static as well as dynamic lung volumes. In addition, chronic infection may have systemic effects, with a stunting of growth in children, and a wasting of lean tissue in adults.

Prevalence of respiratory disease

Earlier in the present century, tuberculosis was rampant in much of the circumpolar region. Johnson (1971) reported that the mortality rate for tuberculosis among Alaskan Amerindians and Inuit was 655 per 100 000 in 1920, and that by 1950 it had increased to an overall prevalence of 1.5–1.8%. In Greenland, the prevalence was even higher, around 7% in the Julianehab District (Grzybowski & Stylbo, 1976), and in the Baffin Zone of the Canadian North West Territories, the annual incidence during the period 1955 to 1957 was 2.9% (Wherrett, 1969). Palva & Finell (1976) noted that in 1953, new cases of tuberculosis in Lapland (403 per 100 000) were twice the average for Finland.

However, in most of these jurisdictions, acculturation to a 'modern' lifestyle has been accompanied by a steady reduction in the prevalence of tuberculosis. The introduction of comprehensive control programmes began around 1950 in Greenland, 1955 in Alaska, and the early 1960s in Canada. By 1972, the new case rate in Lapland had dropped to 72 per 100 000, comparable with the rest of Finland, and in the Canadian Arctic the incidence of primary tuberculosis had decreased from 390 per 100 000 in 1960–62 to 1 per 100 000 in 1970–73. Nevertheless, the annual incidence of new and reactivated cases in Québec in 1981 was still 66.7 per 100 000 Cree Amerindians and 202.4 per 100 000 Inuit (Renaud & Dumont, 1985). In all jurisdictions, the death rate from tuberculosis had dropped to near zero by the early 1970s (Grzybowski & Stylbo, 1976), but many of the population still showed residual effects of tuberculous disease. For instance, among the Inuit of northern Labrador (Nain), 21% had a history of tuberculosis, 12% had suffered pneumonia, and 7% bronchitis (Edwards *et al.*, 1985).

Chronic non-specific respiratory disease has also had an unusually high prevalence rate in some circumpolar settlements. For example, Brody (1965) and Fleshman *et al.* (1968) both reported high chronic bronchitis rates for Alaskan settlements. Hildes *et al.* (1976) examined chest radiographs for 315 adults from the Foxe Basin in 1969/70. In 304 of these individuals, an estimate was also made of the diameter of the descending branch of the right pulmonary artery. More than a quarter of adults over the age of 50 years (a total of about 15 individuals) were judged to have severe radiographic evidence of chronic obstructive pulmonary disease, whereas most of those aged 20–29 years had normal chest radiographs. Hildes *et al.* (1976) were inclined to attribute the changes that they saw in older subjects to the inhalation of cold air (above). They linked these observations to a reduction of dynamic lung volumes, normal static volumes, and an age-related broadening of the pulmonary artery to dimensions that (particularly in the men) exceeded norms for the urban population. A number of the subjects that they examined had incomplete right bundle branch clock, but a multivariate analysis failed to link electrocardiographic abnormalities to the ventilatory changes.

Nikitin (1985) suggested that the indigenous populations of the Soviet arctic had adapted to cold air by increasing their residual volumes, thereby substantially diluting inhaled cold air with warm alveolar air. Nevertheless, he also reported that the incidence of chronic bronchitis and pneumonia was higher in circumpolar settlements than in the southern parts of Siberia. Astakhova *et al.* (1991) noted that as many as 16% of Chukotka natives suffered from chronic non-specific respiratory disease. Reijula *et al.* (1990) reported a similar prevalence of chronic bronchitis (15%) among Finnish reindeer herders, although they attributed much of the pathology that they observed to a reindeer epithelium allergy, rather than to cold exposure.

Morrell *et al.* (1975) and Carson *et al.* (1985) commented that Inuit children had a high incidence of lower respiratory tract infections. Contributing factors here could include dry air, overcrowding, exposure to environmental tobacco smoke, and poorly treated ear infections. However, in most of the cases that they examined, lung function was apparently not impaired relative to their peers from the same settlement.

Physiological consequences

The general physiological consequences of a past history of respiratory disease seem quite small in most instances. Perhaps because the more active and less acculturated members of the community have been particularly vulnerable to chronic respiratory infections, Rode & Shephard (1992b) noted that the body fat content of affected individuals was generally below the average for other indigenous residents of the settlement. However, in a few individuals where disease was advanced, mobility had become impaired. In such circumstances, the body fat content was above average and aerobic power and muscle strength were poor.

Rode & Shephard (1973a) classified the Igloolik Inuit with respiratory problems in terms of the clinical diagnosis. The commonest forms of chronic chest disease in 1969/70 were minor hilar calcification and or Ghon foci associated with primary tuberculous infection. Such findings had little influence upon either pulmonary function or maximal aerobic power (Table 6.8).

Seventeen subjects had a history of successfully treated secondary tuberculosis. In general, such individuals showed a slight stunting of height and a below-average body mass. Particularly in the women there was some

Table 6.8. *The effects of respiratory disease upon pulmonary function and aerobic power. Based on data for the Inuit community of Igloolik (1969/70). All values are expressed as a percentage of values for healthy individuals of the same age*

Diagnosis	Height	Body mass	FEV1.0	FVC	$\dot{D}_{L,co}$	$\dot{V}_{O_2max}$
Hilar calcification						
Male ($n = 8$)	100.4	103.6	106.4	105.4	118.4	109.7
Female ($n = 3$)	99.3	85.0	101.7	97.7	82.0	110.3
Primary tuberculosis						
Male ($n = 9$)	102.3	100.2	106.6	99.6	102.0	101.9
Secondary tuberculosis						
Male ($n = 6$)	99.7	98.2	96.4	104.4	117.3	97.3
Female ($n = 6$)	98.3	91.7	84.3	86.5	85.3	98.5
Advanced tuberculosis						
Female ($n = 5$)	100.4	94.4	103.8	102.0	72.6	118.3
Pulmonary fibrosis						
Male ($n = 7$)	100.3	102.3	93.3	101.0	103.8	102.3
Female ($n = 1$)	101.0	106.0	115.0	108.0		99.0
Emphysema						
Male ($n = 1$)	96	90	85	97	55	100
Bronchiectasis						
Male ($n = 1$)	101	86	49	63	60	82

Source: Rode & Shephard (1973a).
$\dot{D}_{L,co}$ = pulmonary diffusing capacity. $\dot{V}_{O_2max}$ = maximal oxygen intake.

reduction in both one-second forced expiratory volume and forced vital capacity. In cases with more advanced chest disease, there was also a substantial limitation of pulmonary diffusing capacity, but in keeping with our view that maximal oxygen intake is determined by cardiac rather than respiratory factors, none of these changes had more than a marginal effect upon aerobic power.

Radiographic reports of pulmonary fibrosis were linked with a small reduction of forced expiratory volume, but no impairment of oxygen transport. Only one individual was described as having emphysema. This person showed a substantial deficit in both one-second forced expiratory volume and pulmonary diffusing capacity. One person had gross pulmonary disease – a combination of bronchiectasis, emphysema and healed tuberculosis. He showed gross impairment of pulmonary function, with

Table 6.9. *Secular trends in the influence of a history of respiratory disease upon lung function data. Linear regressions against age and standing height fitted to cross-sectional data for Igloolik Inuit examined in 1969/70, 1979/80 and 1989/90. Volumes predicted for a man of 165 cm and a woman of 155 cm at ages of 20 and 60 years, and differences calculated relative to a person with no history of respiratory disease*

	Men		Women	
Year	Age 20	Age 60	Age 20	Age 60
Forced vital capacity (l)				
1969/70	4.95	4.26	3.62	3.30
1979/80	5.48	3.18	4.23	2.83
1989/90	5.65	3.50	4.14	2.44
Delta forced vital capacity (l)				
1969/70	−0.20	−0.26	−0.32	+0.42
1979/80	+0.05	−0.96	−0.07	−0.21
1989/90	+0.09	−0.32	0.0	−0.55
One-second forced expiratory volume (l)				
1969/70	4.10	2.87	2.89	2.49
1979/80	4.68	1.74	3.49	2.04
1989/90	4.67	2.63	3.50	1.79
Delta one-second forced expiratory volume (l)				
1969/70	+0.08	−0.43	−0.30	+0.39
1979/80	+0.22	−1.26	−0.08	−0.16
1989/90	+0.01	−0.44	−0.05	−0.51
Maximal mid-expiratory flow rate (l/s)				
1979/80	4.50	2.09	3.96	1.35
1989/90	4.96	2.77	3.89	1.72
Delta maximal mid-expiratory flow rate (l/s)				
1979/80	−0.21	−0.52	+0.24	−1.34
1989/90	−0.15	−0.49	−0.18	−0.96

Source: Rode & Shephard (1992a).

some reduction of aerobic power and muscle strength and an above-average body fat.

Similar studies were conducted in 1979/80 and 1989/90. This allowed an assessment of the influence of acculturation upon overall function in those with a history of respiratory disease (Rode & Shephard, 1992a). The lung function data can conveniently be summarized by means of linear regressions for those with a history of chronic respiratory disease. Lung volumes are predicted for a 165 cm man and a 155 cm woman at ages 20 and 60 years (Table 6.9). In all three surveys, the effect of a history of respiratory disease upon static and dynamic lung volumes was much larger for the 60-year-old person than for the 20 year old. This reflects in part progression of the disease state with aging, and in part the existence of more serious disease among older individuals who became infected when modern treatment was less accessible. Table 6.9 also emphasizes that in terms of demonstrating impaired respiratory function, the maximal mid-expiratory flow rate is a much more sensitive tool than the measurement of forced vital capacity or one-second expiratory flow rate.

Conclusions

Currently, the lung function of young adults in the circumpolar regions compares favourably with that of city-dwellers, but a secular trend to a more rapid aging of lung function among circumpolar residents also seems to have developed in recent years. Thus, in the latest studies older individuals have demonstrated slightly poorer lung function than sedentary city dwellers.

Possible factors contributing to this adverse trend include an increased use of cigarettes by circumpolar populations, an increased inhalation of cold air when operating snowmobiles, and a progressive loss of physical fitness with adoption of a more sedentary lifestyle. The practical importance of introducing measures to reduce the current prevalence of cigarette smoking among the circumpolar peoples is emphasized by rapidly increasing death rates from lung cancer in many of the arctic settlements.

7 Secular trends in growth and development

The observed pattern of growth in any individual reflects both that person's genetic potential for growth and the extent to which an optimization of environmental factors such as nutrition and mental health has allowed the realization of this potential. The course of growth is also influenced by the timing of puberty. An advance in the age of puberty tends to increase the size of young children, thereby exaggerating any secular trend to an increase of size among the prepubertal segment of a population. But because of earlier epiphyseal closure, an advance of puberty may also limit adult dimensions. Other factors such as increasing kyphosis, the compression of intervertebral discs, osteoporosis and vertebral collapse lead to a progressive decrease in an individual's stature over the span of adult life. These factors, again, may have some inherited component, but are certainly susceptible also to changes of personal environment.

The present chapter examines available information on the growth and development of circumpolar populations, noting limitations of the cross-sectional and semi-longitudinal methodologies that have commonly been employed to date. The influence of acculturation to a 'modern' lifestyle is also explored, looking not only at alterations in growth curves, but also at changes in the age of puberty, and secular trends in adult stature.

Limitations of cross-sectional and semi-longitudinal surveys

Because of the logistic problems of conducting longitudinal surveys in the arctic, the majority of investigators have been content to apply cross-sectional methodologies (Auger *et al.*, 1980; Jamison, 1990; Rode & Shephard, 1973e, 1984b, 1994c; Stewart, 1939; Zammit, 1993).

One assumption implicit in a cross-sectional approach is that growth patterns have remained constant in different cohorts of any given population. This assumption is not strictly valid, even in urban 'white' society, but in settlements that are undergoing rapid 'modernization', the shape of growth curves can become grossly distorted if a cohort that is supposed to represent the typical size of young children ultimately grows several centimetres taller than the teenage cohort that has preceded them. It is also difficult if not impossible to determine either the magnitude or the timing of the pubertal growth spurt accurately from cross-sectional data, since the averaging of measurements that have been obtained on different individuals smooths out and displaces the growth peak. Zemel & Johnston (1994) recently compared estimates of peak height velocity (PHV) and age at PHV obtained from cross-sectional and longitudinal analyses of the same population. Their material was drawn from the third Harvard growth study. In boys, the age at PHV as deduced from the cross-sectional analysis was not grossly incorrect (13.9 ± 0.3 years, versus 14.1 ± 0.9 years in the longitudinal analysis), but in girls the age at PHV was substantially under-estimated from cross-sectional data (11.3 ± 0.8 years, versus 12.2 ± 0.9 years in the longitudinal analysis). In boys, the peak height velocity was significantly underestimated by a cross-sectional analysis of the data (87 ± 23 versus 91 ± 12 mm/year), and in girls the cross-sectional analysis led to a much larger under-estimation of the longitudinal value (69 ± 7 versus 94 ± 12 mm/year). Occasionally, cross-sectional surveys have been repeated by the same investigator, or the standard measurement techniques proposed by the IBP-HAP have been used in two independent studies. However, many of the early cross-sectional data sets were obtained by public health workers, applying relatively crude techniques to small samples of the population. In a few instances (for example, Hrdlicka, 1941), measurements were entrusted to a local schoolteacher, apparently without any supervision.

Rode & Shephard (1973e, 1994c) carried out two semi-longitudinal surveys on the growth of Inuit children in Igloolik in 1969/70 and 1989/90. In each of these surveys, a range of measurements was repeated on each child after an interval of approximately one year. This allowed a crude calculation of growth rates in students of various ages, although the 12 month inter-test interval was sufficiently long that it tended to blunt peak growth rates. Given the relatively small population of Igloolik, the total number of subjects available for examination at any given age was also rather small to draw accurate conclusions about the rate of growth, or indeed to comment upon the extent of errors that others may have introduced by their use of a cross-sectional methodology.

Rode & Shephard (1995) also obtained longitudinal data on the growth of height, sitting height, body mass, triceps skinfold thicknesses and handgrip force in a substantial sample of Igloolik school children, making measurements twice during each year from 1981 through 1989. Participants totalled 281 boys and 266 girls. Some students entered the longitudinal survey subsequent to their growth spurt, and others did not reach puberty over the period of observation. Nevertheless, it was possible to group subjects into three substantial cohorts of children, born in 1970/72, 1973/74, and 1975/76 respectively, and to follow details of the growth and development of each of these cohorts over the pubertal period. The information obtained on these subjects reflected the true course of growth for the individuals examined. After aligning subjects in terms of their individual ages at peak height velocity, it was also possible to plot growth rates accurately for the three cohorts that had been examined. Unfortunately, it is unlikely that the findings can be generalized to subsequent generations of Inuit children, since continuing acculturation to the life of a permanent settlement will probably expose future cohorts to a different environment during their years of growth.

Cross-sectional findings

Height

The World Health Organization has recommended that the growth of all populations be compared with the cross-sectional norms established by the US National Center for Health Statistics (NCHS) in 1977 (Hamill *et al.*, 1979). In order to avoid problems arising from inter-population differences in the age of the pubertal growth spurt, the majority of the values presented here are compared well before puberty (at an age of nine years), and after growth has been largely completed (at an age of 17–19 years).

Historical data (Heller *et al.*, 1967; Hrdlicka, 1941; Schaefer, 1970) and the IBP observations of 1968–70 (Auger *et al.*, 1980; Jamison, 1990) all suggested that children from the various circumpolar populations were quite short relative to NCHS norms. Nevertheless, data collected in Cumberland Sound in 1939 and 1968 (Schaefer, 1970) suggested that over the intervening 30 years, Canadian Inuit children had undergone an increase of standing height and thus a decrease in their weight for height ratio. Schaefer (1970) commented that although a secular trend to an increase of stature was often attributed to an increased intake of dietary protein, the Cumberland Sound Inuit were characterized by a high protein diet both in 1938 and in 1968. He noted that the most obvious dietary change in the intervening 30 years had been an increase of sugar consumption, and he speculated that this might account for the secular trend to greater stature.

At birth, both the height and the body mass of the Inuit child have typically been close to the NCHS norms (Heller *et al.*, 1967; Zachau-Christiansen, 1976), but by 10 months of age, values have dropped to the 10th percentile of NCHS norms (Heller *et al.*, 1967; Jamison, 1990). In older children and adults, the inter-population gradients of height that were observed at the time of the IBP-HAP survey probably reflect differences in either genetic admixture (heterosis, Hulse, 1957) or the extent of acculturation to the lifestyle of 'developed' societies. In the late 1960s, the height of Alaskan Inuit tended to follow the 25th percentile of the NCHS norms throughout the period of growth, the Canadian Inuit were at the 5th–10th percentile, and the height of Greenland Inuit lay between the other two populations.

Jamison (1976a) fitted a polynomial growth curve to his data, claiming that growth of the Alaskan Inuit continued through to an age of 20–25 years. However, since he was fitting an arbitrary curve to a small cross-sectional sample of data, the apparent continuation of growth through into adulthood could reflect either an inappropriate statistical model or an unusual size of subjects in the 20–25 year category, which had already been attained by these individuals during their late teen years. Other authors, ourselves included, have not seen any substantial growth of Evenki, Inuit or nGanasan beyond the age of 20 years in males and 16 years in females (Leonard *et al.*, 1994c; Rode & Shephard , 1973e, 1984b, 1994c; Zammit, 1993).

Petersen & Brant (1984) retested the Inuit boys of Wainwright in 1977. Relative to the 1968 data of Auger *et al.* (1980), they found that boys were significantly taller through to 11 years of age, and girls were also significantly taller between three and nine years of age. They commented that during the intervening period, government authorities had instituted a supplemental nutrition programme, and there had been a marked decrease in the prevalence of anaemia in Wainwright. Postl (1981) examined Keewatin Inuit in 1980, finding that the heights of young children (birth–seven years) lay below the 50th percentile of the NCHS norms (although apparently above the 5–10th percentiles). Zammit (1993) examined 100 Labrador Inuit in 1991. The younger children in this sample again were at the 50th percentile of the NCHS norms, but older individuals had dropped back toward the 10th percentile of norms. A study of Cree Amerindians from the James Bay area of Québec (Moffatt *et al.*, 1985) also suggested little difference from the NCHS norms over the first 30 months of life.

Why are the older children and adults of traditional circumpolar populations so short? Does this finding have a genetic basis? Asians have traditionally been of short stature (Jamison, 1990), and it has thus been

Table 7.1. *Height of children (in cm) at an age of 9.0 years. Comparison of circumpolar populations with the 1977 norms of the US National Center for Health Statistics norms (Hamill et al., 1979)*

Population	Boys	Girls	Author
US NCHS Norms			Hamill *et al.* (1979)
50th percentile	132.2	132.2	
25th percentile	128.2	127.7	
10th percentile	125.2	123.9	
5th percentile	122.9	122.1	
Evenki			
	123	123	Leonard *et al.* (1994c)
Inuit			
Alaska	128.7	128.5	Jamison (1976a)
Fort Chimo, Québec	126.1		Auger (1976)
Igloolik & Hall Beach	125.3	119.7	de Pena (1972)
Labrador	125	132	Zammit *et al.* (1994)
Lapps			
Kautokeino	127.4	126.2	Lewin & Hedegard (1971b)
Inari	128.8	127.3	Lewin *et al.* (1970)

proposed that the Asian heritage of many indigenous circumpolar populations (Chapter 1) could explain why they are small relative to norms for the US population. The 'rules' of Bergmann (1847) and Allen (1877) have also suggested that a compact body form might be a useful genetic adaptation to a cold environment.

Despite the isolation of the populations studied in Table 7.1, in many instances there has been considerable genetic admixture. As might be predicted from a genetic hypothesis (Hulse, 1957; Schreider, 1967), the hybrids have been taller than those of pure ancestry. For instance, Jamison (1970) plotted regression lines of age against height for hybrid and non-hybrid Alaskan Inuit. At the age of nine years, respective values for the hybrids and the non-hybrids were 1.32 and 1.28 m in the boys, and 1.31 and 1.27 m in the girls. However, it is less clear whether the greater growth of the hybrids is due simply to the admixture of European and Inuit genes, or whether it also reflects the fact that in comparison with genetically pure Inuit, individuals of mixed parentage have often become more acculturated to a 'modern' lifestyle, with associated changes in their diet (particularly an increased intake of refined carbohydrates, Martin, 1981; Schaefer, 1970; Suzuki, 1970), and improved levels of general health and hygiene. Certainly, acculturation appears to have increased the average stature at any given age, but the increase has extended to some isolated populations where there has been little genetic admixture (Skrobak-Kaczynski *et al.*, 1974). Moreover, the rapid increase in height of the Japanese and of some circumpolar groups in recent years has argued strongly against the hypothesis that short stature is an invariant genetic characteristic of the Asian and indigenous circumpolar peoples (Matsuura, 1993; Skrobak-Kaczynski *et al.*, 1974).

Leonard *et al.* (1994b) suggested that the energy demands of exposure to severe cold might be an environmental factor restricting the growth of traditional circumpolar residents, particularly during the winter months. However, Icelandic children who also lived in a cold (but subarctic) habitat were found to be at least 50 mm taller than Inuit of similar age (Auger *et al.*, 1980). Moreover, we have seen no evidence that Inuit growth rates differ between the temperate summer and the coldest winter months (Rode & Shephard, 1995).

Another possible complicating factor (J. Gottlieb, pers. commun., 1994) is the administration of methyl phenidate for supposed learning disabilities. In some 'modernized' circumpolar communities, health workers have administered this drug and its analogues widely, but it has a stunting effect, so it could not explain any enhancement in the average growth of the population. By exclusion, the main cause of the increase in stature of indigenous circumpolar children in recent years thus seems a change in dietary patterns and improved standards of general hygiene related to the 'modernization' of their habitat.

In some circumpolar communities, the increase in size of recent cohorts of children has been substantial. Alaskan Inuit children examined during the survey of Jamison (1976a and b) were several centimetres taller than the Inuit from south-western Alaska who had been followed semi-longitudinally from 1928 to 1931 (Hrdlicka, 1941), and data collected in 1977 (Petersen & Brant, 1984) suggested that the secular trend was continuing, at least in prepubertal Alaskan Inuit. Heller *et al.* (1967) examined Inuit children in nine villages in western Alaska between 1957 and 1961. Until puberty, the growth of children in this earlier series was closely comparable with that of the children examined by Jamison (1976a and b), but the postpubertal students seen by Heller *et al.* (1967) were shorter, with heights similar to the Canadian Inuit of 1969/70 (de Pena, 1972).

Over the period from 1969/70 to 1989/90, our data for the Inuit community of Igloolik (Rode & Shephard, 1994c) showed a trend for young children in the more recent cohorts to be taller than their predecessors (Table 7.2); over the two decades, the height of the youngest category of boys (age 10.2 ± 0.9 years) increased from 1.32 ± 0.06 to 1.37 ± 0.06 m, and the height of the girls increased from 1.29 ± 0.05 in 1969/70 to

Table 7.2. *Secular trends in standing height, body mass, sum of three skinfolds, handgrip, leg extension strength, relative maximal oxygen intake and absolute maximal oxygen intake. Repeated measures ANOVA (by age-category, cohort and gender) for data collected at Igloolik in 1969/70, 1979/80, and 1989/90.*

Source of variance	Height	Body mass	Skin folds	Handgrip force	Knee ext.	Rel. $\dot{V}_{O_2max}$ (ml/[kg·min.])	Abs. $\dot{V}_{O_2max}$ (l/min.)
Age category	$p < 0.001$	$p < 0.001$	$p < 0.001$	$p < 0.001$	$p < 0.001$	$p < 0.001$	$p < 0.001$
Gender	$p < 0.001$	ns	$p < 0.001$	$p < 0.001$	$p < 0.001$	$p < 0.001$	$p < 0.001$
Cohort	$p < 0.011$	$p < 0.003$	$p < 0.001$	$p < 0.001$	$p < 0.001$	$p < 0.003$	$p < 0.001$
Age × gender	$p < 0.001$	$p < 0.001$	$p < 0.001$	$p < 0.001$	$p < 0.001$	$p < 0.001$	ns
Age × cohort	$p < 0.025$	ns	ns	ns	$p < 0.001$	ns	ns
Gender × cohort	ns	ns	$p < 0.001$	$p < 0.017$	ns	ns	ns

Source: Rode & Shephard (1994c).

1.36 ± 0.07 m in 1979/80. Given the young age of the students that were being compared, it seems unlikely that the increase in size that we observed in Igloolik could be attributed to earlier maturation, at least in the boys.

The size of the cross-sectional sample was rather small to identify the timing of the growth spurt precisely, but in the boys who were measured in 1989/90, it appeared that rapid growth began at 12–13 years of age, and continued through into the 15th year, whereas in 1969/70, rapid growth did not begin until an age of 13–14 years. In 1989/90, collection of cross-sectional data on the girls began at an age of 11 years, when rapid growth was already established. Inferences regarding timing of the growth spurt are in general supported by the semi-longitudinal and longitudinal data, discussed below.

In postpubertal children, there was little evidence of any secular trend to an increase of height over the period from 1969/70 to 1989/90. It could be that conditions were already optimal for realization of genetic potential in 1969/70, although this is difficult to reconcile with the secular trend in the younger children and the advance in the age of maturation. Alternatively, some other factor (such as the increasing use of powerful snowmobiles) may have been cancelling out secular gains in the older teenagers.

Recent comparisons between the Inuit children of Igloolik and the nGanasan of Volochanka have shown no significant interpopulation differences of standing height (Rode & Shephard, 1994b). A two-way ANOVA (age category × ethnic group) showed that no significant component of variance could be attributed to ethnicity in either boys or girls (Rode & Shephard, 1994b). Likewise, the heights of the children of Evenki pastoralists, as reported by Leonard *et al.* (1994c), were similar to those of Igloolik Inuit students.

Retrospective analysis of successive cross-sectional surveys that included adult subjects between 20 and 50 years of age has provided some further evidence supporting a secular trend to an increase of stature among circumpolar groups in the period immediately preceding the IBP-HAP survey (Auger *et al.*, 1980; Leonard *et al.*, 1994c; Shephard & Rode, 1973), although more recent information has been tainted by uncertainties regarding the extent of any superimposed age-related decrease of stature within a given cohort. As discussed below, there is evidence that aging leads to an unusually rapid shortening of stature in some circumpolar populations. This process may recently have been accelerated by repeated exposure to vibration during the operation of high-speed snowmobiles. At the time of the IBP-HAP survey, Inuit males in western Greenland were only 15 mm shorter at an age of 41–60 years than at 21–40 years, and most of this gradient could be attributed to the normally anticipated decrease of stature with aging (Jorgensen & Skrobak-Kaczynski, 1972). In contrast, Greenlandic

Table 7.3. *Apparent secular trend (cm/decade) in the standing height (cm) of males (M) and females (F), as deduced from comparisons of adults in successive age decades (data for final decade adjusted by 1.5 cm to allow for normal ageing). Data collected in the late 1960s*

Population	Age decades						Author
	20–30 vs 30–40		30–40 vs 40–50		40–50 vs 50–60		
	M	F	M	F	M	F	
Inuit							
Alaska	1	−1	0	2	−2.5	0	Jamison (1976a)
Fort Chimo	2	—	0	—	0	—	Auger (1976)
Igloolik 7 Hall Beach	2	1	2	1.5	−2.5	0	de Pena (1972)
Igloolik							
(1969/70)	0.6	1.2	2.0	1.0	−2.1	2.1	Rode & Shephard (1992a)
(1979/80)	0.8	1.2	−0.5	1.2	1.6	0.1	
(1989/90)	0.6	−0.6	1.0	1.0	−2.4	0.5	
Lapps							
Inari	4	3	0	−1.5	0	1	Lewin et al. (1970)
nGanasan	2.5	3.0	−5.2	−2.3	—	—	Rode & Shephard (1994b)

Source: Based in part on data of Auger et al. (1980).

Inuit women aged 21–40 years were actually some 30 mm shorter than those aged 41–60 years. Comparison of other circumpolar data for successive age decades (Table 7.3) suggests that with the possible exception of the Inari Lapps, any secular trends to an increase of size of the indigenous populations as young adults had slowed or even reversed by the time of the IBP-HAP study.

Direct evidence on the magnitude of the secular trend to a change of stature has been sought in repeated cross-sectional surveys, undertaken either by the same observers or successive teams of investigators. Zammit (1993) found that at an age of 16–18 years, the Labrador Inuit males had a height of 1.64 m, as compared with much lower values reported by Stewart (1939): 1.60 m in 1928 and 1.57 m in 1880–1900; comparable figures for the females of this region were 1.56, 1.50 and 1.48 m. These data suggested a modest but sustained secular trend to an increase of stature, amounting to 7 mm/decade in the males and 8 mm/decade in the females. In the males, the trend appeared to have become somewhat slower in more recent years, but in the females it had accelerated. Skrobak-Kaczynski & Lewin (1976) compared Lapps tested at the time of the IBP-HAP survey with those measured by other observers in 1928 and 1934. Over the 40 year interval, the height of the Skolt Lapps had increased by an average of 106 mm, and the Inari Lapps were on average 50 mm taller. In contrast, the Evenki group studied by Leonard *et al.* (1994c) showed a similar average height to that reported for the same population some 90 years earlier (Boas, 1903).

Recent cross-sectional data for the Inuit of Igloolik (Rode & Shephard, 1994c) have shown not only a cessation of the secular trend to an increase of stature, but even its reversal among older adolescents and adults. These observations underline the fallacy of drawing inferences about secular growth trends from adult heights as seen in a single cross-sectional survey. At an age of 17.0–19.9 years, the male Inuit of Igloolik had respective heights of 1.67 ± 0.07, 1.63 ± 0.05 and 1.64 ± 0.05 m in the surveys of 1969/70, 1979/80 and 1989/90. The corresponding figures for the females were 1.57 ± 0.07, 1.53 ± 0.07 and 1.54 ± 0.04 m. Adult data sets, similarly, showed decreases rather than increases in height across the three surveys (Table 7.4).

Sitting height

The circumpolar peoples have traditionally been regarded as short-legged, with a correspondingly large ratio of sitting to standing height (Auger *et al.*, 1980; Hrdlicka, 1941; Johnston *et al.*, 1982). However, even at the beginning of the IBP-HAP survey, most data sets were not remarkable in

Table 7.4. *Adult height in successive cohorts of Igloolik Inuit examined in 1969/70, 1979/80 and 1989/90*

Age group (years)	1969/70 cohort Males	1969/70 cohort Females	1979/80 cohort Males	1979/80 cohort Females	1989/90 cohort Males	1989/90 cohort Females
20–30	166.1	156.5	164.9	154.5	164.3	153.2
30–40	165.5	155.3	164.1	153.3	163.7	153.8
40–50	163.5	154.3	164.6	152.1	162.7	152.8
50–60	164.1	150.7	161.5	150.5	163.6	150.8

Source: Rode & Shephard (1992a).

Table 7.5. *Sitting height, expressed as a percentage of standing height, in successive age groups of circumpolar residents examined in the late 1960s*

Population	20–30 years M	20–30 years F	31–40 years M	31–40 years F	41–50 years M	41–50 years F	51–60 years M	51–60 years F
Inuit								
Alaska	54.0	53.5	53.8	53.5	53.6	53.4	53.6	52.7
Greenland	54.6	54.1			53.3	53.4		
Igloolik & Hall Beach	54.9	54.3	52.9	54.0	53.3	53.4	53.9	53.6
Lapps								
Inari	54.1	54.3	54.1	53.0	53.9	53.1	53.7	53.4

Source: Based in part on data collected by Auger *et al.* (1980).

this regard (Table 7.5). Currently, the sitting height of the Igloolik Inuit children (Rode & Shephard, 1995) is actually a slightly smaller proportion of stature than in average Canadians (Demirjian, 1980).

Body mass

The body mass of circumpolar residents has typically been large in proportion to standing height (Schaefer, 1970). Johnston *et al.* (1982) noted this trend in St. Lawrence Island Inuit, and blamed it upon their short leg length. Postl (1981) found that the trend to a large mass for height was apparent even in young Keewatin Inuit (1–7 years of age), and in most surveys the same tendency has persisted throughout growth (Table 7.6). At the time of the IBP-HAP study, data for older Inuit children lay between

Table 7.6. *Body mass (kg) in children aged 9.0 years, expressed as a percentage of standing height (cm)*

Population	Male	Female
US NCHS		
50th percentile	20.2	20.1
75th percentile	21.8	22.0
90th percentile	23.8	24.8
Inuit		
Alaska	22.6	23.6
Igloolik & Hall Beach	21.8	21.3
Lapps		
Kautokeino	21.1	20.8
Inari	21.0	20.0

Based in part on data obtained in the late 1960s, and collected by Auger *et al.* (1980). Comparison data taken from US National Center for Health Statistics (Hamill *et al.*, 1979).

the 75th and the 90th percentiles of the cross-sectional norms proposed by the US Center for Health Statistics (Hamill *et al.*, 1979), but the body mass of the Lappish children was somewhat less in relation to their height.

ANOVAs comparing the Igloolik Inuit children of 1989/90 with less acculturated nGanasan children measured in 1992/93 (Rode & Shephard, 1994b) showed that the Igloolik boys were 4–10% heavier than their peers in Volochanka ($p < 0.047$). The girls, also, tended to be heavier in Igloolik than in Volochanka (difference 0–10%, $p < 0.10$).

Cross-sectional data of Rode and Shephard (1994c) for the Igloolik community showed a statistically significant overall tendency for the body mass at any given age to increase across the two decades of observation (Table 7.2). When first examined in 1969/70, the girls aged 13.0–14.9 years were heavier than boys of the same age (47.3 versus 41.7 kg). This was attributed to leg muscle that the girls had developed by carrying babies on their backs over rough ice and snow. In support of their hypothesis, Rode & Shephard (1973e) showed that the girls had a parallel advantage of knee extension strength over the boys. Taking data for the same age-category, the gender difference in body mass was smaller in 1979/80 (46.7 kg in the girls versus 44.9 kg in the boys), and it had been reversed by 1989/90 (46.2 kg in the girls versus 49.5 kg in the boys).

In the Inuit boys of Igloolik, rapid cross-sectional gains of body mass occurred at an age concurrent with the rapid increase in stature, the spurt beginning earlier in 1989/90 (12–13 years) than in 1969/70 (13–14 years). In the girls of this community, the period of rapid increase in body mass had

Table 7.7. *Influence of acculturation on body mass (kg) expressed as a percentage of standing height (cm). Data for Igloolik Inuit obtained in 1969/70, 1979/80 and 1989/90*

Age group (years)	Males			Females		
	1969/70	1979/80	1989/90	1969/70	1979/80	1989/90
9.0–10.9	22.0	22.4	24.2	22.3	25.8	
11.0–12.9	26.2	27.5	27.2	26.8	27.0	28.8
13.0–14.9	28.1	29.6	32.1	31.4	31.4	30.7
15.0–16.9	33.6	34.6	35.3	35.3	34.5	35.5
17.0–19.9	37.6	38.2	38.5	34.3	34.7	35.4

Source: Rode & Shephard (1994c).

already begun at the initial age of sampling, and thus could not be defined precisely for any of our cross-sectional surveys.

The contribution of lean tissue to the unusual body mass of the Inuit child was underlined by Jamison (1976a). He showed that the arm muscle area of St. Lawrence Island Inuit lay between the 50th and the 85th percentiles of US norms.

Cross-sectional data have demonstrated a substantial trend for an increased weight for height ratio among Igloolik Inuit boys of all ages over the period from 1969/70 through 1989/90 (Table 7.7). The younger girls have conformed to the same trend, but in the older age categories the mass for height of the girls has either remained constant or has even decreased across the three surveys. One explanation of this last finding is that girls now less commonly carry babies on their backs. Store-bought garments do not include the traditional amauti for carrying young children, and carpeted homes allow small children to crawl on the floor instead of being carried about. Some girls in the most recent cohort may also have dieted in response to television images of feminine beauty, but, in general, skinfold readings have increased rather than decreased across the three surveys.

Skinfold thicknesses

As in urban societies (Malina & Bouchard, 1988; Shephard & Lavallée, 1993), the Igloolik Inuit have shown greater skinfold thicknesses in the girls than in the boys, this finding being true of even the youngest children that were examined. The trend was already apparent in 1969/70, and it has persisted with continuing acculturation to settlement life (Table 7.8).

Table 7.8. *Triceps skinfold thicknesses (in mm) of selected circumpolar populations, categorized by age group*

Population	Age (years)	Males	Females	Author
Inuit				
Alaska	9–10	7.0	10.0	Jamison (1976a)
	11–12	7.5	11.5	
	13–14	8.0	13.0	
	15–16	6.5	18.0	
	17–19	10.0	18.7	
Fort Chimo	9–10	8.3		Auger (1976)
	11–12	6.8		
	13–14	7.6		
	15–16	6.5		
	17–19	7.2		
Igloolik & Hall Beach	9–10	5.4	4.8	de Pena (1972)
	11–12	5.2	6.2	
	13–14	5.6	6.1	
	15–16	3.7	8.4	
	17–19	4.6	8.9	
Keewatin	3–7	8	9	Postl (1981)
Lapps				
Kautokeino	9–10	8.3	10.4	Lewin & Hedegard
	11–12	9.0	10.6	(1971b)
	13–14	9.4	13.6	
	15–16	8.3	15.1	
Inari	9–10	6.8	9.6	Lewin *et al.* (1970)
	11–12	8.5	9.4	
	13–14	7.6	10.6	
	15–16	6.7	13.7	
	17–19	6.2	13.8	

Cross-sectional comparisons across the three surveys showed a consistent increase of skinfold readings from 1969/70 through 1979/80 to 1989/90, this change being statistically significant for the group as a whole (Table 7.2) and for most of the individual age categories. In each of the three surveys, values also increased with age in both boys and girls.

Whereas the thickness of skinfolds for the Inuit children of Igloolik now approaches that of sedentary city dwellers, the less acculturated nGanasan of Volochanka have continued to show quite low readings (Table 7.9). ANOVAs have shown highly significant differences between the two populations in both boys and girls, with no significant age-category × ethnic group interaction.

Table 7.9. *A comparison of skinfold thicknesses (in mm) between the Inuit children of Igloolik (1989/90) and the nGanasan children of Volochanka (1991). Average of three skinfolds,* ±SD

Age group (years)	Males		Females	
	Inuit	nGanasan	Inuit	nGanasan
11.0–12.9	7.9 ± 3.8	6.2 ± 1.2	11.7 ± 6.9	8.0 ± 2.6
13.0–14.9	9.0 ± 4.2	3.1 ± 1.9	11.6 ± 2.8	10.2 ± 4.1
15.0–16.9	7.9 ± 3.7	6.1 ± 1.4	18.2 ± 6.5	11.2 ± 0.8
17.0–19.9	9.6 ± 3.9	6.4 ± 2.1	16.8 ± 5.3	12.7 ± 6.6

Source: Rode & Shephard (1994g).

Handgrip force

Handgrip force is not easy to measure in small children, and the only available information for circumpolar populations seems that collected by Rode & Shephard (1973e; 1984b; 1994b; 1994c). In the boys of Igloolik, the grip strength was generally maintained from 1969/70 to 1979/80, but it had declined by 1989/90 (Table 7.10). At any given age, the girls showed a small decrease of average values from 1969/70 to 1979/80, and a larger decrease was seen in 1989/90.

In the boys, the rapid pubertal gain of grip strength began earlier in 1989/90 (12–13 years) than in 1969/70 (13–14 years). In 1969/70, the girls aged 13–14 years had a greater grip strength than the boys, but in part because of earlier maturation in the boys and in part because of a loss of strength in the girls, this advantage was not seen in subsequent surveys. By the age of 15.0–16.9 years, the girls had only 83% of the strength of the boys in 1969/70, 68% in 1979/80, and 54% in 1989/90. In the most recent survey, age-related gains in the grip strength of the girls apparently slowed after reaching 15 years, and values were actually lower for the older age categories (Table 7.10).

An ANOVA by age category and ethnic group showed that the grip strength of the Inuit boys tested in 1989/90 did not differ from that of nGanasan boys tested in 1992/3, but in the girls, the nGanasan grip strength values were consistently larger than those seen in the Igloolik Inuit.

Knee extension force

In the Inuit children of Igloolik, the knee extension force for any given age category declined progressively from the 1969/70 survey through that of

1979/80 to 1989/90 (Table 7.10), the inter-survey decrease of strength being more marked in the girls than in the boys.

In the 1969/70 survey, the knee extension force of the young girls increased much more rapidly than that of the boys, so that in the age category 13.0–14.9 years the female values stood at 104% of the male readings. However, by 1979/80, female values for the 13.0–14.9 year age-category had dropped to 95% of the male figure, and in 1989/90 the corresponding readings for the females were only 91% of those for their male peers. At age 15.0–16.9 years, female values from the three surveys were 84%, 81% and 88% of the corresponding male figures.

An ANOVA comparing the 1989/90 data set for the Inuit with the nGanasan of 1992/93 showed that the Volochanka boys had a significantly greater knee extension force than the Igloolik population. However, the Volochanka girls had no advantage over their peers in Igloolik. This reflects the fact that girls in Siberia have never carried their baby brothers and sisters on their backs in the manner adopted by the traditional Inuit of the Canadian arctic.

Aerobic power

In 1969/70, the Inuit children of Igloolik had extremely high levels of aerobic power (Fig. 7.1). In the girls, the largest values relative to body mass were seen in the prepubertal groups (9.0–10.9 years), and in the boys the highest relative values were found between 11.0 and 14.9 years of age. Values declined in the subsequent two surveys, and results for the older age categories are now no higher than would be seen in moderately active 'white' city-dwellers.

At ages 13.0–14.9 years, the nGanasan boys had a 9% advantage of aerobic power over their Inuit peers, but in the remaining categories the Inuit had respective advantages of 16%, 19% and 14%. An ANOVA by age-category and ethnic group showed a significant effect of ethnic group and a significant age-category × ethnic group interaction. The Inuit girls also had a 7–10% advantage of aerobic power relative to the nGanasan girls.

Indices of maturation

Mills (1939) commented on the late onset of menstruation in the Inuit. In the mid 1800s, the age at menarche had reputedly been around 23 years (Weyer, 1932). In the 1880s it was still 19–20 years (Cook, 1893–94), and in the 1930s it was 15–16 years (Bertelsen, 1935; Mills, 1939). By the time of the IBP-HAP survey, cross-sectional comparisons among the Igloolik

Table 7.10. *Handgrip and knee extension force (in N) for Inuit children of Igloolik and nGanasan children of Volochanka.* Mean ± SD

Age group (years)	Igloolik males			nGanasan males (1991)	Igloolik females			nGanasan females (1991)
	1969/70	1979/80	1989/90		1969/70	1979/80	1989/90	
Handgrip								
9.0–10.9	134 ± 19	136 ± 30	113 ± 21	—	112 ± 333	114 ± 30	—	—
11.0–12.9	182 ± 37	188 ± 37	144 ± 38	135 ± 57	180 ± 34	153 ± 38	97 ± 34	109 ± 41
13.0–14.9	234 ± 68	266 ± 79	224 ± 81	210 ± 72	253 ± 38	237 ± 33	153 ± 60	200 ± 58
15.0–16.9	346 ± 75	377 ± 94	368 ± 72	377 ± 125	287 ± 32	256 ± 53	198 ± 51	232 ± 40
17.0–19.9	456 ± 50	470 ± 65	402 ± 60	401 ± 61	272 ± 18	276 ± 33	194 ± 46	220 ± 51
Knee extension								
9.0–10.9	361 ± 41	267 ± 84	407 ± 21	—	315 ± 90	300 ± 82	—	—
11.0–12.9	448 ± 118	352 ± 69	355 ± 104	344 ± 94	428 ± 136	307 ± 101	321 ± 91	299 ± 64
1300–14.9	535 ± 207	442 ± 131	368 ± 120	429 ± 151	555 ± 120	506 ± 142	390 ± 100	367 ± 92
15.0–16.9	655 ± 210	626 ± 186	444 ± 89	598 ± 167	551 ± 208	506 ± 142	390 ± 100	320 ± 114
17.0–19.9	802 ± 154	685 ± 165	509 ± 132	571 ± 190	605 ± 116	528 ± 95	402 ± 56	391 ± 101

Data collected by Rode & Shephard (1994b; 1994c).

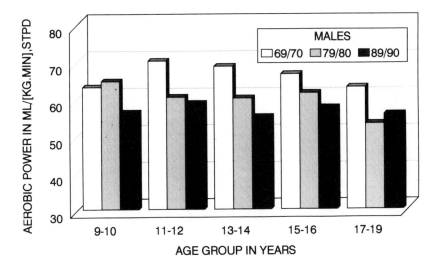

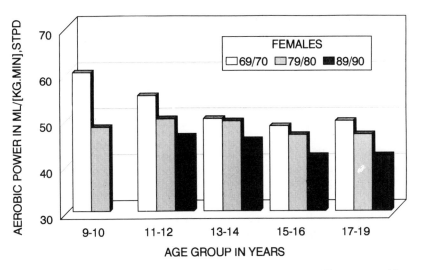

Fig. 7.1. Predicted aerobic power of Inuit children of Igloolik, as surveyed in 1969/70, 1979/80 and 1989/90. Age categories are 9.0–10.9, 11.0–12.9, 13.0–14.9, 15.0–16.9 and 17.0–19.9 years. From Rode & Shephard (1994c).

Inuit (Hildes & Schaefer, pers. commun., 1970) suggested that there was already evidence of further advances in the age of menarche as acculturation to settlement life continued. Women aged 17–29 years reported an average age at menarche of 13.9 years, compared with 14.5 years in women aged 30–40 years, and 14.7 years in those over 50 years of age.

Edgren *et al.* (1976) studied the skeletal maturity of Inari Lapp children between 1968 and 1970. They found an average lag of 11.1 ± 1.4 months relative to Greulich–Pyle standards (although these supposed standards are open to criticism because they were derived from urban children of high socio-economic status; values for average urban children in Canada also lag substantially behind the published curves, Canadian Pediatric Society, 1987; Shephard *et al.*, 1978b).

Observations on Labrador Inuit were undertaken by Zammit *et al.* (1994) in 1991. Again, a substantial 1–2 year lag in bone maturity was observed relative to the Greulich–Pyle standards. The authors of the Labrador study pointed out that normally, an increased weight for height ratio is associated with earlier maturation, and since the Labrador Inuit were extremely thin, they argued that the delayed maturation might have a genetic basis. In keeping with our observations (above) on the age when growth now ceases, Zammit *et al.* (1994) concluded that males started to reach skeletal maturity at an age of 17 years 9 months, and on average females had reached maturity at an age of 15 years 8 months.

Conclusions

In recent years, there has been considerable debate regarding the health of populations that are short (Seckler, 1980). Is short stature of genetic origin, or is it an expression of the unwillingness of the developed world to share food with its poorer neighbours?

The Inuit data illustrate the complexity of the question. If height alone were to be considered, most circumpolar populations would still be regarded as short, but their body mass is normal or even greater than normal in relation to this height. There is thus little evidence to support the idea that the short stature is due to overall malnutrition.

The evidence from cross-sectional surveys of the Igloolik population suggests that in young children acculturation to a 'modern' lifestyle has been associated with a progressive increase of stature at any given age, to the point that in the most recent studies the norms of the NCHS are being approached. The reasons for this secular trend are unclear, but they probably include an increasing genetic pool as contacts with the outside world have increased, changes in composition of the diet, better housing and overall improvements in hygiene. In contrast, the older children and adults in Igloolik remain substantially shorter than the US norms (100–120 mm for males, and around 80 mm for females), suggesting that as individuals become older, additional factors reverse the trend to a greater stature than is evident when they are young children.

As in adults, the acculturation of the children to a 'modern' society has been associated with an increase in body fat, a decrease of muscle strength, and a decrease of aerobic power. In the older children, reasons for these changes include compulsory school attendance (which has reduced the possibility of accompanying parents on hunting trips) and a large reduction in the energy cost of domestic chores (such as preparing food for dog teams in the boys, and carrying children on the back in the girls). However, particularly in young children, one of the most important influences has been the arrival of television; this reached Igloolik by satellite transmission, beginning in the early 1980s. Several previous reports have documented the enormous amount of time that urban children now spend in watching television (Alderson & Crutchley, 1990; Shephard *et al.*, 1975; Sherif & Rattray, 1976), but the Igloolik data are unique in that we have had the opportunity to examine the impact of this cultural change upon the fitness levels of the children concerned. We unfortunately do not have data on the precise number of hours that Inuit children sit in front of the television each week, but casual observation suggests that in the cold winter months, this may even exceed the 20–30 hours per week that is now typical of the North American city-dweller.

Nevertheless, it is encouraging to note that at least some part of the previously observed high levels of aerobic fitness has been conserved by the 18% of male and 12% of female students who have elected to participate regularly in community sport and exercise programmes (Rode & Shephard, 1993; Table 7.11). The current challenge is to make these programmes attractive and culturally relevant to the majority of Inuit adolescents who presently ignore them.

Semi-longitudinal studies

Standing height

There have been two semi-longitudinal studies of the Inuit children living in Igloolik (Rode & Shephard, 1973e, 1994c). Unfortunately, each of these surveys was only able to obtain two measurements on each child, separated by an interval of approximately 12 months. In order to obtain the precision

Table 7.11. *Influence of participation in community sport and exercise programmes upon the fitness of Igloolik Inuit children. A comparison of active (A) and sedentary (S) students. See original text for details of statistics*

Variable	Boys						Girls					
	13-14		15-16		17-19		13-14		15-16		17-19	
	A	S	A	S	A	S	A	S	A	S	A	S
Skinfold (mm)	7.0	9.1	7.5	8.0	8.0	10.1	13.7	11.3	16.3	18.5	15.0	17.2
Handgrip force (N)	319	216	402	363	413	398	142	155	211	196	170	199
Knee ext. force (N)	392	366	461	442	486	517	304	341	388	391	380	406
Aerobic power (ml/[kg·min.])	69.1	53.7	68.3	56.1	66.3	52.1	44.3	45.6	50.0	40.6	52.8	40.0

Source: Rode & Shephard (1993).

Table 7.12. *Growth rates (cm/year) of Inuit children in Igloolik, as seen in semi-longitudinal surveys over a 12-month period; ages shown are average mid-survey values for a given sample*

Age (years)	Boys 1969/70	Age (years)	Boys 1989/90	Age (years)	Girls 1969/70	Age (years)	Girls 1989/90
11.3	3.3			11.1	3.3	11.4	4.6
12.0	4.3	12.3	4.3	11.7	6.5		
12.9	6.9	12.9	6.4	13.3	7.2	12.8	4.4
14.2	6.2	13.9	6.8	14.1	2.5	14.1	2.0
15.2	1.0	14.8	4.0	14.9	1.8	14.9	0.8
16.1	1.3	16.1	1.9	16.1	2.2	16.1	−0.1
16.7	3.3	17.0	1.3	17.1	0.0	16.9	−0.4

Source: Rode & Shephard (1994c).

of data needed for an estimation of growth rates, it was thus necessary to pool data from small samples of children of similar age (Table 7.12). This inevitably masked inter-individual differences in both peak height velocity (PHV) and the age at PHV. Estimates of peak growth rates were further distorted by the relatively long averaging period, and given the asymmetric shape of the velocity curve, the 12-month inter-test interval probably introduced some bias into estimates of the age at PHV.

In the 1969/70 survey, the cohorts of boys with mid-point ages of 12.9 and 14.2 years had the fastest growth rates, each peaking at a rate of some 68 mm/year; the best estimate of age at PHV from this data was 13.6 years. In 1989/90, the fastest growth was seen in those cohorts with mid-point ages of 12.9 and 13.9 years, both peaking at around 69 mm/year. The best estimate of age at PHV for 1989/90 was 13.3 years. In the girls, the 1969/70 data showed the fastest growth in those cohorts with mid-point ages of 11.7 and 13.3 years, peaking at an average figure of 75 mm/year, and the best estimate of age at PHV was 12.6 years. In the 1989/90 sample, there were unfortunately few girls in the critical age range for the calculation of PHV. The fastest growth was seen in cohorts with mid-point ages of 11.4 and 12.8 years, and best estimate of age at PHV for 1989/90 was 12.1 years. The semi-longitudinal data thus confirm the cross-sectional data in suggesting that acculturation to a 'modern' lifestyle has brought about some advancement of the age of PHV over the past two decades.

Body mass

A table analogous to Table 7.12 can be calculated for the two semi-longitudinal sets of data on body mass (Rode & Shephard, 1973e; 1994c). This shows that in the boys, the age of rapid increase of body mass coincides with the age of rapid muscle hypertrophy. In contrast, the period of rapidly increasing body mass in the girls reflects an accumulation of fat in the breast and hips, possibly offset to some extent by a trend to a decrease of muscle mass.

In 1969/70, the fastest increments of body mass in the boys were seen in those with cohort mid-point ages of 14.1 and 15.1 years (respective growth rates of 6.6 and 6.8 kg/year). The best estimate of peak mass velocity for 1969/70 was 14.6 years of age, substantially after the age at PHV. In 1989/90, the fastest increments of body mass were seen at mid-point ages of 12.9 and 13.9 years (respective growth rates of 5.7 and 6.0 kg/year). The best estimate of peak mass velocity for 1989/90 had advanced substantially to 13.4 years. In 1969/70, a peak increase in body mass was less evident among the girls, growth rates being similar (at 5.7–5.9 kg/year) for cohorts with mid-point ages of 11.7, 13.3 and 14.1 years. The best estimate of age at peak mass velocity was 13.0 years. In 1989/90, a peak increase in body mass was even less obvious, with values increasing fairly steadily over the age range 11.4 to 14.9 years.

Skinfold thicknesses

Semi-longitudinal data showed little trend for skinfold thicknesses to increase with age in the boys, thus confirming the view that the observed increments of body mass were attributable to lean tissue, particularly muscle and bone.

However, in the girls, skinfold thicknesses did become larger, with a clear peak of growth. In 1969/70, averaging over three folds, there were respective gains of 2.2 mm/year and 1.9 mm/year at mid-cohort ages of 13.3 and 14.1 years. These values coincided with the period when body mass was increasing particularly rapidly, showing that an accumulation of body fat contributed to the increase of body mass.

Muscle strength

Semi-longitudinal gains in the recorded muscle strength reflect both a true development of muscle tissue from the first to the second test and also some artifactual gains related to a learning of test procedures. The approximate magnitude of such artifacts can be estimated by comparing the overall increments of strength as seen in cross-sectional and semi-longitudinal surveys (Rode & Shephard, 1973e). Over a six year period, the handgrip force increased by an average of 206 N in cross-sectional data, but by 284 N in the summed semi-longitudinal data from six groups of boys, each studied for one year. The average learning artifact was thus a sixth of 76 N, an apparent gain of about 13 N/year. The measurement of knee extension force was rather more susceptible to learning than was the handgrip reading, showing a test/retest artifact of 68 N.

Handgrip force

In 1969/70, the handgrip force of the Igloolik Inuit boys continued to increase relatively late into puberty, the highest annual gains (56, 76 and 62 N/year) being seen at mid-cohort ages of 14.2, 15.2 and 16.1 years. By 1989/90, the peak rate of increase in handgrip force had diminished to 37 N/year, and as might be anticipated from the earlier maturation of those examined, the peak rate of increase in muscle force was seen at a mid-cohort age of 13.9 years; although this was an earlier age than we had found in 1969/70, peak strength velocity was still reached a little later than the age at peak mass velocity.

In the girls who were tested in 1969/70, the greatest increase of handgrip force (54 N/year) occurred in the youngest age category (mid-cohort age of 11.1 years). In 1989/90, the handgrip force of the girls showed only small increments over the ages studied, with no clear age of maximum growth rate.

Knee extension force

In 1969/70, the peak rate of increase of knee extension force for the Igloolik boys was seen rather earlier than for handgrip, with increments of 216, 177, and 138 N/year at mid-cohort ages of 12.9, 14.1 and 15.2 years. By 1989/90, the rates of increase in knee extension force were much smaller (all less than 100 N/year), with no clear peaking of the growth rate.

The girls of 1969/70 also showed a clear peak in the growth of knee extension force, with gains of 164 and 168 N/year at mid-cohort ages of 11.7 and 13.3 years. By 1989/90, the one-year increments of knee extension force were much smaller and more irregular, the largest gains of around 100 N/year being seen in those with mid-cohort ages of 14.9 and 16.1 years.

Aerobic power

The absolute aerobic power of the Igloolik Inuit boys who were tested in 1969/70 showed a sharp peak of growth of 740 ml/min. per year at a mid-cohort age of 15.1 years. This was even later than the age for the peak increase in muscle strength. By 1989/90, the peak of growth in aerobic power had diminished to 450 ml/min. per year, and was occurring much earlier (at a mid-cohort age of 13.9 years).

In the girls, the absolute aerobic power showed little increase over the period of observation, either in 1969/70 or in 1989/90.

Conclusions

Semi-longitudinal data support the evidence of cross-sectional surveys in suggesting that acculturation to the life of a permanent settlement has been associated with an earlier onset of the pubertal growth spurt. Adoption of a 'modern', sedentary lifestyle has also changed the course of increments in body mass, skinfolds, muscle strength and aerobic power. The sharp pubertal increments of strength and aerobic power seen in the Inuit students of 1969/70 have now been replaced by weaker and more variable responses, and in female subjects pubertal increases of body mass now reflect an accumulation of fat rather than a development of muscle.

Longitudinal surveys

The only longitudinal survey of child growth in a circumpolar community is that recently reported by Rode & Shephard (1995) for the Inuit of Igloolik. Their study measured all willing school students every six months during the years 1981–1989. Ideally, accurate measurements should be made on a substantial sample of children four or even six times per year. Some authors have attempted to measure the height of individual students to the nearest 1 mm, but given the 10–20 mm change of stature over the course of a normal day, such attempted precision is unwarranted unless times of laboratory attendance can be rigidly controlled. This is particularly difficult to achieve in a community such as Igloolik, where the sense of time is still not yet strongly implanted. Other practical considerations restricted our measurements to two times per year, and we recognise that this limited the accuracy of our data.

Because growth was averaged over an interval of six months, the apparent peak growth velocities were reduced relative to studies that have made more frequent observations. Inter-measurement differences in heights, as seen in individual study participants, were converted to growth rates for the corresponding six month period. The age of PHV for a given subject was determined by inspection, and growth rates were realigned in terms of peak height velocities. Three substantial cohorts of students were formed, from those born in the years 1970–72, 1973–74, and 1975–76, respectively.

Height and sitting height

Within the limits of the data, ANOVA showed no statistically significant inter-cohort differences of standing height with age category (Fig. 7.2). However, among the boys there was some trend for those born in 1975/76 to mature a little earlier than the two prior cohorts. Thus, the 1975/76 cohort tended to be taller than their peers, particularly at ages 13.25 and 13.75 years.

The standing height was similar for the youngest categories of boys and girls, but over the age range 11–14 years, the height of the girls exceeded that of the boys by some 20 mm. The height of the male subjects tended to an adult asymptote of 1.64 ± 0.05 m, and the females tended to a maximum of 1.53 ± 0.06 m. At all ages, values from each of the three cohorts were consistently a little above the 10th percentile for the urban US population of 1970 (Hamill *et al.*, 1979).

The age of attainment of PHV is summarized in Table 7.13. As in most other communities, the growth spurt in all three cohorts of Igloolik girls occurred more than two years earlier than that of the boys from this settlement. The PHV tended to occur a little earlier in boys born in 1975/76 than in the two earlier cohorts, but any inter-cohort difference was not statistically significant. Prior to puberty, the rate of growth was relatively constant, at about 50 mm/year (Fig. 7.3). However, in both sexes, growth increased at the growth spurt, reaching values of 92 ± 23 mm/year at an age of 11.3 ± 0.7 years in the girls, and 86 ± 37 mm/year at an age of 13.5 ± 0.8 years in the boys. Notice that these rates were substantially higher than those inferred from the semi-longitudinal surveys (where the test/retest interval was 12 months). Thereafter, the longitudinal data showed a slowing of growth as maturity was approached.

The peak growth rates that we saw in the semi-longitudinal data occurred at a somewhat younger age than the corresponding values as obtained from the longitudinal data. However, the agreement was reasonable, given the small sample available for the semi-longitudinal estimates and the limitations inherent in the longitudinal methodology. Both approaches indicated that the Inuit children reached an earlier growth peak than the values of 11.8 years in girls and 14.3 years in boys observed in a longitudinal study of a 'white' urban population living in Saskatoon, a medium-sized city in Western Canada (Mirwald & Bailey, 1986). This suggests that any secular trend to an advance in maturation associated with decreasing isolation of the circumpolar communities may now be approaching its potential nadir.

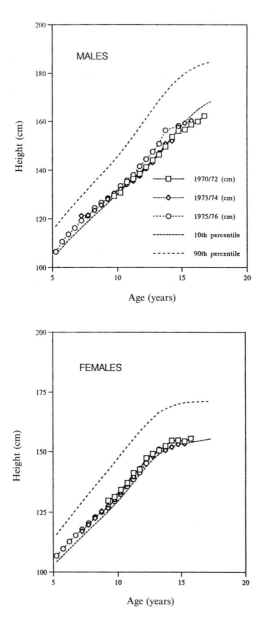

Fig. 7.2. Longitudinal growth curves for three cohorts of Igloolik Inuit children (born in 1970/72, 1973/74, and 1975/76 respectively), compared with NCHS norms (Hamill *et al.*, 1979). From Rode & Shephard (1995).

Table 7.13. *Age (mean ± SD) at peak height velocity in three cohorts of Igloolik Inuit followed longitudinally*

Cohort	Males	Females
Born 1970/72	13.6 ± 0.7	11.3 ± 0.9
Born 1973/74	13.6 ± 1.0	11.2 ± 0.6
Born 1976/76	13.3 ± 0.6	11.3 ± 0.6

Source: Rode & Shephard (1995).

The longitudinal cohorts available to us were rather closely spaced to establish the impact of 'modernization' upon the course of growth and development, but within the limits of our observations, there was relatively little evidence of inter-cohort differences. Nevertheless, repetition of our cross-sectional and semi-longitudinal studies over a longer interval (above) and a comparison between our most recent cross-sectional study and the data of de Pena (1972) obtained on the same community suggest that during the period 1970–1990 there has been a small continuing secular trend to an increase of stature in the younger children – a fact that has modified the shape of both cross-sectional and longitudinal growth curves for the Igloolik community. There also appears to have been a small advance in the age at PHV, but the stature at maturity has shown little change over the two decades of observation.

A comparison of the rate of growth of Igloolik children for warmer and colder seasons (April–October versus October–April) showed closely comparable values. It is thus unlikely that the small size of the Inuit can be attributed to a diversion of food energy from growth processes to the maintenance of body heat during the winter months (Leonard *et al.*, 1994c). Given that children now spend much of their day either in school or at home watching television, it is perhaps not surprising that no growth-retarding effect of the cold winter climate could be demonstrated in Igloolik.

The sitting height showed no statistically significant differences between boys and girls. Nevertheless, the sitting height of the boys tended to be lower relative to Canadian norms (Demirjian *et al.*, 1972) than did that of the girls, particularly in the younger age groups. Our study found no inter-cohort differences of sitting height among the Inuit.

Sitting height is shown as a percentage of standing height in Fig. 7.4. Between the ages of 10 and 14 years, values for Inuit boys and girls tended to be a little lower than in the general Canadian population, the one exception being the cohort of boys born in 1970/72.

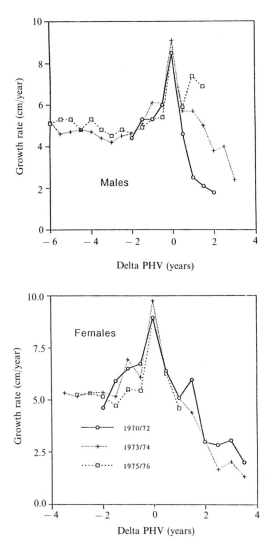

Fig. 7.3. Growth rates of Igloolik Inuit, averaged over six month intervals. Data for three cohorts of subjects (1970/72, 1973/74, and 1975/76) aligned in terms of peak height velocities (PHV). From Rode & Shephard (1995).

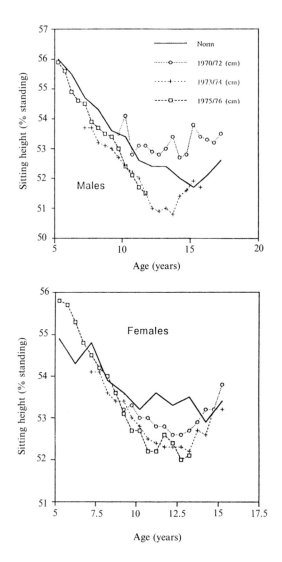

Fig. 7.4. Sitting height in three cohorts of Igloolik Inuit children (born in 1970/72, 1973/74 and 1975/76), expressed as a percentage of standing height. The Canadian cross-sectional norms of Demirjian *et al.* (1972) are also shown. From Rode & Shephard (1995).

Body mass

Body mass data are summarized in Fig. 7.5. Values did not differ between the sexes up to 11 years of age, but thereafter the girls gained weight faster than the boys, remaining heavier throughout the period of observation. At maturity, the male subjects reached a body mass of 65.0 ± 6.8 kg, and female subjects reached values of 54.6 ± 7.6 kg.

The boys showed a slight trend to a higher body mass for the cohort born in 1975/76, but with this exception, there were no significant inter-cohort differences. Despite the short stature of the Inuit children, their average body mass was close to the 50th percentile of the NCHS norms (Hamill *et al.*, 1979) throughout the period of school attendance.

Weight for height curves are plotted in Fig. 7.6. Data for both boys and girls systematically exceeded NCHS norms (Hamill *et al.*, 1979). Three possible hypotheses could explain the substantial ratio of body mass to standing height. The final adult size of the Inuit remains much shorter than that of people living in southern Canada or the United States (Chapter 5). The protein intake remains quite high in Igloolik, as in most other circumpolar populations, and despite the apparent benefit from food supplementation shown in the Alaskan studies of Petersen & Brant (1984), we think it unlikely that there has been any stunting of growth in Igloolik from poor nutrition. Genetic factors are more likely to be involved. The short stature of the Inuit could be enough to distort mass for height ratios relative to 'white' standards. Although the skinfolds are very thin (see below), it is also possible that the Inuit carry an unusually large fraction of their body fat internally, so that the thickness of subcutaneous fat underestimates their body fatness (Rode & Shephard, 1994d; Shephard *et al.*, 1973). Predictions of Inuit body fat content made by applying triceps, subscapular and suprailiac readings to a standard formula for 'white' subjects agree quite well with inferences based upon hydrostatic data (Chapter 4, Rode & Shephard, 1994d), but there remains a need to make a much more detailed analysis of body fat distribution in the Inuit population. Finally, although the grip strength of the Inuit children is not remarkable, the knee extension force is greater than in city-dwellers, so that the Inuit children presumably have an above average lean tissue mass relative to their standing height.

Triceps skinfold thickness

Details of triceps skinfold readings are summarized in Fig. 7.7. As in the adult Inuit of Igloolik (Chapter 5), the body fat content (as indicated by the

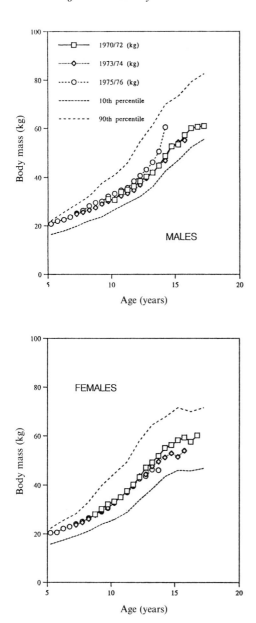

Fig. 7.5. Body mass of three cohorts of Igloolik Inuit children (born in 1970/72, 1973/74 and 1975/76), compared with the NCHS norms of Hamill *et al.* (1979). From Rode & Shephard (1995).

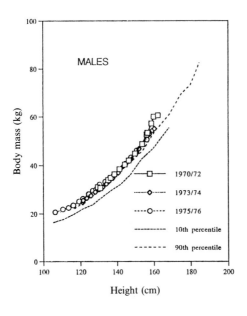

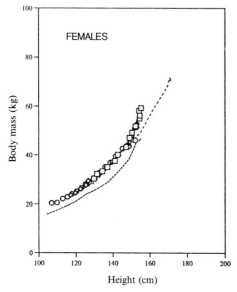

Fig. 7.6. A plot of body mass against height for three cohorts of Igloolik Inuit children (born in 1970/72, 1973/74 and 1975/76), compared with NCHS norms of Hamill *et al.* (1979). From Rode & Shephard (1995).

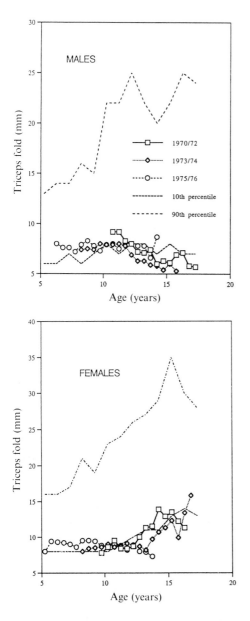

Fig. 7.7. Skinfold thicknesses of three cohorts of Igloolik Inuit, (born in 1970/72, 1973/74 and 1975/76) compared with the cross-sectional norms of Montoye (1975) for the city of Tecumseh, Michigan. From Rode & Shephard (1995).

triceps skinfold) continued to be quite low (particularly in the boys), despite the high body mass for height ratios. Nevertheless, the triceps skinfold readings have increased substantially relative to those measured in the 1969/70 cross-sectional survey of Igloolik, and since there has been little associated increase in body mass, we must presume that lean tissue has been replaced by fat. Following the pattern shown by the children of southern Canada (Demirjian *et al.*, 1972; Shephard & Lavallée, 1993), the girls tended to be slightly fatter than the boys even prior to puberty. With the onset of the growth spurt, all three cohorts of boys tended to become thinner, whereas the girls showed a substantial increase in the thickness of the triceps skinfold at this stage of their development.

Comparing the triceps readings of the Inuit with students living in Tecumseh, Michigan, the preadolescent boys slightly exceeded the 10th percentile of the urban norm (Montoye, 1975), but their readings dropped back to the 5–10th percentile after the growth spurt. Values for the girls remained around the 10th percentile throughout the period of observation.

Handgrip force

Longitudinal data on the growth of hand-grip force are summarized in Fig. 7.8. Values for the Igloolik Inuit are not outstanding relative to those for urban society.

Conclusion

In addition to providing accurate information on PHV and age at PHV, the longitudinal data supports other conclusions drawn from the cross-sectional and semi-longitudinal measurements. In particular, they highlight the problem of using weight for height ratios derived from urban populations as epidemiological tools for examining population health and nutritional status (Waterlow, 1986). It was this approach that led Nutrition Canada (1975) to draw the erroneous conclusion that the Inuit were obese.

The data obtained on children in Igloolik and other circumpolar communities underline the point that a healthy child with a short stature may have an above-average body mass for height. Given the wide availability of skinfold calipers, it is generally better to assess nutritional status from measurements of body mass and skinfold thicknesses than from some manipulation of weight to height ratios. If either pediatricians or nutritionists wish to use height and weight norms, then in principle the curves adopted for evaluating circumpolar populations should be population specific. Because of regional variations in growth and development across

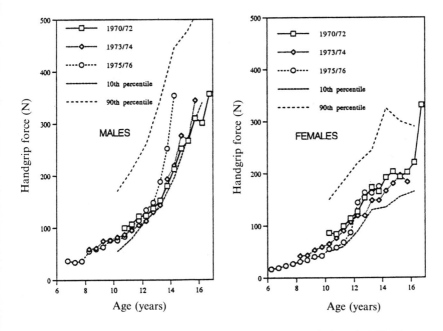

Fig. 7.8. Handgrip force of three cohorts of Igloolik Inuit (born in 1970/72, 1973/74 and 1975/76), compared with the cross-sectional norms of Montoye (1975) for the city of Tecumseh, Michigan. From published data of A. Rode and R.J. Shephard (1995).

the arctic, this may be an unattainable goal (Canadian Pediatric Society, 1987). Nevertheless, if population-specific norms are not available, it is vital that the influence of a short stature be recognised and incorporated into weight for height evaluations.

Changes of adult stature

Our earliest cross-sectional survey of the Igloolik Inuit (1969/70) showed a substantial age-related decrease of height, beginning relatively early in adult life. We suggested that the probable explanation was a substantial secular trend to an increase of adult stature (Shephard & Rode, 1973). However, when the measurements of stature were repeated in 1979/80, we found to our surprise that the average height of the younger categories of adult had decreased rather than increased over the decade (Shephard *et al.*, 1984). This trend was even more obvious when measurements were again made in 1989/90.

Leonard *et al.* (1994c) recently reported a related paradox. Male (but not female) Evenki showed a 70 mm decrease in stature from the third to the sixth decade of life, yet the average height of their subjects did not differ significantly from measurements that had been made on subjects of similar age from the same population some 90 years earlier (Boas, 1903; Comuzzie, 1993).

Such observations could imply either a reversal of the previous secular trend to greater growth during childhood, or the introduction of some adverse influence that was decreasing adult height soon after growth had been completed, for instance by causing a severe kyphosis or vertebral collapse. The younger Igloolik Inuit who were remeasured after an interval of ten years showed larger decreases in stature than could be attributed simply to normal aging. Taking all male villagers over the age of 20 years, stature decreased by an average of 12.5 mm from 1969/70 to 1979/80, and the average decrease had grown to 17.8 mm by 1989/90. Likewise, the women showed an average loss of 22.5 mm from 1969/70 to 1979/80, and a loss of 24.7 mm by 1989/90 (Table 7.14). Such figures show clearly that there had been a loss of stature after maturation rather than a lessening of growth during childhood.

Decreases of height over the adult lifespan were relatively similar in the three cross-sectional surveys, averaging in the men (from 25 to 65 years) 36 mm, 40 mm and 47 mm, and in the women (from 25 to 55 years) 58 mm, 40 mm and 24 mm. However, the tendency of older adults to be shorter than younger members of the community because of a secular trend for children to grow to a greater adult height has now been partially masked by a rapid loss of height among younger adults. In the males, the decrease of stature apparently begins in the third decade of life, and proceeds relatively constantly at a rate of 6–8 mm/decade over the first three decades of adult life, with some possible acceleration from 55–65 years of age. In contrast, the loss of height in the females develops progressively, averaging 6, 11 and 20 mm for the three decades of adult life where detailed figures were available.

In the men, there was a trend for the loss of stature to be related to the extent of snowmobile use. The 10 year change seen in 1979/80 was 24 ± 20 mm in the heaviest snowmobile users, 19 ± 20 mm in moderate snowmobile users and 16 ± 14 mm in light users. Fitting regression equations to data for male adults in 1989/90, the coefficient describing the loss of stature with age (in mm/yr) was much larger for hunters than non-hunters:

Height (hunters, mm) $= 1733 \pm 38 - 2.00 \pm 0.08$ Age ($r = 0.22, p = 0.022$)
Height (others, mm) $= 1652 \pm 12 - 0.40 \pm 1.40$ Age ($r = 0.02$, ns)

Table 7.14. *The decrease in stature of Igloolik Inuit in recent years. Data predicted from linear regressions of height on age for each of three cross-sectional surveys*

Year	Male, age 20 (cm)	Male, age 65 (cm)	Female, age 20 (cm)	Female, age 65 (cm)
1970	165.6	162.5	157.3	150.9
1980	165.5	161.9	155.2	149.8
1990	165.1	160.4	154.1	150.5

Source: Rode & Shephard (1994h).

Could kyphosis be caused by the prolonged adoption of an unusual posture? Crouching is one method of reducing wind exposure, and is also a typical posture of a person who is operating a snowmobile at very high speeds. Furthermore, hunters sometimes spend long hours leaning over seal holes. The trend to kyphosis is exacerbated by the carriage of heavy loads on the back: game for the men, and children for the women. However, there was little obvious evidence of kyphosis in the subjects that we studied. Serial radiographs had been taken on a number of the subjects to monitor the progress of small primary tuberculous lesions, and in these subjects it was possible to demonstrate that there had been a decrease in the length of the thoracic spine over an average observation interval of 11 years. This averaged 22.5 mm over the ten cervical and thoracic vertebrae that were visible. It was difficult to judge how far there had been vertebral collapse and how far the findings could be explained by a compression of the intervertebral discs, but in the two subjects with the largest change, intervertebral disc compression seemed largely responsible.

There are various potential explanations for the early loss of height in adult life. Given the preponderance of change in the hunters, one possibility might be trauma due to the repeated operation of high-powered snowmobiles over rough terrain (Shephard *et al.*, 1984). The decrease in height among the Igloolik population apparently began coincident with the introduction of snowmobiles. Moreover, the prevalence of abnormal spinal radiographs is known to be quite high among the tractor drivers of agricultural communities (Coermann, 1970; Sjoflot, 1982), even though such vehicles move relatively slowly and generate vibrations that have a low frequency (0.5–1.0 Hz) relative to the natural resonant frequencies of the back (4–5 Hz, Coermann, 1970). The potential for all types of vibration injury is much greater during operation of a modern snowmobile (Virokannas *et al.*, 1988), since such machines can move at speeds of up to 140 km/h

(Sundström *et al.*, 1994). In Finland, soft snow and wooded territory restrict the speed of snowmobile operators, but nevertheless measurements of whole-body vibration have shown that more than a half of drivers are exposed to vibration that exceeds the occupational risk limit (Hassi *et al.*, 1984). Hassi and associates (1984) noted that 17% of reindeer herders who used snowmobiles reported neck or shoulder pain, and 51% had back troubles. The Finnish reindeer herders commonly operate their machines from a standing position, as do the Inuit when racing their machines. This posture absorbs some of the vibration, and even greater trauma might be anticipated when the machine is operated from a seated position, as is the Inuit custom when making a long hunting trip.

Female arctic residents are less frequent and less aggressive operators of snowmobiles, and it is thus difficult to attribute their equally large decrease in stature simply to the introduction of snowmobiles. The problem of spinal trauma is likely compounded by other practical consequences of 'modernization'. The traditional 'country' diet was very poor in calcium, but it was rich in vitamin D (Harper *et al.*, 1984; Jeppesen & Harvald, 1985; Nobmann, 1991). Adoption of a 'market' diet may have led to a choice of foods that remain poor in calcium but also contain little vitamin D. Indoor living has also reduced exposure to sunlight, with a consequent reduction in the endogenous synthesis of vitamin D (Loomis, 1967). Another factor modifying bone health in women may be a disappearance of the habit of carrying children on the back. In previous generations, this mechanical loading may have been sufficient to protect the spine against early osteoporosis in the face of a low calcium intake.

To date, there is little evidence that the decrease of adult height has had any adverse consequences, either for cardio-respiratory health or for musculo-skeletal function. However, if the origin is indeed spinal trauma, then in the long term it seems likely that musculo-skeletal pathologies will result. If the cause is calcium deficiency, this also will have important health consequences as the present generation reaches the age when osteoporosis might normally be expected. Plainly, the phenomenon of a premature decrease of adult stature in circumpolar populations merits further examination and close monitoring.

8 Current health status of circumpolar populations

Comment has already been made concerning the influence of acculturation to a 'modern' lifestyle upon certain aspects of health and disease, in particular the prevalence of metabolic abnormalities such as a poor glucose and/or lactose tolerance (Chapter 4), the blood lipid profile, and the prevalence of chronic bronchitis and tuberculosis (Chapter 6). This chapter examines secular trends in positive health and the prevalence of other diseases, including specific comments on acute infections, atherosclerosis, hypertension and stroke, diabetes mellitus, and carcinomas. The paradox is noted that although acculturation brings increased opportunities for health education and access to 'high-tech' medicine, it can also cause an alienation of indigenous populations, with a deterioration of their lifestyle and a reduction of positive health (Chapter 3). The wisdom of traditional healing practices becomes less readily accepted, and growing contacts with the outside world greatly increase the risk of epidemics caused by unfamiliar micro-organisms. Finally, disease may act as a selective agent, reducing the range of genetic variation among circumpolar populations and enhancing the fitness of survivors.

Evidence of positive health

The World Health Organization (1948) defined health not as the mere absence of disease, but as a state of 'complete mental, physical and social well-being'. In essence, the wellness of an individual on any given day was perceived as lying at some point along a continuum that extended from organic illness to a full realization of that person's mental, physical and social potential ('good health', Herzlich, 1973).

There is some evidence that in the recent past, traditional circumpolar peoples have realized their physical potential (aerobic power and muscular strength) to a greater extent than most city-dwellers have done (Chapter 5). However, much of the unusual physical work capacity of the indigenous populations has disappeared as they have become 'acculturated' to the sedentary ways of 'modern' society (Egan, 1991).

It is difficult to apply objective measures of mental and psychological health across cultural boundaries, and perhaps for this reason quantitative assessments of such aspects of well-being are lacking for the indigenous circumpolar peoples. Nevertheless, the high and growing incidence of problems such as substance abuse, family violence and suicide (Chapter 3) point to a progressive decline in mental and social health as circumpolar communities have been exposed to a 'modern' lifestyle.

Acute infections

General considerations

Rausch (1974) suggested that prior to contact with 'developed' societies, the indigenous populations of the circumpolar regions harboured parasites, including lice, pinworms and helminths ingested from raw fish. Conditions such as cystic and alveolar hydatid disease, trichinosis, amebiasis, rabies, botulism and brucellosis were all prevalent, but starvation and accidents were the main factors controlling population growth.

Europeans introduced many additional deadly diseases into the circumpolar regions (Keenleyside, 1990; Marchand, 1943), including smallpox (Fortuine, 1985a), anterior poliomyelitis (Adamson *et al.*, 1949; Rhodes, 1949), tuberculosis (Chapter 6), meningococcal meningitis (Middaugh, 1981), diphtheria (Wilson *et al.*, 1991), syphilis (Fortuine 1985b; Misfeldt, 1991), gonorrhea (Muri & Blackwood, 1981), chlamydiae (Moller *et al.*, 1981; Soininen, 1991a), HIV infection (Craytor, 1991; Imrie, 1991; Misfeldt, 1991; Soininen, 1991a), hepatitis A and B (Larke *et al.*, 1981, 1987; Minuk *et al.*, 1981, 1985; Skinhøj *et al.*, 1977), shigellosis (Ryan, 1981), measles (Brody & Bridenbaugh, 1964; Peart & Nagler, 1954), mumps (Philip *et al.*, 1959a), rheumatic fever (Bender *et al.*, 1981), pneumococcal pneumonia (Davidson *et al.*, 1989), pertussis (D'Aeth *et al.*, 1991; Ireland *et al.*, 1985), rubella (Brody, 1966), and hemophilus influenzae (Bjerregarde, 1983; Nicolle *et al.*, 1981; Ward *et al.*, 1986; Wotton *et al.*, 1981). The acute infections commonly encountered in urban societies have reached isolated arctic settlements by animal and insect vectors, and by direct transmission from the passengers of boats and aircraft (Allen, 1973; Bennike, 1981; Holmes, 1973). Major epidemics have developed because the invading organisms have encountered populations where few if any contacts had any acquired immunity to the disorder in question.

There have been suggestions that resistance to disease is further impaired among indigenous circumpolar populations because of a poor response of the immune system. Evidence in support of this hypothesis has included low salivary levels of IgA (Sayed *et al.*, 1976b) and a poor T cell proliferation rate in response to the mitogens PHA and ConA (Reece & Brotton, 1981). Possible underlying causes of immune deficiency have included (i) a lack of breast feeding and thus failure to ingest the maternal immunoglobulins of the colostrum (Schaefer, 1971), (ii) a dietary deficiency of vitamins and trace minerals important to the immune response (Gross & Newberne, 1980), (iii) an overtaxing of the immune system by repeated bouts of chronic infection, and (iv) genetic influences.

Given a lack of acquired immunity, a poor immune response to novel infections, and communities where the majority of the population are still children (Hall *et al.*, 1973), the mortality rates from outbreaks of many acute infectious diseases have been very high. Moreover, infections have spread very quickly in the closed and crowded winter living accomodation typical of most arctic villages. For instance, 20% of the community died in a single outbreak of influenza among the Inuit of Victoria Island (Nagler *et al.*, 1949), and another outbreak of influenza on St. Lawrence Island affected 87% of local residents (Philip *et al.*, 1959b).

The population of indigenous Alaskans had reached 32 977 by 1880, but the numbers of both Inuit and Amerindians subsequently showed a precipitous drop with the introduction of various European infections. The great influenza epidemic of 1918 alone accounted for some 2000 deaths in Alaska. The population numbers of the 1880s were not regained until the early 1950s (Milan, 1980).

Over the past three decades, the establishment of a chain of nursing stations staffed by nurse-practitioners, tele-satellite contact with major medical centres and the air evacuation of seriously ill patients have greatly curbed deaths from acute infections, but death rates from other causes have remained high in the circumpolar environment (Tables 8.1 and 8.2). Thus, at the time of the IBP-HAP survey, accidental deaths in Alaska were still three times the US national average, at 188/100 000. Diseases of early infancy (61/100 1000) were also about three times the US average, but deaths from influenza and pneumonia, at 36/100 000, had become very similar to those found in urban North American society (Milan, 1980). By 1971, the leading reason for hospitalization in Alaska had become otitis media (Milan, 1980).

Acute infections have shown a parallel decline in Greenland. Thus, in the eastern Greenlandic centre of Angmagssalik, hospital admissions for

Table 8.1. *Age and sex standardized diagnoses for medical contacts in Upernavik over period 1968–85*

Diagnostic group	Cases (%)	Standardized contact ratio
Respiratory diseases	16.0	0.88
Accidents	12.1	2.07
Skin diseases	11.0	1.97
Diseases of CNS, eye, ear	10.0	0.99
Musculo-skeletal & connective tissue	9.5	0.92
Other	9.0	
Symptoms	6.5	2.48
Gastro-intestinal diseases	6.3	1.33
Genito-urinary diseases	5.9	0.62
Infectious diseases	5.8	1.13
Cardiovascular diseases	2.6	0.41
Pregnancy and puerperium	2.6	3.72
Psychiatric diseases	1.6	0.27
Neoplasms	0.8	0.43
Congenital malformation	0.2	0.67
Endocrine, nutritional metabolic diseases	0.1	0.03
Diseases of newborn	0.04	0.91

Source: Adapted from data of Bjerregaard (1988).

Table 8.2. *Secular change in mortality from selected causes in Upernavik, expressed as age-standardized ratio to values for that region in 1968/73*

Condition	Age group (years)	1974/79	1980/85
Measles	0–14	0.25	0.17
Cervical cancer	5–64	0.96	0.84
Rheumatic heart disease	5–44	0.51	0.37
Acute respiratory infections	0–49	0.60	0.39
Chronic bronchitis	0–49	0.84	0.64
Alcohol-related accidents	15+	0.50	0.45
Tuberculosis	5–64	1.58	1.10
Meningitis	0–64	1.34	0.85
Lung cancer	5–64	2.45	3.11
Boat accidents	All	0.55	1.48
Suicides	10+	1.80	3.55
Alcohol-related natural deaths	15+	1.39	1.95

Source: Based on data of Bjerregaard (1988).

infectious diseases dropped from 16.1% of all causes in 1948–60 to 8.0% in 1961–70 and 9.1% in 1971–79 (Helms, 1981).

Diseases of infants

Many of the health problems that contribute to a continuing high infant mortality rate among traditional indigenous arctic populations arise during hunting trips, when families are isolated at remote field camps. Acculturation to the life of larger, permanent settlements has reduced this source of morbidity and mortality, and in many northern communities it has also brought the advantage of professional ante-natal care. On the other hand, in many settlements poverty and a growing prevalence of maternal smoking, alcoholism, and child neglect contribute to infant morbidity and mortality rates that still far exceed national averages.

Green & Kromann (1981) found that during the period 1950–1959, 103 of every 1000 infants in Upernavik died in the first year of life. Statistics had improved only slightly, to 85 infant deaths per 1000 by 1960–1974. Jorgensen *et al.* (1981) noted that between 1975 and 1979, the infant mortality rate was still 72 per 1000 in Eastern Greenland, although the figure for Greenland as a whole had improved to 37 per 1000. Neonatal problems accounted for many of the infant deaths, and if a child's birth weight was less than 2.5 kg, the infant mortality rate was as high as 291 per 1000. By 1983/87, there had been a further small decrease in the overall infant mortality for Greenland, to a rate of around 30/1000 (Bjerregaard & Misfeldt, 1991).

In the Lapp municipalities of northern Finland (Soininen & Akerblom, 1981), the prognosis of the young child was much more favourable than in Greenland. Infant mortality rates were 36 per 1000 in 1950–59, and 21.4/1000 in 1960/69, but had decreased to 13.7/1000 by 1970/79. This last figure was only a little higher than the overall national average for Finland (10.3/1000) during the same period.

Figures for Alaskan natives showed 60 infant deaths per 1000 in 1960, but the average had declined to 31 per 1000 by 1969 (Fleshman, 1972). In the North West Territories of Canada, Brett *et al.* (1976) noted that in 1973 the infant mortality rate was 28.6 per 1000, this average value reflecting rates of 9.1 per 1000 for 'white' children, 44.7 per 1000 for Inuit children, and 46.4 per 1000 for Amerindian children. Among Labrador Inuit, the infant mortality rate was still 60/1000 during the period 1971–82 (Wotton, 1985).

About a half of infant deaths occur during the first seven days of life. Precipitating causes at this stage of development (Brett *et al.*, 1976) include a low birth weight (this being 50% more common in Inuit than in 'white' children, Smith, 1976, and linked in part to maternal smoking, Chapter 7), inadequately utilized ante-natal services, accidents arising during pregnancy, lack of expert medical care (especially while families are away at hunting camps) and a failure to diagnose potentially treatable conditions.

During the post-neonatal period (seven days to one year) factors apparently contributing to deaths include acute infections, malnutrition, lack of medical care, and parental neglect (including alcoholism and child battering).

Diseases of children

Averaging data for the North West Territories over the period 1969–71, the main causes of death in children between birth and 14 years of age (Smith, 1976) were diseases of infancy (35%), accidents and injuries (22%), pneumonia (20%), gastro-intestinal disease (7%) and miscellaneous conditions (15%). Respiratory disorders accounted for 41% of all hospital admissions, and a further 9% were attributable to gastro-intestinal conditions (Smith, 1976). Herbert *et al.* (1976) examined chest radiographs from a group of 11 children aged 8–10 years after they had sustained an attack of adenovirus type III pneumonia. All except two of the 11 children showed residual areas of bronchiectasis and chronic bronchitis, although this seemed as yet to have had little impact on overall respiratory function.

Guenter *et al.* (1991) found that the incidence of hemophilus influenzae meningitis was some twenty times higher among the Keewatin Inuit than in Manitoba Caucasians. Cases peaked around four to six months of age. Ward *et al.* (1986) noted that Alaskan Inuit had a similar vulnerability to H. influenzae. Unfortunately, vaccines against this virus seem to have a low immunogenicity in Inuit (Ward *et al.*, 1988).

In Alaska, type A streptococcal pharyngitis progressing to rheumatic heart disease has been another major cause of morbidity among indigenous children. From 1964 to 1973, the average annual incidence of rheumatic heart disease was 44 cases per 100 000 of the population (Edelen *et al.*, 1976). Gordis *et al.* (1969) set the incidence of first attacks at 55 per 100 000 per year in Alaskan natives, and 84 per 100 000 in Alaskan Inuit, these figures being inversely related to the extent of health care available (Bender *et al.*, 1981). In northern Manitoba and the Keewatin district, the incidence of rheumatic heart disease among the Amerindian and Inuit populations was even higher, at 126 cases per 100 000 for the period 1970–79, although the number of new cases was declining steadily over the period of observation (Longstaffe *et al.*, 1981).

Another important sequel to respiratory disease in early childhood is chronic otitis media. Lupin (1976) noted that the incidence of chronic otitis media was 30–60% in Alaskan natives, 34–54% in British Columbian Amerindians, 30–40% in Baffin region Inuit, and 'high' in Swedish Lapps. Correlates of chronic ear disease were an absence of breast feeding, poverty, over-crowding and lack of personal cleanliness. Tissue scarring and chronic ear infections both seem to be less common in the smaller (and what might be thought the less 'acculturated') settlements. For example, in 1972–73, Baxter (1976) found that only 18% of children in the hamlet of Igloolik had an ear drum abnormality at otoscopy, compared with 60% of children in the larger regional administrative centre of Iqualuit (where there was not only advanced acculturation to an urban lifestyle, but also much poverty). Likewise, only 4.3% of children had active chronic otitis media in Igloolik, compared with 13.2% in Iqualuit.

Hobart (1976) examined the influence of social change on child health, finding that the effects of 'modernization' varied with the size of community. In larger towns, the children of successfully acculturated males with wage employment, uncrowded housing, tanked water, modern toilets and modern heating had good health. In contrast, poverty, often associated with attempts at continuation of a traditional lifestyle in the urban setting, was linked with poor infant and child health, as was wage-employment of the mother, bottle-feeding, over-crowded housing, and substance abuse by one or both parents. In the smaller settlements, well-educated parents again had healthy children, but so did those who were proud to maintain such traditional practices as reliance on 'country' sources of food and carrying young children in the amauti.

The early response of health authorities to the problem of controlling childhood disease in the north was to evacuate infected infants to distant hospitals, thereby adding the trauma of separation from family and a familiar culture to the immediate effects of disease. In the year 1972, a total of 2000 Inuit children were evacuated from the North West Territories to southern hospitals, two thirds of these children being under the age of five years (Smith, 1976). Coulter & Popkin (1976) noted that intractable diarrhoea was the most common reason for children to be transferred from the small hospital at Iqualuit to the more extensive pediatric facilities available in Montreal (17% of 96 transfers 1971–73). Usually, no specific bacterial or viral cause of the diarrhoea was identified, but quite a number of the infants concerned died.

The number of emergency evacuations has declined as better facilities have been developed for the treatment of medical emergencies within local communities.

Sexually transmitted diseases

Gonorrhoea and syphilis became major health problems among 'modernizing' indigenous circumpolar populations, as men moved to larger settlements in search of work, and impoverished women turned to prostitution. Unprotected sexual activity remains a cause for concern in the north, but over the last decade health education and other preventive and therapeutic measures have diminished the prevalence of most sexually transmitted diseases.

A recent study from Greenland suggested that the average number of sexual partners among the indigenous population was as high as 7.5/year for men and 4.4/year for women (Misfeldt, 1991). Some 53% of Greenlandic women reported having a total of more than 20 sexual partners, and 22% more than 40 partners (Olsen *et al.*, 1990).

From (1980) reported that during the late 1970s, the number of new cases of gonorrhoea that were being identified in Greenland could be equated to an incidence of some 15% per year. The incidence apparently peaked at about 38% in 1982 (Misfeldt, 1991). Since that date, control measures have reduced the risk of infection substantially, but in 1988 the prevalence of gonorrhoea was still estimated at 6.9% (Middaugh *et al.*, 1990). In Alaska, the gonorrhoea incidence rate for 1979 was said to be 4.2% per year (Muri & Blackwood, 1981), but it had dropped to 1.08% by 1988 (Middaugh *et al.*, 1990). In the Canadian North West Territories, reported cases of gonorrhoea dropped from 3.76% for the year 1976 to 2.21% for 1987 (Middaugh *et al.*, 1990). Much lower figures (0.06 to 0.10%) were cited for Lapland during the period 1987–89 (Soininen, 1991a).

Retrospective analysis of syphilis cases in Greenland showed an estimated annual incidence of around 2.4% for the year 1976. Shortly thereafter, the incidence declined, this change being associated with the vigorous treatment of all primary contacts and mass examinations of the population (Jorgensen *et al.*, 1988; Misfeldt *et al.*, 1988). But perhaps because these preventive measures gave rise to a sense of false security, or perhaps because of a gradual spread of residual infection, there was then another epidemic, with a second peak incidence of 1.8% during 1987 (Misfeldt, 1991). Moreover, there was some fear that even these alarming figures may have under-estimated the true incidence of the disease, since antibiotic treatment of gonorrhoea may have masked or cured coincidental early cases of syphilitic infection (Aargard-Olsen, 1973).

Syphilis rates among the indigenous population of Alaska were much lower than in Greenland; a peak incidence of 0.035% during 1974 had decreased to 0.011% by 1988 (Middaugh *et al.*, 1990). Only sporadic cases of syphilis have been described among the indigenous populations of the Canadian arctic.

Chlamydiae trachomatis and Mycoplasma hominis are increasing problems among circumpolar populations (Moller *et al.*, 1981; Soininen, 1991a), although the lack of precise diagnostic methods in many nursing stations and damage of specimens in transit to specialized central laboratories have hampered accurate assessments of disease prevalence. Early studies cultured Chlamydiae from 12–15% of patients attending Greenlandic clinics for sexually treated diseases, but antibodies for this organism were present in 80% of female and 25% of male attendees (Mardh *et al.*, 1980; Møller *et al.*, 1981). A more recent report found infection rates of 40% in women and 27% in men (Marschall, 1992).

HIV infections and AIDS are discussed below. Much of the high prevalence of hepatitis B, typical of arctic communities, may also have a sexual origin.

HIV Infections and AIDS

Although the reporting of HIV seropositivity is required in most circumpolar jurisdictions, there remain concerns about an under-reporting of cases by physicians, and a lack of central data collection facilities.

In Alaska, a blinded sero-prevalence study began in 1989, but to date, information has only been made available for prevalence of the disease among blood donors, military recruits and high risk groups who have themselves initiated testing. Despite the high prevalence of sexually transmitted diseases in general, both HIV infections and AIDS as yet seem uncommon among circumpolar groups. One reason is that haemophilia is absent among the Inuit and is uncommon in other arctic societies. There are also few major surgical facilities. Blood transfusions are thus a relatively rare occurrence in the north. In Greenland, there is also little evidence of intravenous drug abuse among the indigenous population, but one study of Alaska found 2.5% of natives sharing needles for intravenous drug administration (Olsen *et al.*, 1990). Homosexual activity is thought to be uncommon among the Inuit, as judged by the rarity of anal sexually transmitted diseases. Nevertheless, other forms of unprotected sexual activity are common among young people.

Occasional cases of HIV infection have now made their appearance among the Alaskan indigenous peoples (Craytor, 1991), Canadian Amerindians (Imrie, 1991), Laplanders (Soininen, 1991a) and Greenlandic Inuit (Misfeldt, 1991), but in all of these communities the prevalence of HIV infection still remains less than in the major urban centres of developed societies. The highest proportion of HIV sero-positive cases among the indigenous Alaskan population was found among the Aleuts (3% prevalence in a biased/high risk sample), with comparison values of 0.58% for Inuit and 0.17% for Amerindians. In Greenland, 22% of the population had been tested for HIV infection by mid 1989; a total of 15 cases of HIV positive serum had been found in 11 893 analyses.

By 1990, only six cases of AIDS had been reported in Alaska; furthermore, a careful review of all death certificates for that state made it unlikely that other fatal cases of AIDS had passed unnoticed. Another report noted that 18 cases of AIDS were now known among all Amerindians, Métis and Inuit in Canada (including some living outside the circumpolar territories); one case had also been discovered in Greenland, but none had been diagnosed in the former Soviet Union (Melbye *et al.*, 1990).

Hepatitis

Infectious hepatitis was common in Alaska at the time of the IBP-HAP survey. Prevalence of the disease tended to peak in the fall months, and it was suggested that mosquitos might be acting as vectors (Barrett *et al.*, 1976). More recently, attention had focused upon other methods of disease transmission, as discussed below.

Five forms of hepatitis virus are now identified (A–E), but only hepatitis A and hepatitis B are problems in the circumpolar territories. Hepatitis A is a highly contagious disease with an acute course. In isolated settlements, it is thought to arise by the importing of cases from lower latitudes where the disease is endemic. Epidemics occur in the arctic every 10–20 years, and the serum of a large proportion of the population becomes positive for hepatitis A virus antigens in the first two to three decades of life (Skinhøj, 1991). In 1982–83, Siebke *et al.* (1986) found that 41% of Inuit in four isolated Alaskan villages had serological evidence of previous hepatitis A infection. Attempts to halt epidemics by the administration of gamma globulins have proven ineffective (McMahon, 1991), and the only recourse seems active immunization.

Hepatitis B, in contrast, may progress to a chronic stage, allowing the build up of a reservoir of infection in closed communities. Environmental factors contributing to spread of the disease include overcrowding, poor hygiene, and a lack of good separation of drinking water from sewage, all conditions that in the past were common in circumpolar settlements. A massive epidemic among the indigenous population of Greenland (Mikkelsen, 1976) was attributed to these causes. Hepatitis B can also be spread through blood, saliva, seminal fluid and breast milk. Transmission of the disease can thus occur from mother to child, between sexual partners or from one drug addict to another. The surface antigen for hepatitis B has been detected on a variety of exposed surfaces in northern households and schools (Petersen *et al.*, 1976). Other, acculturation-related factors, including the traditional practice of premasticating food (Beasley & Hwang, 1983), the sharing of modern toilet equipment such as razors and toothbrushes, sexual promiscuity and a growing prevalence of intravenous drug abuse have been further reasons for a high incidence of hepatitis B among northern populations in recent years.

By the end of the 1970s, rates of infection with hepatitis B (as shown by positive sera) became as high as 80% in some Alaskan and Greenlandic communities (Larke *et al.*, 1981). In 1982–83, Siebke *et al.* (1986) found that 28% of Inuit in four Alaskan villages had antibodies to hepatitis B. Figures for Greenland from the same era showed 4.5% of females and 6% of males with positive sera. Data reported for the Canadian Arctic have included a prevalence of: 31% in the population of Resolute Bay, and 56% in Pond Inlet (Vas *et al.*, 1977); 31.3% in Nasivik and 26% in Arctic Bay (Larke *et al.*, 1981); 40% in the Baffin zone as a whole (Larke *et al.*, 1985); and 27% in Baker Lake (Minuk *et al.*, 1981). In the last three studies, rates were higher in men than in women, and there was a concentration of carriers among the older members of the population.

Minuk *et al.* (1985) speculated that the evacuation of mothers for treatment of tuberculosis and BCG vaccination of infants may have been factors limiting the prevalence of disease in younger residents. The concentration of the disease in men may reflect a gender-specific genetic predisposition to the disease (Schreeder *et al.*, 1983), but it could also reflect the greater number of sexual partners exploited by the average male.

Potential long-term sequelae of hepatitis B include hepatic cirrhosis and hepatocellular carcinoma.

Intestinal parasites

Rausch (1974) reported that up to 30% of Yu'pik Inuit in western Alaska were infected with the fish tapeworm *Diphyllobothrum latum*. Studies from the period immediately following World War II reported similar rates for Inuit of the Igloolik region (Brown *et al.*, 1948), Southampton Island (Brown *et al.*, 1950) and Kuujjuak (Laird & Meerovitch, 1961), with a rate as high as 83% at Ivujjuvik (Port Harrison) (Arch, 1960). However, Curtis & Bylund (1991) pointed out that tapeworm infection was often due to other species of *Diphyllobothrum*. *D. latum* was usually found in pike and perch, but *D. dendriticum* was the commonest tapeworm in arctic char, salmon, trout and whitefish, *D. ursi* and *D. klebanovski* predominated in Pacific salmon, and *D. dalliae* was the dominant species in Alaskan blackfish.

Various fish-eating mammals and birds can serve as intermediate hosts.

Clinical manifestations include diarrhoea, epigastric pain, nausea, vomiting, and (in the case of *D. latum* infections) anaemia. Infection is most readily acquired by eating raw fish, a practice which is diminishing with acculturation. Improved sewage disposal is also interrupting the normal life-cycle of the parasite. Other factors in control are the use of freezers (the tapeworm is killed by the freezing of fish) and vigorous drug therapy of infected individuals.

Trichinella is a parasite carried by bears and some marine mammals. Outbreaks of human infection have been found in both Alaska (Ryan, 1981) and Greenland (Holgersen, 1961). In the early 1960s, *Entamoeba histolytica* was found in 12–17% of the residents of various isolated Greenlandic communities (Babbot *et al.*, 1961), but the prevalence of this infection has diminished with improved living standards.

Freeman & Jamieson (1976) confirmed the wide range of intestinal parasites harboured by the Inuit. One or more of the protozoa, other flagellates, helminths, flukes and pinworms were to be found in about 50% of fecal specimens collected from the indigenous population of Igloolik in 1970/71. Moreover, the prevalence of such organisms was substantially higher than had been reported by Brown *et al.* (1950) in a study of the same community completed some 20 years earlier. Freeman & Jamieson (1976) suggested that one source of parasitic infection was diminishing, as the introduction of portable cooking stoves allowed the cooking of fish even at camps. However, this health gain had seemingly been outweighed by the crowded conditions of many homes, and the then prevalent use of plastic garbage bags as a means of sewage disposal. This last practice predisposed to fecal contamination of the main domestic water tank; the problem has since been corrected in many parts of the Canadian arctic by a combination of intensive instruction in domestic hygiene and the installation of modern toilet facilities.

Animal and insect-borne diseases

Rabies is endemic in Alaskan wildlife (Middaugh & Ritter, 1981). However, Charlton & Tabel (1976) maintained that there had been no rabies in the Canadian North West Territories until 1945, when the disease was introduced by the importing of dogs from Alaska.

Dogs probably play a major role in the transmission of many bacterial, viral and parasitic diseases, including alveolar hydatid disease, since they can so easily contaminate human food and water supplies (Rausch, 1974; Rausch & Wilson, 1985). Greater protection of water supplies and a rapid decline in the dog population of the Alaskan and Canadian arctic should have a favourable influence on the prevalence of a number of diseases as 'modernization' continues.

Much of the tundra is heavily populated with insects in the early summer. In Siberia, insect-borne encephalitis, rabies, rickettsiosis, leptospirosis, tularemia and brucellosis are all well known (Netsky, 1976). Likewise, antibodies for insect-borne encephalitis have been described in the Inari Lapps (Eriksson *et al.*, 1980).

In theory, humans can be vaccinated against a wide range of invading micro-organisms, with the development of antibodies that can attenuate or kill invading micro-organisms, allowing their early neutralization. However, protection is not always complete, and in some instances the theoretical benefit from vaccination is limited by a poor immune response.

Atherosclerosis

Problems of diagnosis

Health statistics for the circumpolar regions lack the sophistication of those for lower latitudes, and an autopsy to confirm the diagnosed cause of death is not always performed even when a medical history is lacking. Prior to 1980, as many as 20% of deaths in Greenland lacked a death certificate (Bjerregaard, 1988). More recently, it has been estimated that as many as 20% of death certificates still represent no more than a qualified guess (Bjerregaard, 1986).

This difficulty is particularly critical when assessing the incidence of ischemic heart disease among circumpolar populations, since often the first intimation of disease is sudden death. Nevertheless, there have been repeated assertions that traditional Inuit and other indigenous inhabitants of the arctic coastline have a low incidence of cardiovascular disease (Dyerberg, 1989; Ehrström, 1951; Rodahl, 1954).

Risk factors

The American Heart Association has recognized four major risk factors that predispose to atherosclerotic heart disease: (1) cigarette smoking, (2) a high blood pressure, (3) high serum cholesterol levels, and (4) a low level of habitual physical activity.

Rode & Shephard (1994b) recently made a detailed comparison of the prevalence of cardiac risk factors for the Inuit coastal community of Igloolik (NWT) and nGanasan living at Volochanka on the Siberian

Table 8.3. *A comparison of blood pressures between Igloolik Inuit and Volochanka nGanasan (mean ± SD of data, in mm Hg)*

Age category (years)	Males		Females	
	Inuit	nGanasan	Inuit	nGanasan
18–29	112 ± 7	127 ± 15	105 ± 8	119 ± 12
	72 ± 7	80 ± 12	69 ± 7	77 ± 11
30–39	110 ± 10	129 ± 17	104 ± 7	133 ± 24
	73 ± 10	88 ± 10	65 ± 6	86 ± 12
40–49	108 ± 13	140 ± 14	106 ± 14	158 ± 19
	69 ± 9	86 ± 11	69 ± 14	94 ± 8
50–59	114 ± 12	120	113 ± 13	160
	71 ± 10	78	72 ± 11	105
60–69	116 ± 14	—	120 ± 36	—
	73 ± 5	—	70 ± 17	—
70–79	138 ± 13	—	—	—
	73 ± 6	—	—	—

Source: Rode & Shephard (1994g).

tundra. In the Inuit, systolic blood pressures were uniformly low normal values in all except the oldest category of men (Table 8.3). A two-way ANOVA by gender and age category showed a significant age effect ($p < 0.011$), but a trend to lower pressures in women than in men was not statistically significant. Diastolic pressures were also low normal values among the indigenous population of Igloolik, with no effects of age category or gender. Only one subject (a women aged 63 years, with a blood pressure of 160/90 mm Hg) was found to be hypertensive, out of a total sample of 145 subjects aged 18–79 years. Average blood pressures were substantially higher in the nGanasan population. Sex-specific ANOVAs showed that ethnic group had a significant influence upon both systolic and diastolic pressures in both men and women ($p < 0.001$). Three of 32 nGanasan men and seven of 30 nGanasan women had systolic pressures in excess of 160 mm Hg, and five nGanasan men and nine nGanasan women had diastolic pressures over 90 mm Hg.

One measure of gender-specific fat distribution, the chest/waist ratio, was (as expected) greater in men than in women (Table 8.4). In both the Inuit and the nGanasan populations, the ratios for both men and women showed a progressive decrease with age category ($p < 0.001$), but there was no significant influence of either ethnic group or age category × ethnic group interaction. Data on central versus peripheral fat deposition (the subscapular/triceps skinfold ratio) (Rode & Shephard, 1994b) showed values for the Inuit that would be regarded as relatively high in an urban

Table 8.4. *Chest/waist ratios for the Igloolik Inuit and the Volochanka nGanasan (mean $\pm$ SD of data)*

Age category (years)	Males		Females	
	Inuit	nGanasan	Inuit	nGanasan
18–29	1.18 $\pm$ 0.05	1.21 $\pm$ 0.05	1.15 $\pm$ 0.05	1.14 $\pm$ 0.05
30–39	1.18 $\pm$ 0.07	1.17 $\pm$ 0.05	1.12 $\pm$ 0.06	1.06 $\pm$ 0.07
40–49	1.15 $\pm$ 0.06	1.16 $\pm$ 0.06	1.09 $\pm$ 0.12	1.03 $\pm$ 0.07
50–59	1.07 $\pm$ 0.06	1.07	1.05 $\pm$ 0.21	1.00
60–69	1.01 $\pm$ 0.05	—	0.99 $\pm$ 0.08	—
70–79	1.07 $\pm$ 0.04	—	—	—

Source: Rode & Shephard (1994g).

Table 8.5. *Ratio of subscapular to triceps skinfold for the Igloolik Inuit (mean $\pm$ SD)*

Age category (years)	Males	Females
20–29	1.74 $\pm$ 0.66	1.06 $\pm$ 0.24
30–39	1.60 $\pm$ 0.52	1.15 $\pm$ 0.31
40–49	1.68 $\pm$ 0.36	1.06 $\pm$ 0.24

Source: Unpublished data of Rode & Shephard (1995).

community (Table 8.5). For instance, Frisancho *et al.* (1994) found average values for the US population of around 0.8 in women and 1.3 rising with age to 1.5 in men. Moreover, the subscapular/triceps ratios of the Inuit has increased as this group has become acculturated to a sedentary lifestyle. The male populations of Siberia are currently much thinner than the Igloolik Inuit, but nevertheless their subscapular/triceps ratios are also close to US norms (Leonard *et al.*, 1994a). It seems that the deterioration of cardiovascular prognosis associated with a high subscapular/triceps ratio is less marked in the arctic than in urban, developed societies, in part because both subscapular and triceps folds are still relatively thin in the northern populations. A further possible factor is an atypical distribution of body fat in the indigenous circumpolar groups. Shephard *et al.* (1973) suggested that the Inuit carried a larger than normal proportion of their body fat internally, and Beall & Goldstein (1992) speculated that even in the absence of marked obesity, a central deposition of body fat might be a 'useful' genetic adaptation to a cold climate. A more spherical body form would minimize heat elimination, and thermogenesis would also be greater

Table 8.6. *Prevalence of risk factors among Igloolik Inuit and Volochanka nGanasan. Smoker = regular smoker of more than one cigarette per day. Hypercholesterolemia = plasma cholesterol >5.7 mmol/l. Hypertension = systolic >160 mm Hg and/or diastolic >90 mm Hg*

Group and sex	Sample size	Number of risk factors (% of sample)			
		None	One	Two	Three
Inuit men	89	20	74	6	0
Inuit women	62	16	77	7	0
nGanasan men	22	9	55	36	0
nGanasan women	10	30	40	30	0

Source: Rode & Shephard (1994g).

with abdominal than with peripheral fat deposition. Rode & Shephard (1993) pointed out that when living in a very cold climate, it was important that fat reserves be carried internally, so that body insulation did not fluctuate with nutritional status and thus utilization of depot fat.

As discussed in Chapter 7, the proportion of male smokers and their daily consumption of cigarettes were both very similar in the Inuit and the nGanasan communities. The proportion of female smokers was higher in Igloolik (80%) than in Volochanka (56%), and among the women, per capita cigarette consumption also tended to be higher for the Inuit than for the nGanasan.

The daily energy expenditures of the men were high in Igloolik at the time of the IBP-HAP survey (Godin & Shephard, 1973a), but habitual physical activity has decreased progressively with the mechanization of both hunting and life within the settlement. Data for Siberia show a continuing high energy expenditure in the men, but relatively low activity levels in the women (Leonard *et al.*, 1994a).

The distribution of the three risk factors for which we have precise and recent data, cigarette smoking, a plasma cholesterol greater than 5.7 mmol/l, and hypertension (a systolic blood pressure greater than 160 mm Hg, or a diastolic pressure greater than 90 mm Hg) is summarized in Table 8.6. Among the Inuit, the cardiac risk would become extremely low if smoking could be eliminated, but among the nGanasan a substantial fraction also have a risk attributable to hypertension.

A few other authors have provided parallel analyses. Alekseev (1991) found that in Yakutia, 2.1% of aboriginals had three major risk factors, 19.9% two risk factors, and a further 48.8% one risk factor. Exercise habits were not included in their analysis, and the relative importance of the other three factors was not identified, but it may be suspected that smoking accounted for much of the total risk. Astakhova *el al.* (1991) found that 63% of aboriginals in Chukotka were smokers, 15% had arterial hypertension, and 6.5% were hypercholesterolemic.

Two important factors protecting traditional coastal communities against cardiovascular disease have been a high intake of ω-3 polyunsaturated fatty acids from fish and marine mammals (Chapter 4) and a high level of daily physical activity. In Greenland Inuit, associations have been demonstrated between plasma levels of ω-3 polyunsaturated fatty acids, prolonged bleeding time and an inhibition of platelet aggregation. Factors reducing the tendency to clotting include an inhibition of thromboxane TXA_2 formation by eicosopentanoic acid ($20:5\omega$-3 fatty acid), and changes in the lipid content of the platelets themselves (Dyerberg & Bang, 1979).

Unfortunately, both habitual physical activity and the consumption of marine food have diminished with 'modernization' of the arctic communities, and the proportion of smokers has also increased.

Ischemic heart disease

In the years 1974–76, Dyerberg & Bang (1981) found that the age-standardized death rate from ischemic heart disease in Greenland was 5.3, compared with 34.7 for Denmark and 40.4 for the United States. Among the Baffin Inuit, the standardized mortality ratio for deaths from circulatory diseases relative to Canadians as a whole was still only 0.6 in 1983–87 (Choinière, 1992). Likewise, Astakhova *et al.* (1991) found ultrasonic evidence of abdominal aortic atherosclerosis in only 2.5% of indigenous Chukotki, but in 13% of immigrants from southern cities. Malyutina *et al.* (1991) found that angina pectoris was equally prevalent in the indigenous population of Chukotka (5.9%) and in Novosibirsk (5.8%), but it was admitted that among the indigenous population, chest pains often proved to be non-specific in origin. Electrocardiographic Q-wave changes were seen 1.5 times more often in indigenous groups than in the general population of Novosibirsk (3.6% versus 2.2%), but other ischemic changes were more common in the general population (6.5%) than in the aboriginal populations (2%).

Acculturation has been associated with an increasing prevalence of ischemic heart disease in a number of northern communities. Bjerregaard (1991) observed that in Greenland, mortality rates from ischaemic heart disease were higher in towns than in settlements. Moreover, in studies from

Yakutia, Alekseev (1991) found that the average area of atherosclerotic lesions in the aortas of the aboriginal population had approximately doubled from 1968 to 1988, although they were still somewhat less marked than in 'white' immigrants to the region. Recent ultrasonic investigations have found no difference in the extent of carotid artery atherosclerosis between Danes and Greenlandic Inuit (Hansen *et al.*, 1991), although the indigenous population may still gain some protection against clinically manifest ischemic heart disease from a lesser tendency to clotting of the blood.

Hypertension and stroke

Although ischemic heart disease has traditionally had a low prevalence in the indigenous populations of the north, many of the ethnic groups concerned have paradoxically shown a substantial prevalence of hypertension and a high incidence of strokes.

In Chukotka, Astakhova *et al.* (1991) found a 15% prevalence of hypertension. Nikitin *et al.* (1981), apparently studying the same population sample, commented that the prevalence of hypertension was lower in coastal settlements where the economy was based upon hunting and fishing (8% in Lorino, 13% in Walen) than in the interior of Siberia, where people worked as reindeer breeders (19% in Kanchalan, 21% in Markova).

Among the Cree Amerindians of Northern Québec, a growing problem of obesity has been associated with an increasing proportion of hypertensive individuals. In 1983–84, Thouez *et al.* (1990) found frank hypertension (a systolic reading > 160 mm Hg, or a diastolic reading > 95 mm Hg) in 17% of male Cree and 22% of female Cree, with another 19% of men and 21% of women reaching borderline values. In contrast, the prevalence of hypertension in (presumably less acculturated) Inuit who were living at higher latitudes was only 5%. Young (1991) found an even higher overall prevalence of hypertension (27%) among Cree and Ojibwa Amerindians of the Keewatin District. High pressures were more common in men, and were concentrated in those with a family history of hypertension, obesity, diabetes, low physical activity levels and little formal education. Other correlates of the disorder included a high total serum cholesterol, a low HDL-cholesterol/total cholesterol ratio, high serum triglycerides, and central fat patterning.

Systolic blood pressures have tended to be low among coastal Inuit, both in Greenland (Jorgensen *et al.*, 1986) and in the North West Territories, but much higher pressures have been found in the nGanasan residents of the Siberian tundra (Rode & Shephard, 1994a). Bjerager *et al.* (1981) described differences in the relationship of blood pressure to age between Inuit living on the west coast of Greenland (Upernavik), the east coast (Scoresbysund) and the north-western coast (Thule). In the last location, systolic pressures were low in young adults, and showed little tendency to rise with age. Early studies of the Inari and Skolt Lapps also found little change of diastolic pressures with age (Sundberg *et al.*, 1975).

Despite low average blood pressures, Kromann & Green (1980) found that the incidence of stroke in the Upernavik district of Greenland was twice that anticipated in Danes. Bjerregaard & Dyerberg (1988) confirmed this in terms of the cerebrovascular disease death rate for Greenland as a whole. Possibly, the 3ω-fatty acids that protect the coastal Inuit against coronary thrombosis also increase the tendency to brain haemorrhage (Kromann & Green, 1980). Bjerregaard (1988) found that the overall death rates for stroke were almost equal in Greenland and in Denmark. Male rates for cerebrovascular disease were also higher in the towns than in the settlements of Greenland (Bjerregaard, 1988).

Diabetes mellitus

As with indigenous populations living in warmer climates, a substantial proportion of Amerindians living in subarctic regions have become obese in recent years. A high prevalence of diabetes mellitus has developed with this trend (Young *et al.*, 1992).

The WHO criterion for the diagnosis of diabetes mellitus is a plasma glucose concentration in excess of 11.1 mmol/l two hours after the oral ingestion of a 75 g glucose load. Available information on the glucose tolerance of circumpolar peoples has already been discussed (Chapter 4). The clinical diagnosis of diabetes mellitus is often complicated by problems in obtaining subjects in a fasting state, and until recently there were also problems in making measurements of blood glucose concentrations in remote locations. Reliance upon clinical diagnoses often yielded fallible data. One study of Dogrib Amerindians found that 13% of male and 7% of female subjects had an impaired glucose tolerance, despite the reputed absence of clinical diabetes mellitus in the Dogrib community (Szathmary & Holt, 1983).

A statewide survey of Alaska in 1985 found that the prevalence of diabetes mellitus was lowest among the Yu'pik Inuit of southwestern Alaska, and highest (2.72%) among the coastal Amerindians (Schraer *et al.*, 1988). Despite the low overall prevalence in the Yu'pik Inuit, the prevalence of diabetes among the pregnant women of this group was twice that for the US as a whole (Murphy *et al.*, 1991), suggesting that many of the Yu'pik may have had an undiagnosed latent glucose intolerance

(Harris, 1988). Middaugh *et al.* (1991) estimated diabetes rates from physician diagnoses at hospital discharge, finding an overall age-adjusted prevalence of 1.36% in Alaskan natives, compared with 2.47% for the US as a whole; however, the rate for the natives excluded those receiving out-patient care at private clinics, and thus may have been an under-estimate of the true prevalence.

A sample of 355 relatively unacculturated Skolt Lapps yielded only one female with diagnosed diabetes mellitus. However, the disease became more common as contacts with urban society increased. Among the more acculturated and semi-urbanized Nellim Lapps, four females and one male cases of diabetes were diagnosed in a population of 154 subjects (Eriksson *et al.*, 1980).

The early data of Schaefer (1968a, b) found a very low prevalence of diabetes mellitus among Canadian Inuit. Young & McIntyre (1985) reported that the physician-diagnosed prevalence of diabetes mellitus for Cree and Ojibwa Amerindians of the Keewatin region over the period 1978–82 was 2.3%. The prevalence rose from 0.06% in those aged 0–14 years to 0.28% in those aged 15–24 years, 4.36% in those aged 25–44 years, 9.88% in those aged 45–64 years, and 8.61% in those 65 years and older. The risk of diabetes was greater in those with a blood pressure greater than 140/90 mm Hg, and in those with a body mass that exceeded the 75th percentile of urban norms. Based on the years when individual cases had been diagnosed, it appeared that the incidence of the disorder had increased substantially from 1958–1962 to 1978–1982. Thouez *et al.* (1990) found blood sugar levels in excess of the WHO ceiling in 4.4% of female and 1.2% of male Cree Amerindians living in northern Québec, compared with 0.38% in female and 0.43% in male Inuit who were living in more isolated settlements at higher latitudes. In a further study of the James Bay Cree Amerindians, Brassard *et al.* (1993) noted that 38.6% of cases of physician-diagnosed diabetes had truncal obesity, and 56% of men and 69% of women had a body mass index of greater than 30 kg/m^2.

Astakhova *et al.* (1991) and Stepanova & Shubnikov (1991) both reported impaired glucose tolerance in 5% of the indigenous population of Chukotka, compared with 16% of newcomers.

Carcinomas

General considerations

There is general agreement that in the indigenous circumpolar populations, some types of carcinoma are more common than in urban populations, and some less so. However, because of technical problems (small total populations, difficulties in detecting and confirming cases in remote locations, and differing reporting periods), details of the prevalence, incidence and mortality for specific types of tumour are disputed.

In some instances, for example, lung cancer, other tumours that are provoked by smoking, and lesions such as cervical cancers that are associated with sexually transmitted viruses, the prevalence (which had originally been low in the arctic) is now increasing rapidly to overtake that found in the urban regions of North America. Acculturation to a 'modern' lifestyle has also changed the prevalence of certain cancers that traditionally were common among indigenous populations (Hildes & Schaefer, 1984; Irvine *et al.*, 1991). In particular, the prevalence of salivary and naso-pharyngeal tumours has diminished in some areas of the Canadian arctic, although the overall incidence still remains quite high (Gaudette *et al.*, 1991).

Detection of tumours tends to occur later among the indigenous circumpolar groups than in 'white' society, and access to specialized treatment is also difficult for those who are living in isolated settlements, so that the survival rates for conditions such as lung cancer remain much poorer than in urban centres (Irvine *et al.*, 1991).

Tumours of the nasopharynx and salivary glands

Several authors have noted a high incidence of cancers of the salivary glands and nasopharynx in traditional circumpolar groups (Freitag *et al.*, 1991; Hildes & Schaefer, 1984; Lanier *et al.*, 1981; Mallen & Shandro 1974; Nielsen, 1986). Relative to North American norms, the prevalence of salivary tumours in the Canadian North West Territories over the period 1970–84 was 9.2 times the expected rate in men, and 20.7 times the expected rate in women. Rates for nasopharyngeal tumours were also 21.5 times and 26.9 times expected values in men and women respectively (Gaudette *et al.*, 1991).

It has been speculated that a genetic abnormality (Simons *et al.*, 1974), and Epstein Barr viral infection (Albeck *et al.*, 1992; de Thé, 1982; Lanier *et al.*, 1981) and some dietary factor such as a high nitrosamine content of dried fish (Hubert *et al.*, 1985; Poirier *et al.*, 1987; Siemssen *et al.*, 1981) may all increase vulnerability to the development of salivary gland and nasopharyngeal tumours in the circumpolar groups.

Lung cancers

In recent years, there has been a rapid increase in cancers of the upper respiratory tract among the indigenous circumpolar populations, probably because the practice of cigarette smoking has become so widespread (Storm *et al.*, 1991). In the Canadian North West Territories, the incidence of lung cancer is now 2.4 times the expected rate in men and 6.4 times the expected rate in women (Freitag *et al.*, 1991; Gaudette *et al.*, 1991). Such figures are almost twice those observed in Alaska and Greenland (Lanier *et al.*, 1990). There is little difference in the prevalence of smoking between the Canadian arctic, Alaska and Greenland, and it may be that in northern Canada the adverse effects of smoking have been supplemented by exposure to extremely cold air and/or earlier exposure to the sooty smoke from oil lamps (Hildes & Schaefer, 1984).

Cancers of the reproductive tract

There is a high risk of both cervical and ovarian cancers among circumpolar women (Freitag *et al.*, 1991; Harvald, 1991; Gaudette *et al.*, 1991; Nielsen *et al.*, 1978). For example, Nielsen *et al.* (1988) found that the incidence of invasive cervical cancer in Greenland in 1981–85 was four times that for Denmark.

This finding has been associated with the prevalence of high-risk sexual behaviour. The age at first intercourse is low in circumpolar communities, there is a high birth rate among teenagers, the rate of many sexually transmitted diseases is high, and a high proportion of youths have multiple sexual partners (Freitag *et al.*, 1991).

Extensive cervical-smear examinations have so far failed to reduce the incidence of invasive cancers, probably because the affected individuals have either escaped screening or have been lost to follow-up (Nielsen *et al.*, 1988).

Other tumours

Among the indigenous populations living in the former Soviet Far East, there is a high incidence of stomach cancer (Kustov *et al.*, 1991). Moreover, the risk of developing a gastric tumour seems to diminish if the individuals concerned move to a more southerly latitude. Similar findings have been reported from Labrador (Thomas & Williams, 1976) and Greenland (Nielsen & Hansen, 1981). Suggested causes include an accumulation of nitrosamines in stored fish, a deficiency of vitamins A and C, and a high incidence of alcoholism.

Infection with hepatitis B virus seems responsible for a high incidence of primary hepatic tumours among the indigenous populations of the Canadian and Alaskan arctic, but, puzzlingly, hepatic tumours are much less common in Greenland (Nielsen & Hansen, 1985).

The incidence of breast cancer in the indigenous Alaskan population is only about a half of that seen in 'white' immigrants (Lanier & Mostow, 1991). In Canada, also, breast cancer has been uncommon in either Inuit (Freitag *et al.*, 1991) or Amerindians (Young & Frank, 1983) until recently, although the rates for these two populations now appear to be rising (Irvine *et al.*, 1991). Possibly, the unacculturated population was protected by a low body fat content and a high level of habitual physical activity.

Perhaps because the skin is covered for much of the year, and perhaps also because the skin has a higher melanin content than in 'white' populations, the incidence of melanomas is low among permanent residents of the arctic regions (Freitag *et al.*, 1991). In Greenland, Krømann *et al.* (1981) noted that the skin cancer rate was only 40% of that expected in men, and only 20% of that expected in women; likewise, the rates for malignant melanomas were 70% of expectation in men, and only 10% of expectation in women.

Disease as a selective agent

One of the original hypotheses of the International Biological Programme was that disease might act as a selective agent, killing those individuals who were poorly adapted to a specific habitat, but allowing the propagation of others better adapted to the milieu. For example, a disease such as tuberculosis might prove fatal before the period of reproduction and child-rearing had been completed, or a condition such as ischemic heart disease might influence the course of natural selection because antecedents of the disorder (for example, a lack of physical fitness) made the susceptible individual less attractive to members of the opposite sex, and therefore less likely to procreate. A further possibility might be the survival of a particular variant in an isolated community, because the disease(s) that would otherwise prove fatal to that individual were rarely encountered in the chosen habitat.

The organizers of the IBP-HAP survey hoped that a detailed study of genetic markers would allow them to establish correlations between the carriage of specific genes and susceptibility to or resistance against particular diseases. A wealth of data of this type was collected (Eriksson *et al.*, 1980), but in general it proved hard to interpret. Given the small number of people living in most settlements, it was difficult to be certain whether unusual genetic variants had arisen through selective pressures, or as a consequence of a random sexual encounter between a circumpolar resident and a traveller from some distant part of the outside world. Further, the apparent limited range of alleles in many circumpolar communities might reflect the fact that investigators had made a less exhaustive study of blood and tissue specimens from these populations rather than being a sign of true homozygosity. If a real phenomenon, the limited extent of variation could reflect a population that had been kept small by a challenging habitat and that had experienced only limited contact with conquering invaders. Alternatively, again because of a small total population, there could have been a 'genetic drift', with some normally encountered variants dying out merely by chance (Harvald, 1976).

The main historic pressures upon survival in most circumpolar communities have been early childhood disease, starvation and accidental death. Shephard (1980) speculated that because of the high infant mortality (an average of 232 per 1000 in Igloolik from 1959 to 1968) there may have been a selective survival of infants with an above average level of cardio-respiratory function. Provision of adequate food supplies for a growing family depended mainly on the skills of the hunter, not all of which were physical in nature. But in particularly difficult times, those members of the population with a large mobilizable reserve of intra-abdominal fat may have had a selective advantage in resisting starvation. In some instances, a well-developed muscular strength or aerobic power may also have helped to avoid accidental death, but the survival of catastrophe often depended more upon intelligence than on unusual physical prowess.

Detailed investigation has revealed more specific instances of selective pressures. Many circumpolar groups, particularly those living away from the coastline, have had narrow pelvises. One reason for this was ultra-violet deprivation and a resulting rachitic deficiency of Vitamin D. The narrow pelvis often caused difficulties in labour, thus exerting a specific selective influence favouring women who were able to maintain their vitamin D reserves (Lewin & Hedegard, 1971a, b; Eriksson *et al.*, 1980). The vitamin-D binding alpha globulin seems identical with group-specific component (Gc) protein, and it is interesting that the Gc^2 allele tends to be most frequent in populations where exposure to sunshine is low (Mourant *et al.*, 1975). However, interpretation of the predominance of this characteristic as a genetic adaptation to birthing problems in the arctic is weakened by the fact that the Lapps, the group most vulnerable to rickets before the introduction of vitamin supplementation, do not show an unusually high frequency of the Gc^2 allele. Another possible genetic adaptation to lack of sunlight and a resulting deficiency of Vitamin D is a fair complexion.

Certainly, the Lapps have less pigmentation than the coastal dwelling Inuit and Ainu (who have access to fish oils rich in Vitamin D).

Other evidence suggesting an inherited susceptibility to particular diseases has emerged from genetic studies of circumpolar populations. For example, Harvald (1976) found a high correlation between the histocompatibility antigen (HL-A) W27 and the development of certain rheumatic diseases. More than 90% of patients with ankylosing spondylitis were W27 positive, and this particular antigen was four times more common in Greenlandic Inuit than in the Danish population. A number of other antigens, such as HL-A8, are rare in both Inuit and Lapps, possibly because the circumpolar populations have been exposed historically to a much narrower range of diseases than have urban societies (Simpson, 1981). One practical consequence of the limited range of HL antigens is that the circumpolar groups rarely suffer from auto-immune diseases (Svejgaard *et al.*, 1975). The limited number of such antigens may also explain why there is a low incidence of juvenile diabetes in the arctic (Thomsen *et al.*, 1975). Tissue grafts also survive for longer in the Inuit than in large urban populations (Dossetor *et al.*, 1976; Kissmeyer-Nielsen *et al.*, 1971), and the incidence of congenital disorders is low (Simpson, 1981).

However, one negative consequence of their unusual immune characteristics is that many circumpolar groups have a low incidence of antibodies to major infections. It remains unclear how far this reflects a lack of exposure to the infections in question, and how far there are peculiarities of the immune system that limit the production of such antibodies. Some of the studies of vaccination discussed above suggest that the Inuit immune system lacks the capacity to respond to certain pathogens, and several authors (Nielsen *et al.*, 1971; Persson *et al.*, 1972; Sayed *et al.*, 1976b) have reported an absence or low levels of salivary Immunoglobulin A among the Inuit.

Conclusions

The nature of the arctic habitat and the lifestyle of traditional circumpolar populations protect them against certain chronic medical disorders that are common in the urban environment. However, contact with the bacteria and viruses of developed societies has brought devastating epidemics to populations that previously lacked immunity to the diseases in question. Currently, acculturation to certain adverse habits of 'modern' society (over-eating, lack of physical activity and cigarette smoking) and adoption of a poorly chosen 'market' diet are also increasing the prevalence of such chronic diseases as ischemic heart disease, diabetes, and several types of cancer among the circumpolar populations. Life in isolated settlements has apparently permitted the survival of groups that have little immunity to diseases that are common in large cities. A high infant death rate and the problems of childbirth for women with narrow rachitic pelvises may have had some influence upon the course of natural selection, but as yet there is little objective evidence to support the hypothesis that disease has had a major influence in impressing particular genotypic and phenotypic characteristics upon the circumpolar populations.

9 Postscript: lessons from traditional circumpolar life and options for the future

In most parts of the arctic, the traditional lifestyle of indigenous populations can no longer be sustained in the face of a rapid natural increase in the population and commercial exploitation of the circumpolar habitat by the multinational corporations of 'modern' society. This final chapter will highlight a few valuable characteristics of the traditional northern heritage that merit preservation. It will review desirable future adaptations of indigenous circumpolar society, and will draw some lessons for sedentary city-dwellers, health professionals and indigenous populations that have colonized other habitats.

The heritage of traditional circumpolar life

A heavy use of imported technology, fossilized carbon energy and other material resources allows the city-dweller to exploit the coldest parts of the arctic, regions that destruction of the natural flora and fauna has made uninhabitable for the first nations. Perhaps for this reason, temporary arctic sojourners and immigrants from lower latitudes often suggest or imply that the ways of their society are in some fashion superior to those of the indigenous circumpolar residents. As a corrective to such a judgment, it seems useful to list some important lessons that 'developed' societies could learn from indigenous populations. Accumulated knowledge and behavioural adaptations have allowed peoples with no great inherent biological advantages to colonize one of the earth's least promising regions, and to succeed in their endeavour without recourse to the technological props that the 'modern' city-dweller finds so essential when venturing into the arctic.

We will focus upon the overall philosophy and theology of the first nations, their attitudes to education, diet, and physical activity, and other favourable features of their traditional lifestyle.

Overall philosophy

One important element in the traditional philosophy and theology of the indigenous circumpolar peoples was a profound respect for the fruits of the earth and the oceans. For the Amerindian, the earth, and for the Inuit, the sea, were seen as sacred sources of life-giving food and material resources (Chapter 1). Humans were recognised as an integral part of the local ecosystem, profoundly at one with nature and depending for their very survival upon a preservation of the richness of their immediate habitat. The local flora and fauna were viewed as gifts, to be used only as needed, and to be shared freely with others in need – particularly the older and the weaker members of the community.

At the same time, indigenous populations did not have the modern fear of death that sometimes leads people in urban society to spend vast sums in prolonging the life of a friend or relative by a few painful weeks. If a winter was hard, and an elderly member of a small circumpolar settlement felt that their useful contribution to the community had been completed, they were willing to sacrifice their life so that the available stores of food might be enough to ensure the survival of the younger members of the group (Chapter 3).

Such a holistic world view stands in stark contrast with the mores of our modern commercial society. The 'white' doctrine is to subjugate the natural environment, rather than to work in cooperation with it. Too often today human society applauds a materialistic emphasis that maximizes exploitation of the natural world in a minimum of time, and at the lowest immediate cost. The ever-expanding consumer economy pays little heed to the amount of pollution that it may create, the disruption of the traditional indigenous economy that it causes, the need for ultimate renewal of the natural resources that are being exploited, or the cumulative costs that the raping of an environment may impose upon future generations. The citizens of 'developed' society aggressively claim territory and accumulate products not because of any personal need, but merely because such acquisitions are seen as a means of gaining status. The idea of sharing possessions with others has been largely abandoned in the relentless quest for material wealth, and in many instances it is hard to persuade people even to care for the older members of their own families.

Thus, the intrinsic global burden of an expanding world population is enormously exaggerated by a small group of the world's wealthy: 'modern' people in 'developed' societies who consume an excessive fraction of natural resources and account for much of the currently unsustainable burden of air, land and sea pollution.

Education

Many casual visitors from 'developed' societies have considered indigenous residents of the circumpolar territories to be ill-educated. In fact, the main difference between traditional methods of education and those subsequently imposed upon the first nations by 'white' administrators seems that the former approach was better adapted to the immediate needs of the local people.

It is instructive to peruse a modern Inuktittuk dictionary. The brevity of the 44 letter syllabic alphabet introduced by nineteenth-century missionaries has now undergone transliteration into Roman letters. Beyond this external change, notice the 23 types of snow that are still described, and ask how many of these variants the average 'well-educated' 'white' person can recognise. Observe also the humorous literal translation of the words for a white administrator – the person who arrives with a suitcase, a ludicrous piece of equipment to bring to the arctic habitat!

It is equally interesting to watch how the indigenous people learn from their environment. The skilled hunter, for example, can follow the movements of animals, birds and aircraft from a distance where the inexperienced city-dweller cannot even detect them (Chapter 1). And in the home, the Inuit woman demonstrates skills that her 'white' counterpart could not emulate even with much technological support. Compare the 11 CLO of insulation provided by the parka that she fashions for her husband with the 4 CLO of protection found in the military clothing that 'modern' tailors have crafted for Canadian arctic regiments (Chapter 1).

The traditional basis of education in circumpolar communities was experiential – an approach that the wisest of our 'modern' educators are beginning to realize is much more effective than book-learning. The skills of hunting, domestic economy and overall arctic survival were learned as children listened to and discussed their experiences of the natural world with the elders of the community. This approach gave grandparents and other senior citizens an important role in the overall economy, something which is sadly lacking in 'developed' societies. The continued and successful colonization of an unforgiving habitat bears eloquent testimony to the practical success of the indigenous approach to learning.

Against these documented achievements, we may set the approaches of our urban educationalists, the 'learned' people with doctorates in education. Their earliest policy, both in North America and in Siberia (Tumnaettuvge, 1991) was to send children to distant residential schools, sometimes against the wishes of the parents (Chapter 3). This may have been the only educational option in that era, and it produced many of the current leaders of the northern peoples. However, the residential schools generally prohibited use of any language other than that of the dominant culture, and the ideas of 'white' society were inculcated by rote learning from textbooks. Subsequently, the scholars, by then young adolescents, were shipped back to the small settlements from whence they had been removed. They returned as strangers: people who could not speak the local language, who were alienated from their parents by a foreign culture, and who lacked even the basic skills needed for survival in the arctic habitat.

The next tactic of 'modern' educationalists – a government-imposed concentration of whole families in large settlements – allowed education to proceed without breaking up the family, but it still showed little understanding of the place of the circumpolar residents in the local ecology. In many areas, natural resources within a reasonable travelling distance of the enlarged settlement quickly became depleted of game and unable to support the traditional way of life.

Pupils may now emerge from a 'modern' 12 grade school with a substantial knowledge of computers, science and world religions, but too often they have been taught by teachers who know little about either the arctic or the rich heritage of indigenous traditions. Graduates cannot pursue the lifestyle of their parents, but educators have given little thought to ways of equipping them for any alternative employment. It is thus not surprising that many indigenous students see little reason to complete their enforced high school education. Instead, they turn to alcohol, drugs or even suicide as an escape from what they regard as an unenviably bleak future.

Diet, activity patterns and lifestyle

One of the dominant concerns of modern society is that many adults are now falling victim to ischaemic heart disease and various types of cancer at a relatively early age. A commonly suggested reason for the growing prevalence of these conditions is that city-dwellers are eating the wrong sorts of food and are not undertaking sufficient physical activity. As a reaction to this concern, some of the urban population engage in over-rigorous dieting or become obsessed with exercise, to the point of developing anorexia nervosa or precipitating a heart attack during a session of heavy exercise (Shephard, 1993b).

Again, there seems much that 'modern' society can learn from the wisdom of the traditional circumpolar inhabitant. To our knowledge, the traditional Inuit never adopted a deliberate 'reducing diet', but because of a lifestyle where levels of physical activity were adequate, the energy was derived more from protein and fat than from refined carbohydrates, there was no significant accumulation of sub-cutaneous fat as the individual became older (Rode & Shephard, 1971). Again, in part because of the absence of television and magazine images of the (supposedly) sexually attractive young women with a very low body fat content, there were no instances of anorexia nervosa and bulimia among traditional circumpolar residents.

The lifestyle of the coastal peoples of the arctic was marked by a very low incidence of ischaemic heart disease and hypertension (Chapter 8). This was linked to such favourable lifestyle practices as the choice of a diet rich in polyunsaturated fish oils (Chapter 4) and a high overall daily energy expenditure reached through sustained, moderate intensity activity (Chapter 1). The typical city-dweller is so harried by 'lack of time' that any occasional bouts of exercise must be pursued at an exhausting rate. But in contrast, the arctic resident seeks to avoid sweating, and adopts a pace that can be sustained over much of the day (Chapter 1). In 'developed' societies, the consumption of refined carbohydrates has also risen steeply over the past century. But finally, scientists are beginning to commend the fish oils that are rich in ω-3 fatty acids (Simopoulos, 1991), and to recognize that sustained low intensity activity may be the optimum exercise prescription for overall health (American College of Sports Medicine, 1991; Shephard, 1993b).

Other favourable practices

Lactation and infant care

In traditional Inuit society, lactation continued for several years (Chapter 3). This provided a close bonding between mother and child, a bond that was reinforced by carrying the infant on the back throughout the day in the amauti. 'Modern' societies are increasingly recognising the value of breast feeding, and parents are also experimenting with pouches to carry small children in a manner analogous to that adopted by the Inuit.

The prolonged period of lactation provided arctic populations with a relatively effective form of natural contraception. In contrast, we may note the potentially carcinogenic contraceptive preparations used by women in many 'educated' societies.

Gender equity

To some observers, circumpolar society has seemed patriarchal in its structure, with rigidly ascribed gender roles. Such a practice has been determined in part by the high infant mortality rate, and thus the need for a high birth rate to sustain population numbers. Nevertheless, when the territory of Nunavut achieves independence in 1999, its government will be the first Canadian legislative body to consider definitive measures to assure an equal representation of women and men in its assembly. Moreover, at least initially it will avoid the extravagant posturing of opposing parties that has limited the effectiveness of political endeavour in southern Canada in recent years.

Social behaviour

The traditional social behaviour of the circumpolar peoples is marked by a sharing of material possessions. This is well-characterized by the willingness to adopt orphaned children (Chapter 3, Egan, 1991), but it is seen also when hunting, as members of an expedition share information and work toward common goals in the hunting of their prey.

The Inuit in general have had little taste for the affected conversation that has marked the social behaviour of 'developed' society. For example, when a meeting with a friend or colleague is completed, the visitor will end the discussion simply by saying 'Taima', rather than presenting a large number of specious excuses for leaving.

Communication depends to a large extent upon listening rather than talking. Information is transmitted by facial expression, body language and a change in the tone of the voice, rather than an excessive use of words.

Attitudes to time

In some settlements, the use of clocks was virtually unknown prior to the IBP investigations. This attitude is in marked contrast with urban society, where time has become a tyrant.

In 'developed' society, perceived time pressures often cause people to work in opposition to nature. For instance, they may try to accomplish self-imposed tasks on schedule, despite adverse natural conditions. A typical Torontonian may drive several hours in a blizzard in a fruitless attempt to reach his or her place of employment at the expected time of duty.

However, the extreme environment of the arctic has taught the indigenous peoples of this region the importance of working with rather than against the environment. No rigid nine-till-five timetable is imposed – if the weather conditions appear favourable for hunting, other planned projects are abandoned, and game may be pursued throughout much of a 24-hour day (Chapter 1). But if conditions are unfavourable for outdoor pursuits, people will rest and/or find alternative indoor employment until the weather improves.

Capacity for improvization

The absence of a ready access to 'market' supplies has allowed the circumpolar resident to develop a brilliant capacity for improvization. This talent continues to be of service even now, when much mechanical equipment is available.

For instance, if a snowmobile breaks down, the indigenous arctic resident still knows how to fashion an effective overnight snow shelter, and how to make a functional repair to equipment using a minimum of tools and supplies; with the patience and persistence learned on the land and the hunt, the Inuit will invest many hours in fashioning a necessary part from other parts or implements. In contrast, the city-dweller can think of little alternative but to find a colleague who can drive him or her to a dealer to purchase some new parts for the machine. Such a plan is not compatible with survival in the arctic winter, and for this reason at least one city (Point Barrow) restricts snowmobile use to the indigenous population during the coldest winter months.

Future adaptations of circumpolar society

The burden of acculturation

The extent of the changes that have already occurred in arctic society seems overwhelming even to the external observer. The older circumpolar residents have faced the trauma of moving from the clean, open spaces of the tundra to squalid prefabricated villages. Here, they have encountered an unfamiliar type of dwelling, a greatly increased population density, and a contamination of their previously pristine air and water resources. Their diet has changed from an almost total dependence upon 'country' foods to a reliance upon 'market' purchases as a means of satisfying daily energy needs (Chapter 4). The defences of the body have been challenged by new and unfamiliar diseases. The virtual autonomy of the nomad has been exchanged for a regimented life, imposed by petty government bureaucrats. The traditional occupations of hunter, trapper and reindeer-herder have given place to work as a menial assistant of the southerner: an operator of heavy machinery, a government clerk, or an unemployed dependent. Traditional culture has been shaken to its roots by an imported language, religion, educational system and technology. Family structures have collapsed as menfolk have migrated to urban centres in search of work (Chapter 3), and an aggressively dominant southern culture has dictated major changes in attitudes, behaviour and values.

Many of the pressures of acculturation are experienced, although to a much smaller degree, by all migrants to a new society. But in the case of 'white' settlers in the north, knowledge that the move was self-imposed, with the option to return to their previous environment, facilitates acceptance of the situation and adjustment to it. In most cases, the indigenous populations of the arctic did not seek to be 'modernized' – rather, change was thrust upon them by the greed of entrepreneurs and the paternalism of state and church.

Given that the process of acculturation seems likely to continue in most regions of the arctic, what can be done to facilitate the coping strategies of the indigenous populations? Practical suggestions include a facilitation of higher education, and a shift of responsibility for government, health care, religion and the environment from immigrants to the indigenous populations.

Education

Currently, very few indigenous residents of the arctic have the educational background that would allow them to assume a managerial role in 'modern' society. Most teachers in northern high schools are 'white' immigrants who only remain in the arctic for one to three years. Few of them speak the local language or understand the local culture. It is thus not surprising that many students leave school before completing their diplomas and that few teenagers have the educational background necessary to enter university or to pursue careers in medicine, education, civil administration or political science.

Unfortunately, the economic realities of serving a very small population militate against the production of textbooks in local languages and dialects, and a good knowledge of English remains necessary for a student to progress through the higher school grades (Boessen, 1991). The long-term remedy is to increase the number of indigenous teachers, individuals who are familiar with the local language and culture, able to communicate with the students and to serve as role models for them. Many school boards have now adopted such a policy for the lower school grades. But in order to teach high school students in a culturally sensitive and relevant manner, a cadre of well-educated indigenous people is needed. This is currently lacking. For example, a 1991 study found fewer than 200 Canadians of native ancestry with qualifications in medicine, nursing, physiotherapy, occupational therapy or laboratory technology (Thomlinson *et al.*, 1991).

The most promising method of increasing the formal education of the indigenous population is an affirmative action education programme. This allows mature students with inherent promise to complete high school diplomas at an accelerated rate (Stephens, 1991) and gain facilitated entrance to university programmes. There is a related need for support services that are operated by indigenous people within schools and universities. These can help the passage of students through the unfamiliar educational milieu. Necessary assistance may include subsidies to cover the costs of accomodation, books and supplies, transportation from remote locations, baby-sitting services, and dental and optometry bills (Stephens, 1991). In some instances, it may also be appropriate to allow longer than normal to complete a given course requirement (Thomlinson *et al.*, 1991).

In the past, one serious objection to the provision of greater educational opportunities for indigenous populations was that such a process increased the pace of acculturation. However, if it be accepted that acculturation has become inevitable, then plainly it is important that an appropriate percentage of the local population receives the benefits of higher education, so that they can play their appropriate role as physicians, nurses, teachers and administrators in northern communities. In some instances (such as the preparation of nurses), a combination of audio-conferences, video-tapes and extensive use of FAX machines may enable students to complete much of their education while remaining within small and remote communities (Pflaum *et al.*, 1991). There is then less disruption of previous lifestyle, and perhaps for this reason the educational accomplishments of such programmes may match those realized by centrally operated courses in major urban centres. An important bonus of community-based higher education is that it reduces the risk that the person who is being educated will be permanently seduced by the comforts of life in the south.

Government

Currently, most governments have established a multiplicity of agencies to administer services in the circumpolar territories. In Canada, governmental expenditures on a typical indigenous family of five people now exceed $60 000 per year, yet often the end result of such a large expenditure is disappointing in terms of health, education, and social adjustment.

As with education, the system of government administration has tended to be paternalistic, policies being determined by bureaucrats who sometimes had only a limited understanding of local customs, culture, or needs (Clayton, 1991). Moreover, much of the available funding was squandered as a result of inter-departmental bickering and the provision of overlapping services. For instance, in Canada the federal government had assumed responsibility for the provision of health services to indigenous populations, but the Province or Territory was responsible for providing the same services to southern immigrants. Often, there were prolonged inter-agency disputes about who had responsibility for indigenous people who had moved to larger centres or had established relationships with 'white' partners.

The transfer of services to the indigenous populations, now beginning as the result of the settlement of land claims, should eliminate many of the barriers of language, culture, and paternalism that have previously alienated the indigenous populations from the territorial administration. Such policies should also facilitate the direction of services to areas where they are really needed.

Land claim agreements provide an opportunity not only to open the path for the indigenous peoples to participate fully in new industrial developments in the north, as business executives, lawyers, doctors, and nurses, but also to conserve and strengthen the natural fauna and flora on which a traditional lifestyle depends, through the establishment of national parks such as the Northern Yukon National Park (Berger, 1991).

Health care

A similar pattern of medical services has developed in many circumpolar jurisdictions, a three-tiered system being operated by the federal government concerned (Chapter 3). Within a given settlement, a small hospital with a few beds and staff of nurse-practitioners has been built; this has provided a base for out-patient treatment, uncomplicated childbirth, emergencies and the holding of patients awaiting air evacuation. In larger towns, a 20–30 bed hospital has been established. This is usually staffed by four to five general practitioners and a general surgeon, but more serious cases are still evacuated to major urban centres.

In terms of curative medicine, the system is probably as good as can be developed for such remote regions without going to exorbitant expense. However, services have been presented in a paternalistic manner, with little recognition of the input and insights that could be gained from indigenous nursing aides. Moreover, as in 'developed' societies, many of the local population have come to accept 'high-tech' medical services in a passive manner, not recognizing their own ability to contribute to the optimization of personal health through changes in their own lifestyle.

A further important criticism of current medical programmes is their continued preoccupation with curative medicine. Too often, physicians do no more than treat the symptoms of a stressful acculturation to a 'modern' society, with its accompanying social and economic deprivation. It seems likely that community health could be much improved through a redirection of funding into projects of health education and preventive medicine. To be effective, such efforts would require a strong input from those familiar with the local language and culture. In Alaska, this is now being accomplished through Native Health Boards (Ivey & Duyan, 1991). In Canada, also, there have been moves to democratize the delivery of health care (Connell *et al.*, 1991; O'Neil, 1991; Read & Watts, 1991), with the establishment of Regional Boards of Health and community health centres that offer a wide range of health services. Likewise, in Greenland, negotiations are currently proceeding to transfer the Health Service from Danish control to home rule (Misfeldt, 1991).

Finally, there is scope not only to encourage the formal medical education of representatives of the indigenous populations, but also to incorporate their traditional healing practices into the 'white' medical system – for example, the 'sweat lodge' and herbal remedies of the Amerindian shaman (Wheatley, 1991), the Toullmos pressure points of the Skolts (Soininen, 1991b), and the services of local midwives or Sundhed-smedhjaelper in Greenland (Boessen, 1991).

Religion

Religious institutions in the past played a major role in enforced acculturation by shipping children (sometimes against the wishes of their parents) to distant residential schools (Chapter 1) where the language, culture and religion of the dominant society were required, and native customs were belittled (Read & Watts, 1991). On occasion, it appears that the children were also physically and sexually abused by their instructors.

The dark consequences of this policy are now becoming recognized, and at least one major Canadian denomination (the United Church of Canada) has recently issued a formal apology to the native peoples for the treatment that they have received from 'white' immigrants. Currently, this church has established a separate Conference of indigenous congregations, with a theological school staffed by indigenous people, and there has been an increasing recognition of and acceptance of native spirituality. One measure of this new attitude is that the immediate past-moderator of the United Church of Canada is of indigenous origin.

Environment

Although a major fraction of the food needs of the indigenous population is now met from 'market' sources, 'country' foods still provide an important source of protein and ω-3 omega fatty acids to many of the circumpolar peoples, particularly those living in the smaller coastal settlements. For this reason, it is important that the indigenous populations be given early control over the long-term management of wildlife and the overall exploitation of the arctic environment (Berger, 1991).

Self-government seems to be the most effective method of achieving this goal.

Lessons for other groups

Indigenous populations

Many other indigenous populations have undergone, to an equal or even a larger extent, the changes of lifestyle that we have seen among the indigenous populations of the arctic. Moreover, they show proportionately larger increases in the prevalence of those chronic diseases that we regard as typical of 'modern' society, such as diabetes and atherosclerosis.

However, in many parts of the world there are still small isolated communities that continue in the role of hunter/gatherer, pastoralist or primitive agriculturalist. If the encroachments of 'modern' civilization permit, they should be encouraged to continue in this course, and if it appears that they also will undergo enforced acculturation to an urban lifestyle, they should be counselled to continue with the diet and the pattern of moderate sustained activity that has so far protected them from chronic disease.

Health-care professionals

City-dwellers are unlikely to adopt many facets of traditional circumpolar life directly. But it is important that we acknowledge the wisdom inherent in traditional practices and that we communicate our appreciation of this wisdom to the indigenous groups concerned. For too long, they have suffered the cultural trauma imposed by the vaunting of our supposedly superior knowledge.

In some instances – for example, the consuming of moderate amounts of fish oils, and the taking of prolonged bouts of moderate-intensity exercise – we may improve our health and well-being by adopting appropriately adapted variants of the traditional circumpolar lifestyle. In a longer-term perspective, if we are to achieve the skilful adaptation to our environment that circumpolar groups previously accomplished on a regional scale, we will also need to emulate other important values of arctic society – respect for nature, sharing rather than competing with our fellow citizens, and the consumption of no more of the world's resources than we need for our daily survival.

Finally, we must recognize that the acceptance of support from a highly technological and specialized society has cost us much of the capacity for innovation that previously allowed survival in a variety of hostile environments. If technological support is withdrawn because of war or some natural disaster, our easy rejection of this cultural inheritance could prove extremely costly to our prospects for survival.

References

Aargard-Olsen, G. (1973). Consumption of antibiotics in Greenland, 1964–1970. II. Effect of coincidental administration of antibiotics on early syphilitic infections. *Br. J. Ven. Dis.*, **49**, 27–29.

Abbey, S.E., Hood, E., Young, L.T. & Malcolmson, S. (1991). New perspectives on mental health problems in Inuit women. In *Circumpolar Health 90*, ed. B.D. Postl *et al.*, pp. 285–287. Winnipeg: Canadian Society for Circumpolar Health.

Adams, D. (1978). Inuit recreation and cultural change: a case study of the effects of acculturative change on Tununirmuit lifestyle and recreation patterns. University of Alberta, Edmonton, Unpublished MA thesis.

Adams, T. & Covino, B.G. (1958). Racial variations to a standardized cold stress. *J. Appl. Physiol.*, **12**, 9–12.

Adamson, J.D., Moody, J.P., Peart, A.F., *et al.* (1949). Poliomyelitis in the Arctic. *Can. Med. Assoc. J.*, **61**, 339–348.

Aidaraliyev, A.A. & Maximov, A. (1991). Specifications of adaptation in aboriginal and newcomer populations of Alaska and Chukotka in the process of intensive physiological loads. In *Circumpolar Health 90*, ed. B.D. Postl *et al.*, pp. 537–538. Winnipeg: Canadian Society for Circumpolar Health.

Albeck, H., Nielsen, N.H., Hansen, H.E., Bentzen, J., Ockelmann, H.H., Bretlau, P. & Hansen, H.S. (1992). Epidemiology of nasopharyngeal and salivary gland carcinoma in Greenland. *Arct. Med. Res.*, **51**, 189–195.

Alderson, J. & Crutchley, D. (1990). Physical education and the national curriculum. In *New Directions in Physical Education*, ed. N. Armstrong, pp. 37–62. Leeds, UK: Human Kinetics Publishing.

Alexseev, V.P. (1991). Coronary atherosclerosis and ischemic heart disease in aboriginal and newcomer populations of Yakutia. In *Circumpolar Health 90*, ed. B.D. Postl *et al.*, pp. 406–407. Winnipeg: Canadian Society for Circumpolar Health.

Allen, J.A. (1877). The influence of physical conditions in the genesis of species. *Radical Rev.*, **1**, 108–140.

Allen, T.R. (1973). Common cold epidemiology in Antarctica. In *Polar Human Biology*, ed. O.G. Edholm & E.K.E. Gunderson, pp. 123–124. London: Heinemann Medical Books.

Aluli, N.E. (1991). Prevalence of obesity in a native Hawaiian population. *Am. J. Clin. Nutr.*, **53**, 1556S–1560S.

American College of Sports Medicine (1991). *Guidelines for Graded Exercise Testing and Prescription*, 4th edn. Philadelphia: Lea & Febiger.

Andersen, K.L. (1963). Physical working capacity of Arctic people. *WHO Public*

Health Paper, **18**, 159–169.

Andersen, K.L. (1969). Racial and inter-racial differences in work capacity. *J. Biosoc. Sci.*, Suppl. **1**, 69–80.

Andersen, K.L. & Hart, J.S. (1963). Aerobic working capacity of Eskimos. *J. Appl. Physiol.*, **18**, 764–768.

Andersen, K.L., Bølstad, A., Loyning, Y. & Irving, L. (1960a). Physical fitness of Arctic Indians. *J. Appl. Physiol.*, **15**, 645–648.

Andersen, K.L., Loyning, Y., Nelms, J.D., Wilson, O., Fox, R.H. & Bolstad, A. (1960b). Metabolic and thermal response to a moderate cold exposure in nomadic Lapps. *J. Appl. Physiol.*, **15**, 649–653.

Andersen, K.L., Elsner, R.E., Saltin, B. & Hermansen, L. (1962). *Physical Fitness in Terms of Maximal Oxygen Intake of Nomadic Lapps*. USAF Arctic Aeromed. Lab. Alaska Tech. Rept. AAL-TDR-61-53. Alaska: Fort Wainwright.

Andersen, K.L., Hart, J.S., Hammel, H.T. & Sabean, H.B. (1963). Metabolic and thermal response of Eskimos during muscular exertion in the cold. *J. Appl. Physiol.*, **18**, 613–618.

Andersen, K.L., Shephard, R.J., Denolin, H., Varnauskas, E. & Masironi, R. (1971). *Fundamentals of Exercise Testing*. Geneva: World Health Organization.

Anderson, T.W., Brown, J.R., Hall, J.W. & Shephard, R.J. (1968). The limitations of linear regressions for the prediction of vital capacity and forced vital capacity. *Respiration*, **25**, 140–158.

Andrew, B. & Sarsfield, P. (1985). Innu health: the role of self-determination. In *Circumpolar Health 84*, ed. R. Fortuine, pp. 428–430. Seattle: University of Washington Press.

Arch, I. (1960). Fish tapeworms in Eskimos in the Port Harrison area. *Can. J. Publ. Health*, **51**, 268–270.

Aromaa, A., Björkstén, F., Eriksson, A.W., Maatela, J., Kirjarinta, M., Fellman, J. & Tamminen, M. (1975). Serum cholesterol and triglyceride concentrations of Finns and Finnish Lapps. I. Basic data. *Acta Med. Scand.*, **198**, 13–22.

Arthaud, J.B. (1970). Causes of death in 339 Alaskan Eskimos as determined by autopsy. *Arch. Pathol.*, **90**, 433–438.

(1972). Anaplastic parotid carcinoma (malignant lymphothelial lesion) in seven Alaskan natives. *Am. J. Clin. Pathol.*, **57**, 275–412.

Asmussen E. & Christensen, E.H. (1967). *Kompendium: Legemsölvelsernes Specielle Teori*. Copenhagen: Universitets Fond til Tilverbringelse of Läremidler.

Astakhova, T., Rjabikov, A., Astakhov, V., Bondareva, Z., Lutova, F. & Bulgakov, Y. (1991). Risk factors and non-communicable disease in native residents and newcoming population of Chukotka. In *Circumpolar Health 90*, ed. B.D. Postl *et al.*, pp. 408–409. Winnipeg: Canadian Society for Circumpolar Health.

Åstrand, I. (1960). Aerobic work capacity in men and women, with special reference to age. *Acta Physiol. Scand.*, **49**, Suppl. 169, 1–92.

Aubrey, L.P., Langner, N., Lawn, J., Sainnawap, B. & Beardy, B. (1991). Nutrient intake of adults aged 15 to 65 years in two Northern Ontario communities. In *Circumpolar Health 90*, eds. B.D. Postl *et al.*, pp. 774–777. Winnipeg: Canadian Society for Circumpolar Health.

Auger, F. (1974). Poids et plis cutanés chez les Esquimaux de Fort Chimo (Nouveau Québec). *Anthropologica*, **16**, 137–169.

(1976). Growth patterns of Fort Chimo and Spotted Island Eskimos. In

Circumpolar Health, ed. R.J. Shephard & S. Itoh, p. 266. Toronto: University of Toronto Press.

Auger, F., Jamison, P.L., Balsev-Jorgensen, J., Lewin, T., De Pena, J.F. & Skrobak-Kaczynski, J. (1980). Anthropometry of circumpolar populations. In *The Human Biology of Circumpolar Populations*, ed. F.A. Milan, pp. 213–255. London: Cambridge University Press.

Babbot, F.L., Fry, W.W. & Gordon, J.E. (1961). Intestinal parasites for man in Arctic Greenland. *Am. J. Trop. Med. Hyg.*, **10**, 185–190.

Baikie, M., Ratnam, S., Bryant, D.G., Jong, M. & Bokhout, M. (1989). Epidemiologic features of hepatitis B virus infection in northern Labrador. *Can. Med. Assoc. J.*, **141**, 791–795.

Balikci, A. (1967). Female infanticide on the Arctic coast. *Man*, **2**, 615–625.

Bang, H.O. & Dyerberg, J. (1972). Plasma lipids and lipoproteins in Greenlandic west coast Eskimos. *Acta Med. Scand.*, **192**, 85–94.

(1980). Lipid metabolism and ischemic heart disease in Greenland Eskimos. *Adv. Nutr. Res.*, **3**, 1–22.

Bang, H.O., Dyerberg, J. & Hjorne, N. (1976). The composition of food consumed by Greenlandic Eskimos. *Acta Med. Scand.*, **200**, 69–73.

Bang, H.O., Dyerberg, J. & Sinclair, H.M. (1980). The composition of the Eskimo food in northwestern Greenland. *Am. J. Clin. Nutr.*, **33**, 2657–2661.

Barrett, D.H., Burks, J.M., McMahon, B., Berquist, K.R. & Maynard, J.E. (1976). Hepatitis B in western Alaska. In *Circumpolar Health*, ed. R.J. Shephard & S. Itoh, p. 315 (abstr.). Toronto: University of Toronto Press.

Bass, D.E. & Henschel, A. (1956). Responses of body fluid compartments to heat and cold. *Physiol. Rev.*, **36**, 128–144.

Baugh, C.W., Bird, G.S., Brown, G.M., Lennox, C.S. & Semple, R.E. (1958). Blood volumes of Eskimos and white men before and during acute cold stress. *J. Physiol.*, **140**, 347–358.

Baxter, J.D. (1976). Ear disease among the Eskimo population of the Baffin zone. In *Circumpolar Health*, ed. R.J. Shephard & S. Itoh, pp. 384–389. Toronto: University of Toronto Press.

Beall, C.M. & Goldstein, M.C. (1992). High prevalence of excess fat and central fat patterning among Mongolian pastoral nomads. *An. J. Hum. Biol.*, **4**, 747–756.

Beasley, R.P. & Hwang, L.Y. (1983). Post-natal infectivity of hepatitis B surface antigen-carrier mothers. *J. Infect. Dis.*, **147**, 185–190.

Beaudry, P.H. (1968). Pulmonary function of the Canadian Eastern Arctic Eskimo. *Arch. Env. Health*, **17**, 524–528.

Beckner, G.L. & Winsor, T. (1954). Cardiovascular adaptation to prolonged physical effort. *Circulation*, **9**, 835–846.

Bell, R.R., Draper, H.H. & Bergan, J.G. (1973). Sucrose, lactose and glucose tolerance in northern Alaskan Eskimos. *Am. J. Clin. Nutr.*, **26**, 1185–1190.

Bell, R.R. & Heller, C.A. (1978). Nutrition studies: an appraisal of the modern North Alaskan Eskimo diet. In *Eskimos of Northwestern Alaska: A Biological Perspective*. ed. P.L. Jamison, S.L. Zegura & F.A. Milan, pp. 145–156. Stroudsberg, PA: Dowden, Hutchinson & Ross.

Bender, T.R., Jones, T.S., DeWitt, W.E., Kaplan, G.J., Saslow, A.R., Nevius, S.E., Clark, P.S. & Gangarosa, E.J. (1972). Salmonellosis associated with whale meat in an Eskimo community: serologic and bacteriologic methods as adjuncts to an epidemiologic investigation. *Am. J. Epidemiol.*, **96**, 153–160.

Bender, T.R., Brant, L.J. & Marnell, R.W. (1981). Factors affecting streptococcal colonization of children in a developing remote area of the United States. In *Circumpolar Health*, 81, ed. B. Harvald & J.P. Hart-Hansen, pp. 415–419. Oulu: Nordic Council for Arctic Medical Research, Report Series 33.

Bennike, T. (1981). Disease and health aspects under equator and on the polar circle. In *Circumpolar Health*, 81, ed. B. Harvald & J.P. Hart-Hansen, pp. 229–240. Oulu: Nordic Council for Arctic Medical Research, Report Series 33.

Bennington, E. (1978). Sports programs as a vehicle of acculturation for indigenous northern youth. University of Alberta, Edmonton: Unpublished MA thesis.

Berger, T. (1991). Key note address. In: *Circumpolar Health 90*, ed. B.D. Postl *et al.*, pp. 2–7. Winnipeg: Canadian Society for Circumpolar Health.

Bergmann, C. (1847). Uber die Verhältnisse der Wärmeökonomie des Thiere zu ihrer Grosse. *Gottinger Studien*, **3**, 595–708.

Berkes, F. & Farkas, C.S. (1978). Eastern James Bay Cree Indians: changing patterns of wild food use and nutrition. *Ecol. Food Nutr.*, **7**, 155–172.

Berry, J.W. (1976). Acculturative stress in North Canada-Ecological, cultural and psychological factors. In *Circumpolar Health*, ed. R.J. Shephard & S. Itoh, pp. 490–497. Toronto: University of Toronto Press.

Berry, J.W., Wintrob, R.M., Sindell, P.S. & Mawhinney, T. (1981). Culture change and psychological adaptations among the James Bay Cree. In *Circumpolar Health* 81, ed. B. Harvald & J.P. Hart-Hansen, pp. 481–489. Nordic Council for Arctic Medical Research, Report Series 33.

Bertelsen, A. (1935). Meddelelser om Gronland 117 (1). Cited by C.A. Mills (1939).

Bjerager, P., Kromann, N., Thygesen, K. & Harvald, B. (1981). Blood pressure in Greenland Eskimos. In *Circumpolar Health* 81, ed. B. Harvald & P.J. Hart-Hansen, pp. 317–320. Oulu: Nordic Council for Arctic Medical Research, Report Series 33.

Bjerregaard, P. (1983). Housing standards, social group, and respiratory infections in children of Upernavik, Greenland. *Scand. J. Soc. Med.*, **11**, 107–111.

(1986). Validity of Greenland mortality statistics. *Arct. Med. Res.*, **42**, 18–24.

(1988). Causes of death in Greenland, 1968–1985. *Arct. Med. Res.*, **47**, 105–123.

(1990). Fatal accidents in Greenland. *Arct. Med. Res.*, **49**, 132–141.

(1991). Disease patterns in Greenland. Studies on morbidity in Upernavik 1979–1980 and mortality in Greenland 1968–1985. *Arct. Med. Res.*, **50**, Suppl. 4, 1–62.

Bjerregaard, P. & Dyerberg, J. (1988). Mortality from ischemic heart disease and cerebrovascular disease in Greenland. *Int. J. Epidemiol.*, **17**, 514–519.

Bjerregaard, P. & Misfeldt, J.C. (1991). Infant and child mortality in Greenland is too high. In *Circumpolar Health* 90, ed. B.D. Postl *et al.*, pp. 558–560. Winnipeg: Canadian Society for Circumpolar Health.

Björkstén, F., Aromaa, A., Eriksson, A.W., Maatela, J., Kirjarinta, M., Fellman, J. & Tamminen, M. (1975). Serum cholesterol and triglyceride concentrations of Finns and Finnish Lapps: interpopulation comparisons and the occurrence of hyperlipidemia. *Acta Med. Scand.*, **198**, 23–33.

Bleiberg, F., Brun, T.A. & Goihman, S. (1980). Duration of activities and energy expenditure of female farmers in dry and rainy seasons in Upper Volta. *Br. J. Nutr.*, **43**, 71–82.

Bligh, J. & Chauca, D. (1981). Effects of hypoxia, cold exposure and fever on pulmonary artery pressure, and their significance for arctic residents. In *Circumpolar Health* 81, ed. B. Harvald & J.P. Hart-Hansen, pp. 606–607. Oulu: Nordic Council for Arctic Medical Research, Report Series 33.

Blondin, B. (1990). Traditional use of tobacco among the Dene. *Arct. Med. Res.*, **49**, Suppl. 2, 51–53.

Blonk, M.C., Bilo, H.J.G., Nauta, J.J.P., Popp-Snidjers, C., Mulder, C. & Donker, A.J.M. (1990). Dose-response of fish oil supplementation in healthy volunteers. *Am. J. Clin. Nutr.*, **52**, 120–127.

Boas, F. (1903). The Jesup North Pacific Expedition. *Am. Mus. J.*, **3**, 72–119. (1964). *The Central Eskimo*. Lincoln: University of Nebraska Press.

Bodey, A.S. (1973). The role of catecholamines in human acclimatization to cold: a study of 24 men at Casey, Antarctica. In *Polar Human Biology*, ed. O.G. Edholm & E.K.E. Gunderson, pp. 141–149. London: Heinemann Medical Books.

Boessen, E.M. (1991). Training/education of local people for posts in the health service – Neocolonialism? In *Circumpolar Health* 90, ed. B.D. Postl *et al.*, pp. 149–150. Winnipeg: Canadian Society for Circumpolar Health.

Bogardus, C., Lillioja, S. & Ravussin, E. (1990). The pathogenesis of obesity in man: results of studies of Pima Indians. *Int. J. Obesity*, **14**, 5–15.

Bøllerud, J.J.E. & Blakeley, R.A. (1950). Survey of the basal metabolism of the Eskimo. Point Barrow, Alaska: *Arctic Aeromedical Laboratory, USAF, Project Report* 21-01-020.

Bonaa, K.H., Bjerve, K.S. & Nordoy, A. (1992). Habitual fish consumption, plasma phospholipid fatty acids and serum lipids: the Tromso study. *Am. J. Clin. Nutr.*, **55**, 1126–1134.

Bone, R.M. (1992). *The Geography of the Canadian North: Issues and Challenges*. Toronto: Oxford University Press.

Booyens, J., Louwrens, C.C. & Katzeff, I.E. (1986). The Eskimo diet: prophylactic effects ascribed to the balanced presence of natural cis unsaturated fatty acids, and to the absence of unnatural trans and cis isomers of unsaturated fatty acids. *Med. Hypotheses*, **21**, 387–408.

Borre, K. (1991). Seal blood, Inuit blood and diet: a biocultural model of physiology and cultural identity. *Med. Anthropol. Quart.*, **51**, 48–62.

Bossé, R., Sparrow, D., Garvey, A.J., Costa, P.T., Weiss, S. & Rowe, J.W. (1980). Cigarette smoking, aging and decline in pulmonary function: a longitudinal study. *Arch. Environ. Health*, **35**, 247–252.

Bouchard, C. (1992). Genetic determinants of endurance performance. In *Endurance in Sport*, ed. R.J. Shephard & P.O. Åstrand, pp. 149–159. Oxford: Blackwell Scientific.

Bouchard, C. & Pérusse, L. (1994). Heredity, activity level fitness and health. In *Physical Activity, Fitness and Health*, ed. C. Bouchard, R.J. Shephard & T. Stephens, pp. 106–118. Champaign, IL: Human Kinetics Publishers.

Bowman, J.D., Mala, T.A., Segal, B. & McElvey, J.G. (1985). Recreational drug use by Alaska's youth: a report on secondary school students in two regions. In *Circumpolar Health* 84, ed. R. Fortuine, pp. 348–351. Seattle: University of Washington Press.

Brassard, P., Robinson, E. & Dumont, C. (1993). Descriptive epidemiology of non-insulin dependent diabetes mellitus in the James Bay Cree population of

Québec, Canada. *Arct. Med. Res.*, **52**, 47–54.

Brett, B., Taylor, W.C. & Spady, D.W. (1976). The North West Territories perinatal and infant mortality study: infant mortality in the North West Territories, 1973. In *Circumpolar Health*, ed. R.J. Shephard & S. Itoh, pp. 435–440. Toronto: University of Toronto Press.

Brocklehurst, J.C. (1957). Fatal outbreak of botulism among Labrador Eskimos. *Br. Med. J.*, **2**, 924.

Brody, J.A. (1965). Lower respiratory illness among Alaskan Eskimo children. *Arch. Environ Health*, **11**, 619–623.

(1966). The infectiousness of rubella and the possibility of reinfection. *Am. J. Publ. Health*, **56**, 1082–1087.

Brody, J.A. & Bridenbaugh, E. (1964). Prophylactic gamma-globulin and live measles vaccine in an island epidemic of measles. *Lancet*, **2**, 811–813.

Brown, G.M. (1954). Metabolic studies of the Eskimo: In *Cold Injury. Transactions of the Third Conference, Feb. 22–25, 1954. Fort Churchill, Manitoba.* ed. M.I. Ferrer. New York: Josiah Macey Foundation.

(1957). Vascular physiology of the Eskimo. *Rev. Canad. Biol.*, **16**, 279–292.

Brown, G.M. & Page, J. (1952). The effect of chronic exposure to cold on temperature and blood flow of the hand. *J. Appl. Physiol.*, **5**, 221–227.

Brown, G.M., Bird, G.S., Boac, L.M., Delahaye, D.J., Green, J.E., Hatcher, J.D. & Page, J. (1954). Blood volume and basal metabolic rate of Eskimos. *Metabolism*, **3**, 247–254.

Brown, G.M., Semple, R.E., Lennox, C.S., Bird, G.S., Baugh, C.W. & Gasmann, E. (1955). Physiological adjustments to acute cold exposure in Eskimos and white men. *Fed. Proc.*, **14**, 322–323 (abstr.).

Brown, G.M., Semple, R.E., Lennox, C.S., Bird, G.S. & Baugh, C.W. (1963). Responses to cold of Eskimos of the eastern Canadian Arctic. *J. Appl. Physiol.*, **18**, 970–974.

Brown, J.R. (1966). The metabolic cost of industrial activity in relation to weight. *Med. Serv. J. Can.*, **22**, 262–272.

Brown, M., Sinclair, R.G., Cronk, L.B., Clark, G.C. & Kuitunen-Ekbaum, E. (1948). Intestinal parasites of Eskimos on Southampton Island, North West Territories. *Can. J. Publ. Health*, **41**, 508–512.

Brown, M., Green, J.E., Boag, T.J. & Kuitunen-Ekbaum, E. (1950). Parasitic infections in the Eskimos at Igloolik, N.W.T. *Can. J. Publ. Health*, **41**, 508–512.

Brun, T.A., Bleiberg, F. & Goihman, S. (1981). Energy expenditure of male farmers in dry and rainy seasons in Upper Volta. *Br. J. Nutr.*, **45**, 67–75.

Burch, E.S. (1986). The Caribou Inuit. In *Native Peoples: The Canadian Experience*, ed. R.B. Morrison & C.R. Wilson, pp. 106–133. Toronto: McLelland & Stewart.

Burrows, B., Lebowitz, M.D., Camilli, A.E. & Knudson, R.J. (1986). Longitudinal changes in forced expiratory volume in one second in adults: methodological considerations and findings in healthy non-smokers. *Am. Rev. Resp. Dis.*, **133**, 974–980.

Burton, A. & Edholm, O.G. (1969). *Man in a Cold Environment*. New York: Hafner.

Butschenko, L.A. (1967). *Das Rühe- und Belastungs-EKG bei Sportlern*. Leipzig: Johann Ambrosius Barth.

Cameron, T.W. (1968). Northern sylvatic helminthiasis. *Arch. Env. Health*, **17**, 614–621.

Canadian Pediatric Society (1987). Growth charts for Indian and Inuit children. *Can. Med. Assoc. J.*, **136**, 118–119.

Carrier, R., Landry, F., Potvin, R. & Bouchard, C. (1972). Comparison between athletes, normal and Eskimo subjects from the point of view of selected biochemical variables. In *Training: Scientific Basis and Application*, ed. A.W. Taylor, pp. 180–185. Springfield, IL: C.C. Thomas.

Carson, J.B., Postl, B.D., Spady, D. & Schaefer, O. (1985). Lower respiratory tract infections among Canadian Inuit children. *Circumpolar Health 84*, ed. R. Fortuine, pp. 226–228. Seattle: University of Washington Press.

Cavalli-Sforza, L.L., Piazza, A., Menozzi, P. & Mountain, J.L. (1988). Reconstruction of human evolution: bringing together genetic, archaeological and linguistic data. *Proc. Natl. Acad. Sci.*, **85**, 6002–6006.

Chagnon, YY.C., Allard, C. & Bouchard, C. (1984). Red blood cell genetic variation in Olympic athletes. *J. Sport Sci.*, **2**, 121–129.

Chance, N.A. (1968). Implications of environmental stress. Strategies of developmental change in the north. *Arch. Environ. Health*, **17**, 571–577.

Chapman, M. (1980). Infanticide and fertility among Eskimos: a computer simulation *Am. J. Phys. Anthropol.*, **53**, 317–327.

Charlton, K.M. & Tabel, H. (1976). Epizootiology of rabies in Canada. In *Circumpolar Health*, ed. R.J. Shephard & S. Itoh, pp. 301–305. Toronto: University of Toronto Press.

Childs, M.T., King, I.B. & Knopp, R.H. (1990). Divergent lipoprotein responses of fish oils with various ratios of eicosapentanoic acid and docosohexanoic acid. *Am. J. Clin. Nutr.*, **52**, 632–639.

Choinière, R. (1992). Mortality among the Baffin Inuit in the mid 80s. *Arct. Med. Res.*, **51**, 87–93.

Clark, A.M. (1974). *Development of Caribou Eskimo Culture*. Ottawa: National Museums of Canada, 169 pp.

Clayton, H.J. (1991). Indian Health and community development: notes for an address. In *Circumpolar Health 90*, ed. B.D. Postl *et al.*, pp. 42–43. Winnipeg: Canadian Society for Circumpolar Health.

Cobiac, L., Clifton, P.M., Abbey, M., Belling, G.B. & Nestel, P.J. (1991). Lipid, lipoprotein, and hemostatic effects of fish vs fish oil ω-3 fatty acids in mildly hyperlipemic males. *Am. J. Clin. Nutr.*, **53**, 1210–1216.

Coermann, R.R. (1970). Mechanical vibrations. In *Ergonomics and Physical Environmental Factors*, ed. L. Parmeggiani, Geneva, ILO Occupational Health and Safety Series, **21**, 17–41.

Colbert, M.J., Mann, G.V. & Hursh, L.M. (1978). Nutrition studies: clinical observations on nutritional health. In *Eskimos of North Western Alaska*, ed. P.L. Jamison, S.L. Zegura & F.A. Milan, pp. 162–173. Stroudsberg, PA: Dowden.

Collins, K.J. & Roberts, D.F. (1988). *Capacity for Work in the Tropics*. London: Cambridge University Press.

Comuzzie, A.G. (1993). *Genomic, genetic and morphological variation in a sample of modern Evenki, and their relationship with other indigenous Siberian populations.* PhD dissertation. Lawrence, KS: University of Kansas.

Connell, G., Flett R., & Stewart, P. (1991). Implementing primary health care through community control: the experience of Swampy Cree Tribal Council. In *Circumpolar Health 90*, ed. B.D. Postl *et al.*, pp. 44–46. Winnipeg: Canadian Society for Circumpolar Health.

Connor, W.E. & Connor, S.L. (1990). Diet, atherosclerosis, and fish oil. *Adv. Int. Med.*, **35**, 139–171.

Consolazio, C.F., Johnson, R.E. & Pecora, J.L. (1963). *Physiological Measurements of Metabolic Function in Man.* New York: McGraw Hill.

Cook, F.A. (1893–94). Gynecology and obstetrics among the eskimos. *Trans. N.Y. Obstetr. Soc.*, **8**, 154–169.

Corcoran, A.C. & Rabinowitch, I.M. (1937). A study of the blood lipids and blood proteins in Canadian Eastern Arctic Eskimos. *Biochem. J.*, **31**, 343–348.

Cotes, J. (1965). *Lung Function*, Oxford: Blackwell Scientific Publications.

Coulter, D.M. & Popkin, J.S. (1976). Intractable diarrhea in Baffin Island Eskimos. In *Circumpolar Health*, ed. R.J. Shephard & S. Itoh, p. 468. Toronto: University of Toronto Press.

Couture, L., Chagnon, M., Allard, C. & Bouchard, C. (1986). More on red blood cell genetic variation in Olympic athletes. *Can. J. Appl. Sport Sci.*, **11**, 16–18.

Covino, B.G. (1961). Temperature regulation in the Alaskan Eskimo. *Fed. Proc.*, **20**, 209 (abstr.).

Crawford, M.H. & Duggirala, R. (1992). Digital dermatoglyphic patterns of Eskimo and Amerindian populations: relationships between geographic, dermatoglyphic, genetic and linguistic distances. *Hum. Biol.*, **64**, 683–704.

Craytor, W.S. (1991). HIV infection and AIDS in Alaska: Epidemiology and prevention strategies. In *Circumpolar Health 90*, ed. B.D. Postl *et al.*, pp. 356–359. Winnipeg: Canadian Society for Circumpolar Health.

Criqui, M. (1985). The problem of response bias. In *Behavioral Epidemiology and Disease Prevention*, ed. R.M. Kaplan & M.H. Criqui, pp. 15–30. New York: Plenum Press.

Crowe, K.J. (1969). *A cultural geography of Northern Foxe Basin, NWT.* Ottawa: Northern Science Research Group, Dept. of Indian Affairs and Northern Development.

Curtis, M.A. & Bylund, G. (1991). Diphyllobothriasis: Fish tapeworm disease in the circumpolar north. *Arct. Med. Res.*, **50**, 18–25.

D'Aeth, R.G., Grauwiler, A., Sinclair, M.A., Pearson, R., Penman, D., Szabo, G. & Tremblay, L. (1991). A community response to an outbreak of pertussis in the Yukon. In *Circumpolar Health 90*, ed. B.D. Postl *et al.*, pp. 384–387. Winnipeg: Canadian Society for Circumpolar Health.

Dahlberg, A.A. (1980). Craniofacial studies. In *The Human Biology of Circumpolar Populations*, ed. F.A. Milan, pp. 169–192. London: Cambridge University Press.

Damas, D. (1984). *Arctic. Handbook of North American Indians*, Vol. 5. Washington, DC: Smithsonian Institute.

Darwin, C. (1859). *On the Origin of Species by means of Natural Selection or the Preservation of Favoured Races in the Struggle for Life.* London: Watts & Co.

Davidson, M., Schraer, C.D., Parkinson, A.J., *et al.* (1989). Invasive pneumococcal disease in an Alaskan native population, 1980 through 1986. *J. Am. Med. Assoc.*, **261**, 715–718.

Davies, A.G. (1973). Effects of season of sledging on waking palmar sweating. In *Polar Human Biology*, ed. O.G. Edholm & E.K.E. Gunderson, pp. 240–245. London: Heinemann Medical Books.

Davies, L.E.C. & Hanson, S. (1965). The Eskimos of the North West passage: a survey of dietary composition and various blood and metabolic measurements. *Can. Med. Assoc. J.*, **92**, 205–216.

Dawber, T.R., Kannel, W.B. & Friedman, G.D. (1966). Vital capacity, physical activity and coronary heart disease. In *Prevention of Ischemic Heart Disease*, ed. W. Raab, pp. 254–265. Springfield, IL: C.C. Thomas.

DeLany, J.P., Vivian, V.M., Snook, J.T. & Anderson, P.A. (1990). Effects of fish oil on serum lipids in men during a controlled feeding trial. *Am. J. Clin. Nutr.*, **52**, 477–485.

Demirjian, A. (1980). *Anthropometry Report: Height, Weight and Body Dimensions*. Ottawa: Health & Welfare Canada.

Demirjian, A., Jenicek, J. & Dubuc, M.B. (1972). Les normes staturo-pondérales de l'enfant urbain Canadien–Francais d'âge scolaire. *Can. J. Publ. Health*, **63**, 14–30.

Dempsey, J.A. & Manohar, M.A. (1992). Pulmonary system. In *Endurance in Sport*, ed. R.J. Shephard & P.O. Åstrand, pp. 61–71. Oxford: Blackwell Scientific Publications.

de Pena, J. (1972). Growth and Development. In *IBP Annual Report #4*, ed. D.R. Hughes, pp. 47–69. Toronto: University of Toronto Dept. of Anthropology.

Desai, I.D. & Lee, M. (1974a). Nutritional status of Canadian Indians. I. Biochemical studies at Upper Liard and Ross River, the Yukon Territory. *Can. J. Publ. Health*, **65**, 369–374.

(1974b). Vitamin E status of Indians of Western Canada. *Can. J. Publ. Health*, **65**, 191–196.

Dessypris, A., Kirjarinta, M., Miettinen, A. & Lamberg, B-A. (1981). Seasonal variations of serum TSH, T_3 and autoantibodies in adult inhabitants of Lapland. In *Circumpolar Health 81*, ed. B. Harvald & J.P. Hart-Hansen, pp. 603–605. Oulu: Nordic Council for Arctic Medical Research, Report Series 33.

de Thé, G. (1982). Epidemiology of Epstein–Barr virus and associated diseases. In *The Herpes Viruses*, ed. B. Roizman, pp. 25–100. New York: Plenum Press.

Dilley, M. (1985). Alaska's improved pregnancy outcome project: an evaluation. In *Circumpolar Health 84*, ed. R. Fortuine, pp. 384–387. Seattle: University of Washington Press.

Dippe, S.E., Bennett, P.H., Dippe, D.W., Humphry, T., Burks, J. & Miller, M. (1976). Glucose tolerance among Aleuts on the Pribilof Islands. In *Circumpolar Health*, ed. R.J. Shephard & S. Itoh, p. 156. Toronto: University of Toronto Press.

Dockery, D.W., Ware, J.H., Ferris, B.G., Glicksberg, D.S., Fay, M.E., Spiro, A. & Speizer, F.E. (1985). Distribution of forced expiratory volume in one second and forced vital capacity in healthy white, adult never smokers in six U.S. cities. *Am. Rev. Resp. Dis.*, **131**, 511–520.

Doolan, N., Appavoo, D. & Kuhnlein, H.V. (1991). Benefit–risk considerations of traditional food use by the Sahtu (hare) Dene/Metis of Fort Good Hope, N.W.T. In *Circumpolar Health 90*, ed. B.D. Postl *et al.*, pp. 747–751. Winnipeg: Canadian Society for Circumpolar Health.

Dossetor, J.B., McConnachie, P. R., Stiller, C.R., *et al.* (1973). The major histocompatibility complex in Eskimos. *Transplant Proc.*, **5**, 209–213.

Dossetor, J.B., Schlaut, J.W., Olson, L., Alton, J.M., Kovithathongs, T. & McConnachie, P.R. (1976). Histocompatibility studies on Canadian Inuit (Eskimos). In *Circumpolar Health*, ed. R.J. Shephard & S. Itoh, pp. 220–221. Toronto: University of Toronto Press.

Draper, H.H. (1976). A review of nutritional research. In *Circumpolar Health*, ed. R.J. Shephard & S. Itoh, pp. 120–129. Toronto: University of Toronto Press.

(1980). Nutrition. In *The Human Biology of Circumpolar Populations*, ed. F.A. Milan, pp. 257–284. London: Cambridge University Press.

Dreyer, G. (1920). *The Assessment of Physical Fitness*. London: Cassell.

Dumont, C. & Wilkins, R. (1985). Mercury surveillance in several Cree Indian communities of the James Bay region, Québec. In *Circumpolar Health 84*, ed. R. Fortuine, pp. 88–91. Seattle: University of Washington Press.

Duncan, J.W. & Scott, E.M. (1972). Lactose intolerance in Alaskan Indians and Eskimos. *Am. J. Clin. Nutr.*, **25**, 867–868.

Durnin, J.V.G.A. & Passmore, R. (1967). *Energy, Work and Leisure*. London: Heinemann Medical Books.

Durnin, J.V.G.A. & Womersley, J.A. (1974). Body fat assessed from total body density and its estimation from skinfold thickness: measurements on 481 men and women aged from 16 to 72 years. *Br. J. Nutr.*, **32**, 77–97.

Duval, B. & Thérien, F. (1981). Demography, mortality and morbidity of the Northern Quebec Inuit. In *Circumpolar Health 81*, ed. B. Harvald & J.P. Hart-Hansen, pp. 206–211. Oulu: Nordic Council for Arctic Medical Research, Report Series 33.

Dyerberg, J. (1989). Coronary heart disease in Greenland Inuit: a paradox. Implications for western diet patterns. *Arct. Med. Res.*, **48**, 47–54.

Dyerberg, J. & Bang, H.O. (1979). Hemostatic function and platelet polyunsaturated fatty acids in Eskimos. *Lancet*, **2**, 433–435.

(1981). Factors influencing morbidity of acute myocardial infarction in Greenlanders. In *Circumpolar Health 81*, ed. B. Harvald & J.P. Hart-Hansen, pp. 300–303. Oulu: Nordic Council for Arctic Medical Research, Report Series 33.

Dyerberg, J., Bang, H.O. & Hjorne, N. (1975). Fatty acid composition of the plasma lipids in Greenlandic Eskimos. *Amer. J. Clin. Nutr.*, **28**, 958–966.

Dyerberg, J., Bang, H.O., Stoffersen, E., Moncada, S. & Vane, J.R. (1978). Eicosopentaenoic acid and prevention of thrombosis. *Lancet*, **2**, 117–119.

Eagan, C.J. (1966). Biometeorological aspects in the ecology of man at high latitudes. *Int. J. Biometeorol.*, **10**, 293–304.

Eaton, S.B. & Konner, M.J. (1985). Paleolithic nutrition: a consideration of its nature and current implications. *N. Engl. J. Med.*, **312**, 283–289.

Edelen, J.S., Burks, J.M., Barrett, D.H. & Steer, P. (1976). Rheumatic fever and rheumatic heart disease among Alaskan natives, 1964–73. In *Circumpolar Health*, ed. R.J. Shephard & S. Itoh, pp. 318–319. Toronto: University of Toronto Press.

Edgren, J., Bryngelsson, C., Lewin, T., Fellman, J. & Skrobak-Kaczynski, J. (1976). Skeletal maturation of the hand and wrist in Finnish Lapps. In *Circumpolar Health*, ed. R.J. Shephard & S. Itoh, pp. 248–254. Toronto: University of Toronto Press.

Edwards, A.C., Martin, J.R., Johnson, G.J. & Green, J. (1985). Chronic disease survey of a Labrador community. In *Circumpolar Health 84*, ed. R. Fortuine, pp. 249–253. Seattle: University of Washington Press.

Egan, C. (1991). Perceived and observed health status of Inuit receiving social assistance. In *Circumpolar Health 90*, ed. B.D. Postl *et al.*, pp. 39–41. Winnipeg: Canadian Society for Circumpolar Health.

Ehrström, M.C. (1951). Medical studies in North Greenland 1948–1949. VI. Blood pressure, hypertension and atherosclerosis in relation to food and mode of living. *Acta Med. Scand.*, **140**, 416–422.

Eisenberg, M.S. & Bender, T.R. (1976). Botulism in Alaska, 1947 through 1974. *J. Am. Med. Assoc.*, **235**, 35–38.

Ekoé, J.M., Thouez, J.P., Petitclerc, C., Foggin, P.M. & Ghadrian, P. (1990). Epidemiology of obesity in relationship to some chronic medical conditions among Inuit and Cree Indian populations in New Québec, Canada. *Diab. Res. Clin. Pract.*, **10**, s17–s27.

Ellis, G. (1971). *Units, Symbols, and Abbreviations. A Guide for Biological and Medical Authors*. London: Royal Society of Medicine.

Elsner, R.W., Nelms, J.D. & Irving, L. (1960). Circulation of heat to the hands of Arctic Indians. *J. Appl. Physiol.*, **15**, 662–666.

Engström, I., Eriksson, B.O., Karlberg, P., Saltin, B. & Thorén, C. (1971). Preliminary report on the development of lung volumes in young girl swimmers. *Acta Paediatr. Scand.*, Suppl. **217**, 73–76.

Erb, B.D. (Chairman) (1970). *Physician's Handbook for Evaluation of Cardiovascular and Physical Fitness*. Nashville, TN: Tennessee Heart Association Physical Exercise Committee.

Eriksson, A.W., Lehmann, W. & Simpson, N.E. (1980). Genetic studies on circumpolar populations. In *The Human Biology of Circumpolar Populations*, ed. F.A. Milan, pp. 81–168. London: Cambridge University Press.

Eriksson, H. (1958). The respiratory response to acute exercise of Eskimos and whites. *Acta Physiol. Scand.*, **41**, 1–11.

Ervasti, O., Virokannas, H. & Hassi, J. (1991). Frostbite in reindeer herders. *Arct. Med. Res.*, **50**, (Suppl. 6), 89–93.

Etzel, R.A., Jones, D.B., Schliffe, C.M., Lyke, J.R., Dunaway, C.E. & Middaugh, J.P. (1991). Saliva cotinine concentrations in young children in rural Alaska. In *Circumpolar Health* 90, ed. B. D. Postl *et al.*, pp. 566–567. Winnipeg: Canadian Society for Circumpolar Health.

Fagan, B.M. (1987). *The Great Journey: The Peopling of Ancient America*. New York: Thames & Hudson.

Feldman, S.A., Ho, K., Lewis, L.A., Mikkelson, B. & Taylor, B.C. (1972). Lipid and cholesterol metabolism in Alaskan Arctic Eskimos. *Arch. Pathol.*, **94**, 42–58.

Feldman, S.A., Rubenstein, A., Ho, K., Taylor, C.B., Lewis, L. & Mikkelson, B. (1975). Carbohydrate and lipid metabolism in the Alaskan Arctic Eskimo. *Am. J. Clin. Nutr.*, **28**, 588–594.

Ferrell, R.D., Chakraborty, R., Gershowitz, H., Laughlin, W.S. & Schull, W.J. (1981). The St. Lawrence Island eskimos: genetic variation and genetic distance. *Am. J. Phys. Anthropol.*, **55**, 351–358.

Ferris, B. G., Chen, H., Puleo, S. & Murphy, R.L. (1976). Chronic non-specific respiratory disease in Berlin, New Hampshire, 1967–1973. A further follow-up study. *Am. Rev. Resp. Dis.*, **113**, 475–485.

Fitzgerald, G.W.N. & Ehrenkranz, J.R.L. (1985). The seasonal occurrence of peptic ulcer disease among the Inuit of Northern Labrador. In *Circumpolar Health* 84, ed. R. Fortuine, pp. 285–287. Seattle: University of Washington Press.

Fleshman, J.K. (1972). Disease prevalence in the Alaskan Arctic and subarctic. *Acta Soc. Med. Scand.*, Suppl. **6**, 217.

Fleshman, J.K., Wilson, J.F. & Cohen, J.J. (1968). Bronchiectasis in Alaska native children. *Arch. Env. Health*, **17**, 517–523.

Fletcher, C., Peto, R., Tinker, C. & Speizer, F.E. (1976). *The Natural History of Chronic Bronchitis and Emphysema. An Eight-year Study of Early Chronic Bronchitis.* Oxford: Oxford University Press. 272 pp.

Folk, G.E. (1966). *Introduction to Environmental Physiology*, Philadelphia: Lea & Febiger.

Food and Agricultural Organization (1957). *Calorie Requirements.* Nutritional Studies #15. Rome: FAO.

(1973). *Energy and Protein Requirements.* WHO Technical Report Series #522. Geneva: World Health Organization.

Foote, D.C. & Greer-Wootten, B. (1966). Man–environment interactions in an Eskimo hunting system. In *Symposium on Man–Animal Linked Cultural Sub-systems.* Washington, DC: American Association for the Advancement of Science.

Forbes, A.L. (1974). Nutritional status of Indians and Eskimos as revealed by Nutrition Canada. In *Proc. 3rd Int. Symp. Circumpolar Health*, Yellowknife, NWT, 8–11 July, 1974. p. 17.

Forsius, H. (1980). Behavior. In *The Human Biology of Circumpolar Populations*, ed. F.A. Milan, pp. 339–358. London: Cambridge University Press.

Forsius, H., Damsten, M. & Fellman, J. (1981). Sexual maturity, childbirth and breast-feeding among Skolt Lapp women. In *Circumpolar Health*, 81, ed. B. Harvald & J.P. Hart-Hansen, pp. 185–186. Oulu: Nordic Council for Arctic Medical Research, Report Series 33.

Forsius, H., Kokkonen, E.-R. & Hirvonen, P. (1985). Social environment, personality structure and school avoidance: a study in a northern Finnish rural district. In *Circumpolar Health 84*, ed. R. Fortuine, pp. 320–321. Seattle: University of Washington Press.

Fortuine, R. (1971). The health of the Eskimos, as portrayed in the earliest written accounts. *Bull. Hist. Med.*, **45**, 97–114.

(1985a). Communicable disease control in the early history of Alaska. I. Smallpox. In *Circumpolar Health 84*, ed. R. Fortuine, pp. 187–190. Seattle: University of Washington Press.

(1985b). Communicable disease control in the early history of Alaska. II. Syphilis. In *Circumpolar Health 84*, ed. R. Fortuine, pp. 191–194. Seattle: University of Washington Press.

Fournelle, H.J., Rader, V. & Allen, C. (1966). A survey of enteric infections among Alaskan Indians. *Publ. Health Rep.*, **81**, 797–803.

Freeman, M.M. (1971a). The significance of demographic changes occurring in the Canadian Eastern Arctic. *Anthropologica*, **13**, 215–236.

(1971b). A social and ecological analysis of systematic female infanticide among the Netsilik Eskimos. *Am. Anthropol.* **73**, 1011–1018.

(1984). Arctic ecosystems. In *Arctic*, ed. D. Damas, pp. 36–48. Washington, DC: Smithsonian Institute.

Freeman, R.S. & Jamieson, J. (1976). Parasites of Eskimos at Igloolik and Hall Beach. In *Circumpolar Health*, ed. R.J. Shephard & S. Itoh, pp. 306–315. Toronto: University of Toronto Press.

Freitag, S.C., Pim, C.P. & Wideman, M. (1991). Cancer registration and trends in cancer incidence in the North West Territories. In *Circumpolar Health 90*, ed.

B.D. Postl *et al.*, pp. 459–461. Winnipeg: Canadian Society for Circumpolar Health.

Friesen, B. (1985). Haddon's strategy for prevention: application to native house fires. In *Circumpolar Health* 84, ed. R. Fortuine, pp. 105–109. Seattle: University of Washington Press.

Frisancho, A.R., Smith, S. & Albalak, R. (1994). Relationship of serum cholesterol and truncal fat distribution among Mexican Americans is accentuated by obesity. *Am. J. Hum. Biol.*, **6**, 51–59.

From, E. (1980). Some aspects of venereal diseases in Greenland. *Br. J. Vener. Dis.*, **56**, 65–68.

Galster, W.A. (1976). Mercury in Alaskan mothers and infants. *Environ. Health Persp.*, **15**, 135–140.

Gassaway, A.R. (1969). Diet and environment in Finnmark. *Geogr. Rev.*, **59**, 440–442.

Gaudette, L.A., Dufour, R., Freitag, S. & Miller, A. (1991). Cancer patterns in the Inuit population of Canada, 1970–1984. In *Circumpolar Health* 90, ed. B.D. Postl *et al.*, pp. 443–446. Winnipeg: Canadian Society for Circumpolar Health.

Gerasimova, E., Perova, N., Ozerova, I., Polesky, V., Metelskaya, V., Sherbakova, I., Levachev, M., Kulakova, S., Nikitin, Yu. & Astakhova, T. (1991). The effect of dietary ω-3 polyunsaturated fatty acids on HDL cholesterol in Chukot residents vs Muscovites. *Lipids*, **26**, 261–265.

Gessain, R. (1968). *Ammasalik, ou la civilisation obligatoire*. Paris: Flammarion.

Giddings, J. (1960). The archaeology of the Bering Strait. *Current Anthropol.*, **1**, p. 130.

Glaser, E.M. & Shephard, R.J. (1963). Simultaneous acclimatization to heat and cold in man. *J. Physiol.*, **169**, 592–602.

Glassford, R.G. (1970a). Games of the traditional Canadian Eskimo. In *Proceedings of the First Canadian Symposium on the History of Sport and Physical Education*, Edmonton, Alberta, ed. M.L. Howell, pp. 133–152. Edmonton: University of Alberta.

(1970b). Application of a theory of games to the traditional Eskimo culture. University of Illinois, Urbana, unpublished PhD thesis.

Glindmeyer, H.W., Diem, J.E., Jones, R.N. & Weil, H. (1982). Non-comparability of longitudinally and cross-sectionally determined annual change in spirometry. *Am. Rev. Resp. Dis.*, **125**, 544–548.

Godin, G. & Shephard, R.J. (1973a). Activity pattern of the Canadian Eskimo. In *Polar Human Biology*. ed. O.G. Edholm & E.K.E. Gunderson, pp. 193–215. London: Heinemann Medical Books.

(1973b). Body weight and the energy cost of activity. *Arch. Env. Health*, **27**, 289–293.

Gordis, L., Lilienfeld, A.M. & Rodriguez, R. (1969). A Community-wide study of acute rheumatic fever in adults. Epidemiologic and preventive factors. *JAMA* **210**: 862–865.

Gottschalk, C.W. & Riggs, D.S., Protein bound iodine in the serum of soldiers and of Eskimos in the Arctic. *J. Clin. Endocrinol. Metab.*, **12**, 235–243.

Green, A. & Kromann, N.P. (1981). Fertility and mortality rate in the Upernavik district, Greenland, 1950–1974. In *Circumpolar Health* 81, ed. B. Harvald & J.P. Hart-Hansen, pp. 163–165. Oulu: Nordic Council for Arctic Medical Research, Report Series 33.

Greenberg, J.H., Turner, C.G., & Zegura, S.L. (1986). The settlement of the Americas: a comparison of the linguistic, dental and genetic evidence. *Curr. Anthropol.*, **27**, 447–497.

Greksa, L.P. & Baker, P.T. (1982). Aerobic capacity of modernizing Samoan man. *Hum. Biol.*, **54**, 777–799.

Gross, R.L. & Newberne, P.M. (1980). Role of nutrition in immunologic function. *Physiol. Rev.*, **60**, 188–302.

Grzybowski, S. & Stylbo, K. (1976). The relevance of studies of tuberculosis in Eskimos to antituberculous program planning. In *Circumpolar Health*, ed. R.J. Shephard & S. Itoh, pp. 334–341. Toronto: University of Toronto Press.

Grzybowski, S., Stylbo, K., & Dorken, E. (1976). Tuberculosis in Eskimos. *Tubercle*, **57**, (Suppl. 4), S1–S58.

Guenter, D., Siber, G., Law, B. & Moffatt, M. (1991). Antibody to H. Influenzae Type B capsular polysaccharide in maternal and cord sera from Inuit, native Indian and Caucasian subjects in the North West Territories and Manitoba. In *Circumpolar Health* 90, ed. B.D. Postl *et al.*, pp. 344–345. Winnipeg: Canadian Society for Circumpolar Health.

Haglin, L. (1991). Nutrient intake among Saami people today compared with an old traditional Saami diet. In *Circumpolar Health* 90, ed. B.D. Postl *et al.*, pp. 741–746. Winnipeg: Canadian Society for Circumpolar Health.

Hall, C.E., Cooney, M.K. & Fox, J.P. (1973). The Seattle Virus-Watch. IV. Comparative epidemiological observations of infections with influenza A and B viruses, 1965–1969, in families with young children. *Am. J. Epidemiol.*, **98**, 365–380.

Hamill, P.V., Drizd, T.A., Johnson, C.L., Reed, R.B., Roche, A. F. & Moore, W.M. (1979). Physical growth: National Centre for Health Statistics. *Am. J. Clin. Nutr.*, **3**, 607–629.

Hannigan, J. (1991). *Statistics on the Economic and Cultural Development of Northern Aboriginal People of the USSR (for the period 1980–1989)*. Ottawa: Indian and Northern Affairs Canada.

Hansen, J.C., Jensen, T.G. & Tarp, U. (1991). Changes in blood mercury and lead levels in pregnant women in Greenland, 1983–1988. In *Circumpolar Health* 90, ed. B.D. Postl *et al.*, pp. 605–607. Winnipeg: Canadian Society for Circumpolar Health.

Hansen, J.C., Pedersen, H.S. & Mulvad, G. (1994). Fatty acids and antioxidants in the Inuit diet. Their role in ischemic heart disease (IHD) and possible interactions with other dietary factors. A review. *Arct. Med. Res.*, **53**, 4–17.

Haraldson, S. (1974). Evaluation of the Alaska Native Health Service. *Alaska Medicine*, **16**, 51–60.

Harper, A.B., Laughlin, W.S. & Mazess, R.B. (1984). Bone mineral content in St. Lawrence Island Eskimos. *Hum. Biol.*, **56**, 63–78.

Harris, M.I. (1988). Gestational diabetes may represent discovery of pre-existing glucose intolerance. *Diabetes Care*, **11**, 402–411.

Harrison, G.A. (1979). *Population Structure and Human Variation*. London: Cambridge University Press.

(1982). *Energy and Effort*. London: Taylor & Francis.

Hart, J.S., Sabean, H.B., Hildes, J.A., Depocas, F., Hammel, H.T., Andersen, K.L., Irving, L. & Foy, G. (1962). Thermal and metabolic responses of coastal Eskimos during a cold night. *J. Appl. Physiol.*, **17**, 953–960.

Hart-Hansen, J.P. (1976). Criminal homicide in Greenland. In *Circumpolar Health*, ed. R.J. Shephard & S. Itoh, pp. 548–554. Toronto: University of Toronto Press.

Hart-Hansen, J.P., Hancke, S. & Møller-Petersen, J. (1991). Atherosclerosis in

Greenland. In *Circumpolar Health* 90, ed. B.D. Postl *et al.*, pp. 400–403. Winnipeg: Canadian Society for Circumpolar Health.

Harvald, B.J. (1976). Current genetic trends in the Greenlandic population. In *Circumpolar Health*, ed. R.J. Shephard & S. Itoh, pp. 165–169. Toronto: University of Toronto Press.

Harvald, B. (1991). Fighting cancer in Greenland: the Danish Cancer Society strategies. In *Circumpolar Health* 90, ed. B.D. Postl *et al.*, pp. 457–458. Winnipeg: Canadian Society for Circumpolar Health.

Harvey, E.B. (1976). Psychiatric consultation and social work at a secondary school for Eskimo, Indian and Aleut students in Alaska. In *Circumpolar Health*, ed. R.J. Shephard & S. Itoh, pp. 517–525. Toronto: University of Toronto Press.

Hassi, J., Virokannas, H., Anttonewn, H., & Järvenpää, I. (1984). Health hazard in snowmobile usage. Paper presented at 6th International Congress of Circumpolar Health, Anchorage, Alaska, May 1984.

Hasunen, K. & Pekkarinen, M. (1975). Nutrient intake of adult Finnish Lapps. *Nordic Council for Arctic Medical Research, Rept.*, **13**, 15–32.

Hatcher, J.D., Page, J. & Brown, M. (1950). Study of the peripheral circulation of the Eskimo. *Rev. Canad. Biol.*, **9**, 76–77.

Hauschild, A.H. & Gavreau, L. (1985). Food-borne botulism in Canada, 1971–84. *Can. Med. Ass. J.*, **133**, 1141–1146.

Heinbecker, P. (1928). Studies on the metabolism of Eskimos. *J. Biol. Chem.*, **80**, 461–475.

Heller, C.A., Scott, E.M. & Hammes, L.M. (1967). Height, weight and growth of Alaskan Eskimos. *Am. J. Dis. Child.*, **113**, 338–344.

Helms, P. (1981). Changes in disease and food patterns in Angmagssalik, 1949–1979. In *Circumpolar Health* 81, ed. B. Harvald & J.P. Hart-Hansen, pp. 243–251. Oulu: Nordic Council for Arctic Medical Research, Report Series 33.

Herbert, F.A., Burchak, E.R. & Wilkinson, D. (1976). Bronchiectasis and bronchitis in Indian children following adeno III virus pneumonia in infancy: a long-term follow-up. In *Circumpolar Health*, ed. R.J. Shephard & S. Itch, pp. 467–468. Toronto: University of Toronto Press.

Herzlich, C. (1973). *Health and Illness*. London: Academic Press.

Hildes, J.A. (1963). Comparison of coastal Eskimos and Kalahari Bushmen. *Fed. Proc.*, **22**: 843–845.

Hildes, J.A. (1966). The circumpolar people-health and physiological adaptations. In *The Biology of Human Adaptability*, ed. P.T. Baker & J.S. Weiner, pp. 497–508. Oxford: Clarendon Press.

Hildes, J.A. & Schaefer, O. (1973). Health of Igloolik Eskimos and change with urbanization. *J. Hum. Evol.*, **2**, 241–246.

(1984). The changing picture of neoplastic disease in the western and central Canadian Arctic. *Can. Med. Assoc. J.*, **130**, 25–33.

Hildes, J.A., Irving, L. & Hart, J.S. (1961). Estimation of heat flow from hands of Eskimos by calorimetry. *J. Appl. Physiol.*, **16**, 617–623.

Hildes, J.A., Schaefer, O., Sayed, J.E., Fitzgerald, E.J. & Koch, E.A. (1976). Chronic lung disease and cardiovascular consequences in Iglooligmuit. In *Circumpolar Health*, ed. R.J. Shephard & S. Itoh, pp. 327–331. Toronto: University of Toronto Press.

Hirota, K., Asami, T., Toyoda, H. & Shimazu, D. (1969). Aerobic and anaerobic work capacity of the Ama. *Res. J. Phys. Educ.*, **13**, 260–265.

Hirvonen, J. (1982). Accidental hypothermia. *Arctic Medical Res. Rep.*, **30**, 15–19.

Hobart, C.W. (1976). Socio-economic correlates of mortality and morbidity among Inuit infants. In *Circumpolar Health*, ed. R.J. Shephard & S. Itoh, pp. 452–461. Toronto: University of Toronto Press.

Holgersen, P.B. (1961). Trikinose I Upernavik vinteren 1959–60. *Nord. Med.*, **66**, 1089–1093.

Holmes, M.J. (1973). Respiratory virus disease in the Antarctic: Immunological studies. In *Polar Human Biology*, ed. O.G. Edholm & E.K.E. Gunderson, pp. 125–134. London: Heinemann Medical Books.

Hong, S.K., Rennie, D.W. & Park, Y.S. (1986). Cold acclimatization and deacclimatization of Korean women divers. *Ex. Sport Sci. Rev.*, **14**, 231–268.

Honigmann, J. & Honigmann, I. (1966). *Eskimo Townsmen*. Ottawa: Canadian Centre for Anthropology, St. Paul's University.

Hopkins, D.M. (1967). *The Bering Land Bridge*. Stanford, CA: Stanford University Press.

Hoygaard, A. (1941). Studies on the Nutrition and Physiopathology of the Eskimos. *Dept. Norske Videnskapsakedemis Skrifter, Mat. naturv.* Klasse 9. Cited by Rodahl (1952).

Hrdlicka, A. (1941). Height and weight in Eskimo children. *Am. J. Phys. Anthropol.*, **28**, 331–341.

Hubert, A., Robert-Lamblin, J. & Hermann, M. (1985). Environmental factors, dietary behaviors, and nasopharyngeal carcinoma: anthropological approach. In *Circumpolar Health* 84, ed. R. Fortuine, pp. 261–265. Seattle: University of Washington Press.

Huertas, J.R., Mataix, F.J., Manas, M., Bargossi, A.M. & Battino, M. (1994). Dietary polyunsaturated fatty acids and peroxidative risks in sport practice. *J. Sports Med. Phys. Fitness*, **34**, 101–108.

Hughes, D. & Milan, F.A. (1980). Introduction. In *The Human Biology of Circumpolar Populations*, ed. F.A. Milan, pp. 1–11. London: Cambridge University Press.

Hulse, F.A. (1957). Exogamie et heterosis. *Arch. Suisses Anthrop. Gen.*, **22**, 103–113.

Huttunen, P., Hirvonen, J. & Kinnula, V. (1981). The occurrence of brown adipose tissue in outdoor workers. *Eur. J. Appl. Physiol.*, **46**, 339–345.

Ikai, M., Ishii, K., Miyamura, M., Kusano, K., Bar-Or, O., Kollias, J. & Buskirk, E.R. (1971). Aerobic capacity of Ainu and other Japanese on Hokkaido. *Med. Sci. Sports*, **3**, 6–11.

Imrie, R.A. (1991). Establishing the joint national committee on aboriginal AIDS education and prevention. In *Circumpolar Health* 90, B.D. Postl *et al.*, pp. 363–367. Winnipeg: Canadian Society for Circumpolar Health.

Innis, S.M. & Kuhnlein, H.V. (1987). The fatty acid composition of northern Canadian marine and terrestrial mammals. *Acta Med. Scand.*, **222**, 105–109.

Ireland, B., Knutson, L., Alward, W. & Hall, D.B. (1985). Pertussis: A study of incidence and mortality in a Yukon-Kuskokwim delta epidemic. In *Circumpolar Health* 84, ed. R. Fortuine, pp. 229–234. Seattle, University of Washington Press.

Irvine, J., Gillis, D.C., Tan, L., Chiu, S., Liu, L. & Robson, D. (1991). Lung, breast and cervical cancer incidence and survival in Saskatchewan northerners and registered Indians (1967–1986). In *Circumpolar Health* 90, ed. B.D. Postl *et al.*, pp. 452–456. Winnipeg: Canadian Society for Circumpolar Health.

Irving, L. (1972). *Arctic Life of Birds and Mammals including Man*. New York: Springer Verlag.

Ishiko, T. (1967). Aerobic capacity and external criteria of performance. *Can. Med. Assoc. J.*, **96**, 746–749.

Itoh, S. (1974). *Physiology of Cold-adapted Man*. Sapporo: Hokkaido University School of Medicine.

—— (1980). Physiology of circumpolar peoples. In *Human Biology of Circumpolar Populations*, ed. F.A. Milan, pp. 285–304. London: Cambridge University Press.

Itoh, S., Shirato, H., Hiroge, T., Kuroshima, A. & Doi, K. (1969). Cold pressor responses of Hokkaido inhabitants. *Jap. J. Physiol.*, **19**, 198–211.

Itoh, S., Doi, K. & Kuroshima, A. (1970a). Enhanced sensitivity to noradrenaline of the Ainu. *Int. J. Biometeorol.*, **14**, 195–200.

Itoh, S., Kuroshima, A., Hiroshige, T. & Doi, K. (1970b). Finger temperature responses to local cooling in several groups of subjects in Hokkaido. *Jap. J. Physiol.*, **20**, 370–380.

Ivey, G.H. & Duyan, K.R. (1991). The Alaska area native health service: improving the health status of the native people of Alaska. In *Circumpolar Health* 90, ed. B.D. Postl *et al.*, pp. 66–69. Winnipeg: Canadian Society for Circumpolar Health.

Jacobs, I., Romet, T., & Kerrigan-Brown, D. (1985). Muscle glycogen depletion during exercise at 9 °C and 21 °C. *Eur. J. Appl. Physiol.*, **46**, 47–53.

Jamison, P.L. (1970). Growth of Wainwright Eskimos: stature and weight. *Arct. Anthropol.*, **7**, 86–94.

—— (1976a). Growth of Eskimo children in northwestern Alaska. In *Circumpolar Health*, ed. R.J. Shephard & S. Itoh, pp. 223–229. Toronto: University of Toronto Press.

—— (1976b). Growth and development of children as an indicator of the health of Arctic populations. *Collegium Anthropol.*, **10**, 179 (abstr.).

—— (1990). Secular trends and the pattern of growth in arctic populations. *Soc. Sci. Med.*, **30**, 751–759.

Jeppesen, B.B. & Harvald, B. (1985). Low incidence of urinary calculi in Greenland Eskimos as explained by a low calcium/magnesium ratio. In *Circumpolar Health* 84, ed. R. Fortuine, pp. 288–290. Seattle: University of Washington Press.

Jetté, M. (1983). *Anthropometric characteristics of the Canadian Population. Nutrition Canada 1970–72*. Ottawa: University of Ottawa.

Johansson, S.R. (1982). The demographic history of the native peoples of North America: a selective bibliography. *Yearbook Phys. Anthropol.*, **25**, 133–152.

Johnson, B.D. & Dempsey, J.A. (1991). Demand versus capacity in the ageing pulmonary system. *Ex. Sport Sci. Rev.*, **19**, 171–210.

Johnson, M.W. (1971). Tuberculosis in Alaska: Experience with twenty year control program, 1950–1970. Paper presented at Second International Symposium on Circumpolar Health, Oulu, June, 1971.

Johnston, F.E., Laughlin, W.S., Harper, A.B. & Ensroth, A.E. (1982). Physical growth of St. Lawrence Island Eskimos: Body size, proportions and composition. *Am. J. Phys. Anthropol.*, **58**, 397–401.

Jones, N.L. & Kane, M. (1979). Inter-laboratory standardization of methodology. *Med. Sci. Sports*, **11**, 368–372.

Jorgensen, B.B., Misfeldt, J. & Larsen, S.O. (1988). Syphilis in Greenland

1970–1987. The number of cases and their distribution. *Arct. Med. Res.*, **47**, 54–61.

Jorgensen, J.B. & Skrobak-Kaczynski, J. (1972). Secular changes in the Eskimo community of Aupilagtoq. *Z. Morph. Anthrop.*, **64**, 12–19.

Jorgensen, K.A., Nielsen, A.H. & Dyerberg, J. (1986). Hemostatic factors and renin in Greenland Eskimos on a high eicosapentaenoic acid intake. Results of the fifth Umanak expedition. *Acta Med. Scand.*, **219**, 473–479.

Jorgensen, P., Møller, J. & Zachau-Christensen, B. (1981). Live born in Greenland, birth weight, neonatal and infant mortality. In *Circumpolar Health* 81, ed. B. Harvald & J.P. Hart-Hansen, pp. 164–166. Oulu: Nordic Council for Arctic Medical Research, Report Series 33.

Karlsson, J. (1970). Maximal oxygen intake in Skolt Lapps. *Arctic Anthropol.*, **7**, 19, 1970.

Kattus, A.A. (1972). *Exercise Testing and Training of Apparently Healthy Individuals: A Handbook for Physicians.* New York: American Heart Association.

Katzmarzyk, P.T., Leonard, W.R., Crawford, M.H. & Sukernik, R.I. (1994). Resting metabolic rate and daily energy expenditure among two indigenous Siberian populations *Am. J. Hum. Biol.*, **6**, 719–730.

Kawahata, A. & Sakamoto, H. (1951). Some observations on sweating of the Ainu. *Jap. J. Physiol.*, **2**, 166–169.

Keenleyside, A. (1990). Euro-American whaling in the Canadian Arctic: its effects on Eskimo health. *Arctic Anthropol.*, **27**, 1–19.

Kemp, W.H. (1971). The flow of energy in a hunting society. *Scientific Amer.*, **224**, 104–115.

Kestin, M., Clifton, P., Belling, G.B. & Nestel, P. (1990). ω-3 fatty acids of marine origin lower systolic blood pressure and triglycerides but raise LDL cholesterol compared with ω-3 and ω-6 fatty acids from plants. *Am. J. Clin. Nutr.*, **51**, 1028–1034.

Kimble, G.H.T. & Good, D. (1954). *Geography of the Northland.* New York: John Wiley.

Kirjarinta, L. & Eriksson, A.W. (1976). Preliminary studies on Diabetes Mellitus in Finnish Lapps. *Arct. Med. Res. Rep.*, **15**, 41–47.

Kirk, R., Szathmary E.J. (1985). *Out of Asia: Peopling the Americas and the Pacific.* Canberra: Journal of Pacific History Special Publication (cited by Young, 1993).

Kissmeyer-Nielsen, F., Andersen, H., Hauge, M., Kjerbye, K.E., Mogensen, B. & Svejgaard, A. (1971). HL-A types in Danish Eskimos from Greenland. *Tissue Antigens*, **1**, 74–80.

Klausner, S.Z., Foulks, E.F. & Moore, M. (1980). *Social change and the Alcohol Problem on the Alaskan North Slope.* Philadelphia: University of Pennsylvania Center for Research on the Acts of Man, p. 3.

Knowler, W.C., Pettitt, D.J., Saad, M.F., Charles, M.A., Nelson, R.G., Howard, B.V., Bogardus, C. & Bennett, P.H. (1991). Obesity in the Pima Indians: its magnitude and relationship with diabetes. *Am. J. Clin. Nutr.*, **53**, 1543S–1551S.

Kodama, S. (1970). *Ainu: Historical and Anthropological Studies.* Sapporo, Japan: Hokkaido University School of Medicine (Cited by Milan, 1980).

Koishi, H., Okuda, T., Matsudaira, T., Takaya, S. & Takemura, K. (1975). Nutrition and cold tolerance. In *Anthropological and Genetic Studies on the Japanese. Japanese IBP Synthesis Volume 2 on Human Adaptability.* ed. S. Watanabe, S. Kondo & E. Matsunaga, pp. 309–319. Tokyo: Tokyo University Press.

Kollias, J., Boileau, R.A., Bartlett, R.H. & Buskirk, E.R. (1972). Pulmonary function and physical condition in lean and obese subjects. *Arch. Environ. Health*, **25**, 146–150.

Kosatsky, T. & Dumont, C. (1991). Determinants of exposure to methylmercury among the James Bay Cree, 1987–1989. In *Circumpolar Health* 90, ed. B.D. Postl *et al.*, pp. 693–695. Winnipeg: Canadian Society for Circumpolar Health.

Kraus, R. & Buffler, P. (1976). Suicide in Alaskan Natives: A preliminary report. In *Circumpolar Health*, ed. R.J. Shephard & S. Itoh, pp. 556–557. Toronto: University of Toronto Press.

Kriska, A.M., LaPorte, R.E., Pettitt, D.J., Charles, M.E., Nelson, R.G., Kuller, L.H., Bennett, P.H. & Knowler, W.C. (1993). The association of physical activity with obesity, fat distribution and glucose intolerance in Pima Indians. *Diabetologia*, **36**, 863–869.

Krog, J. & Wika, M. (1978). Studies of hand blood flow of the Igloolik Eskimo. *Med. Biol.*, **56**, 148–151.

Krog, J., Folkow, B., Fox, R.H. & Andersen, K.L. (1960). Hand circulation in the cold of Lapps and North Norwegian fishermen. *J. Appl. Physiol.*, **55**, 1811–1817.

Krømann, N. & Green, A. (1980). Epidemiological studies in the Upernavik district, Greenland. *Acta Med. Scand.*, **208**, 401–406.

Krømann, N.P., Nielsen, N.H. & Hart-Hansen, J.P. (1981). Skin cancer in Greenland. In *Circumpolar Health* 81, ed. B. Harvald & J.P. Hart-Hansen, pp. 278–279. Oulu: Nordic Council for Arctic Medical Research, Report Series 33.

Kuhnlein, H.V. (1984a). Traditional and contemporary Nuxalk foods. *Nutr. Res.*, **4**, 789–809.

(1984b). Nutritional value of traditional food practices. *Res. Food Nutr. Sci.*, **4**, 67–71.

Kuhnlein, H.V. (1991). Nutrition of the Inuit: A brief overview. In *Circumpolar Health* 90, ed. B.D. Postl *et al.*, pp. 728–730. Winnipeg: Canadian Society for Circumpolar Health.

Kuno, Y. (1956). *Human Perspiration*. Springfield, IL: C.C. Thomas.

Kuroshima, A., Itoh, S., Azuma, T. & Agishi, Y. (1972). Glucose tolerance test in the Ainu. *Int. J. Biometeorol.*, **16**, 193–197.

Kustov, V.I., Roslyakov, A.G. & Marochko, A.U. (1991). Stomach cancer among the aboriginal population of the Soviet Far East: an epidemiological study. In *Circumpolar Health* 90, ed. B.D. Postl *et al.*, pp. 468–469. Winnipeg: Canadian Society for Circumpolar Health.

Laird, M. & Meerovitch, E. (1961). Parasites from Northern Canada. I. Entozoa of Fort Chimo Eskimo. *Can. J. Zool.*, **39**, 63–67.

Lamb, H.H. (1965). Britain's changing climate. In *The Biological Significance of Climatic Changes in Britain*, ed. C.G. Johnson & L.P. Smith, pp. 3–31. London: Academic Press.

Lammert, O. (1972). Maximal aerobic power and energy expenditure, of Eskimo hunters in Greenland. *J. Appl. Physiol.*, **33**, 184–188.

Landsberg, H.E. (1970). *Survey of World Climatology*. Paris: Elsevier.

Langaney, A., Gessain, R. & Robert, J. (1974). Migration and genetic kinship in Eastern Greenland. *Hum. Biol.*, **21**, 272–278.

Langner, N.R. & Steckle, J.M. (1991). National data base on breast feeding among

Indian and Inuit women: Canada 1988. In *Circumpolar Health* 90, ed. B.D. Postl *et al.*, pp. 563–565. Winnipeg: Canadian Society for Circumpolar Health.

Lanier, A.P. & Mostow, E.N. (1991). Screening for cancer in remotely populated regions – lessons from mammography and breast cancer. In *Circumpolar Health* 90, ed. B.D. Postl *et al.*, pp. 462–464. Winnipeg: Canadian Society for Circumpolar Health.

Lanier, A.P., Henle, W., Bornkamm, G. & Bender, T.R. (1981). The association of Epstein–Barr virus and nasopharyngeal carcinoma: an update of studies among Alaskan natives. In *Circumpolar Health* 81, ed. B. Harvald & J.P. Hart-Hansen, pp. 268–271. Oulu: Nordic Council for Arctic Medical Research, Report Series 33.

Lanier, A.P., Bulkow, L.R., Novotny, T.E., Giovino, G.A. & Davis, R.M. (1990). Tobacco use and its consequences in northern populations. *Arct. Med. Res.*, **49**, Suppl. 2, 17–22.

Larke, R.P.B., Eaton, R.D.P. & Schaefer, O. (1981). Epidemiology of hepatitis B in the Canadian arctic. In *Circumpolar Health* 81, ed. B. Harvald & J.P. Hart-Hansen, pp. 401–406. Oulu: Nordic Council for Arctic Medical Research, Report Series 33.

Larke, R.P.B., Froese, G.J., Devine, R.D.O. & Lee, V.P. (1985). Hepatitis B in the Baffin region of Northern Canada. In *Circumpolar Health* 84, ed. R. Fortuine, pp. 199–202. Seattle: University of Washington Press.

Larke, R.B., Froese, G.J., Devine, R.D. & Petruk, M.W. (1987). Extension of the epidemiology of hepatitis B in circumpolar regions through a comprehensive serologic study in the North West Territories of Canada. *Science*, **142**, 633–645.

Larsen, H. & Rainey, F. (1948). *Ipiutak and the Arctic Whale Hunting Culture. Anthropological Papers* 42. New York: American Museum of Natural History.

Larson, L.A. (1974). *Fitness, Health and Work Capacity: International Standards for Assessment*. New York: MacMillan.

Larsson, K., Ohlsén, P., Larsson, L., Malmberg, P., Rydström, P-O. & Ulriksen, H. (1993). High prevalence of asthma in cross-country skiers. *Br. Med. J.*, **307**, 1326–1329.

Laughlin, W.S. & Harper, A.B. (1979). *The First Americans: Origins, Affinities and Adaptations*. New York: Gustav Fischer.

(1988). Peopling of the continents: Australia and America. In *Biological Aspects of Human Migration*, ed. C.G.N. Mascie-Taylor & G.W. Lasker, pp. 14–40. London: Cambridge University Press.

Lavallée, C. & Robinson, E. (1991). Physical activity, smoking and overweight among the Cree of Eastern James Bay. In *Circumpolar Health* 90, ed. B.D. Postl *et al.*, pp. 770–773. Winnipeg: Canadian Society for Circumpolar Health.

LeBlanc, J. (1975). *Man in the Cold*. Springfield, IL: C.C. Thomas. 195 pp.

Lederman, J.M., Wallace, A.C. & Hildes, J.A. (1962). Arteriosclerosis and neoplasms in Canadian Eskimos. In *Biological Aspects of Aging: Proceedings of the Fifth International Congress on Gerontology*, pp. 201–207. New York: Columbia University Press.

Lee, J.F. (1985). The effects of a smoking prevention program for Alaska youth. In *Circumpolar Health* 84, ed. R. Fortuine, pp. 357–360. Seattle: University of Washington Press.

Lee, R.B. (1969). !Kung Bushman subsistence: an input output analysis. In

Environment and Cultural Behavior, ed. A.P. Vayda, pp. 47–79. New York: Natural History Press.

Leighton, A.H. & Hughes, C.C. (1955). Notes on Eskimo patterns of suicide. *Southwest J. Anthropol.*, **11**, 327–328.

Leith, D.E. & Bradley, M. (1976). Ventilatory muscle strength and endurance training. *J. Appl. Physiol.*, **41**, 508–516.

Leonard, W.R., Crawford, M.H., Comuzzie, A.G. & Sukernik, R.I. (1994a). Correlates of low serum lipid levels among the Evenki herders of Siberia. *Am. J. Hum. Biol.*, **6**, 329–338.

Leonard, W.R., Katzmaryzyk, P.T., Comuzzie, A.G., Crawford, M.H. & Sukernik, R.I. (1994b). Growth and nutritional status of the Evenki Reindeer herders of Siberia. *Am. J. Hum. Biol.*, **6**, 339–350.

Leonard, W.R., Katzmarzyk, P.T. & Crawford, M.H. (1994c). Energetics and population ecology of Siberian herders. *Am. J. Hum. Biol.*, in press.

Levine, V.E. (1937). The basal metabolic rate of the Eskimo. *J. Biol. Chem.*, **119**, Proc. LXI-LXII.

Lewin, T. (1971). History of the Skolt Lapps. In *Introduction to the Biological Characteristics of the Skolt Lapps*, ed. T. Lewin, *Proc. Finnish Dental Soc.*, **67**, Suppl. 1. (cited by Milan, 1980).

Lewin, T. & Hedegard, B. (1971a). Human biological studies among Skolt Lapps and other Lapps. In *Introduction to the Biological Characteristics of the Skolt Lapps*, ed. T. Lewin, *Proc. Finnish Dental Soc.*, **67**, Suppl. 1, 63–70.

(1971b). Anthropometry among Skolts, other Lapps and other ethnic groups in northern Fennoscandia. In *Introduction to the Biological Characteristics of the Skolt Lapps*, ed. T. Lewin, *Proc. Finnish Dental Soc.*, **67**, Suppl. 1, 71–98.

Lewin, T., Jürgens, H.W. & Louekari, L. (1970). Secular trend in the adult height of Skolt Lapps. *Arctic Anthropol.*, **7**, 53–62.

Ling, D. (1976). Audiological problems of the Eskimo population in the Baffin zone. In *Circumpolar Health*, ed. R.J. Shephard & S. Itoh, pp. 409–412. Toronto: University of Toronto Press.

Livingstone, S.D., Grayson, J., Reed, L.D. & Gordon, D. (1978). Effect of a local cold stress on peripheral temperatures of Inuit, Oriental and Caucasian subjects. *Can. J. Physiol. Pharmacol.*, **56**, 877–881.

Lloyd, E.L.L. (1985). Environmental cold may be a major factor in some respiratory disorders. In *Circumpolar Health 84*, ed. R. Fortuine, pp. 66–69. Seattle: University of Washington Press.

Long, T.P. (1978). The prevalence of clinically treated diabetes among Zuni reservation residents. *Am. J. Publ. Health*, **68**, 901–902.

Longstaffe, S., Postl, B.D., Kao, H., Nicolle, L. & Ferguson, C.A. (1981). Rheumatic fever in Manitoba Indian and Caucasian children, 1970–1979. In *Circumpolar Health 81*, ed. B. Harvald & J.P. Hart-Hansen, pp. 420–424. Oulu: Nordic Council for Arctic Medical Research, Report Series 33.

Loomis, W.F. (1967). Skin-pigment regulation of Vitamin D biosynthesis in man. *Science*, **157**, 501–506.

Lupin, A.J. (1976). Ear disease in western Canadian natives – a changing entity – and the results of tympanoplasty. In *Circumpolar Health*, ed. R.J. Shephard & S. Itoh, pp. 389–398. Toronto: University of Toronto Press.

Lynge, I. (1976). Alcohol problems in western Greenland. In *Circumpolar Health*, ed. R.J. Shephard & S. Itoh, pp. 543–547. Toronto: University of Toronto Press.

(1981). Psychical problems of the Inuit woman in industrial confrontation. In *Circumpolar Health* 81, ed. B. Harvald & J.P. Hart-Hansen, pp. 461–463. Oulu: Nordic Centre for Arctic Medical Research, Report Series 33.

McAlpine, P.J., Chen, S.H., Cox, D.W., Dossetor, J.B., Giblett, E., Steinberg, A.G. & Simpson, N.H. (1974). Genetic markers in blood in a Canadian Eskimo population with a comparison of allele frequencies in circumpolar populations. *Hum. Hered.*, **24**, 114–142.

MacArthur, R.S. (1974). *Cognitive Abilities of Eskimos of Igloolik, Canada and Upernavik District, Greenland.* IBP-HA Project, Igloolik, North West Territories, Annual Report no. 6: University of Toronto: Department of Anthropology.

McDougall J.D., Wenger, H.A. & Green, H.J. (1983). *Physiological Testing of the Elite Athlete.* Ottawa: Canadian Association of Sport Sciences.

Mackey, M.G.A. & Orr, R.D. (1985). 'Country Food' use in Makkovik, Labrador, July 1980 to June 1981. In *Circumpolar Health* 84, ed. R. Fortuine, pp. 143–150. Seattle: University of Washington Press.

McMahon, B.J. (1991). Prevention of viral hepatitis in circumpolar populations. *Arct. Med. Res.*, **50**, 187–190.

McNair, A., Gudmand-Hoyer, E., Jarnum, S. & Orrild, L. (1972). Sucrose malabsorption in Greenland. *Br. Med. J.*, **2**, 19–21.

Mailhot, J. (1968). *Inuvik Community Structure–Summer 1965.* Ottawa: Department of Indian Affairs and Northern Development.

Malina, R.M. & Bouchard, C. (1988). Subcutaneous fat distribution during growth. In *Fat Distribution during Growth and Adolescence*, ed. C. Bouchard & F.E. Johnston, pp. 63–84. New York: A.R. Liss.

Mallen, R.W. & Shandro, W.G. (1974). Nasopharyngeal carcinoma in Eskimos. *Can. J. Otolaryngol.*, **3**, 175–179.

Malyutina, S.K., Ryabikov, A.N., Astakhova, T.I., Serova, N.V. & Nikitin, Y.P. (1991). IHD in male population of Novosibirsk and Chukotka. In *Circumpolar Health* 90, ed. B.D. Postl et al., pp. 429–432. Winnipeg: Canadian Society for Circumpolar Health.

Mann, G.V., Scott, E.M., Hursh, L.M., Heller, C.A., Youmans, J.B., Consolazio, C.F., Bridgforth, E.B., Russell, A.L. & Silverman, M. (1962). The health and nutritional status of Alaskan Eskimos: a survey of the interdepartmental committee on Nutrition for National Defence-1958. *Am. J. Clin. Nutr.*, **11**, 31–75.

Marachev, A.G. & Matveev, L. (1978). Morpho-functional manifestations of the respiratory tract adaptation in north inhabitants. In *Scientific and Technical Progress and Circumpolar Health*, ed. V. Kaznacheev, pp. 104–105. Novosibirsk: USSR Academy of Medical Sciences, Siberian Branch.

Marchand, J.F. (1943). Tribal epidemics in the Yukon. *J. Am. Med. Assoc.*, **123**, 1019–1020.

Mardh, P-A., Lind, I., From E. & Andersen, A-L. (1980). Prevalence of chlamydia trachomatis and neisseris gonorrhea infections in Greenland. *Br. J. Ven. Dis.*, **56**, 327–331.

Marschall, B. (1992). Chlamydia Trachomatis infections in a Greenlandic district. *Arct. Med. Res.*, **51**, 23–28.

Martin, J.D. (1981). Health care in Northern Canada – an historical perspective. In *Circumpolar Health* 81, ed. B. Harvald & J.P. Hart-Hansen, pp. 80–87. Oulu:

Nordic Council for Arctic Medical Research, Report Series 33.

Martin, R.B., Burr, D.B. & Schaffler, M.B. (1985). Effects of age and sex on the amount and distribution of mineral in Eskimo tibiae. *Am. J. Phys. Anthropol.*, **67**, 371–380.

Mascie-Taylor, C.G.N. & Lasker, G.W. (1988). *Biological Aspects of Human Migration*. London: Cambridge University Press.

Matsuura, Y. (1993). Analysis on annual changing trend of stature and body weight of Japanese children and youth; from 1929 to 1988 (Syowa 4th to 63rd). *Bull. Inst. Health & Sport Sci., Univ. of Tsukuba*, **16**, 121–144.

Maynard, J.E. (1976). Coronary heart disease risk factors in relation to urbanization in Alaskan Eskimo men. In *Circumpolar Health*, ed. R.J. Shephard & S. Itoh, p. 294. Toronto: University of Toronto Press.

Mazess, R.B. & Mather, W. (1974). Bone mineral content of northern Alaskan Eskimos. *Am. J. Clin. Nutr.*, **27**, 916–925.

(1975). Bone mineral content in Canadian Eskimos. *Hum. Biol.*, **47**, 45–63.

Meehan, J.P. (1955). Body heat production and surface temperatures in response to a cold stimulus. *J. Appl. Physiol.*, **7**, 537–541.

Meehan, J.P., Stoll, A.M. & Hardy, J.D. (1954). Cutaneous pain threshold in the native Alaskan Indian and Eskimo. *J. Appl. Physiol.*, **6**, 397–400.

Melbye, M., Lanier, A., Erben, R. & Gromyko, A. (1990). AIDS and HIV epidemiology in the circumpolar regions. *Arct. Med. Res.*, **49**, Suppl. 3, 9–16.

Melgaerd, J. (1960). Prehistoric culture sequences in the Eastern Arctic as elucidated by stratified sites at Igloolik. In *Man and Culture*, ed. F.C. Wallace. Selected papers of the 5th International Conference of Anthropological and Thenological Sciences. Philadelphia: University of Pennsylvania.

Mellerowicz, H. (1966). Bericht uber die Sitzung der Internationales Ergometrie-Standardisierungskommission. *Sportarzt und Sportmedizin* **8**: 405–408.

Mercier, J., Varray, A., Ramonatxo, M., Mercier, B. & Préfaut, C. (1991). Influence of anthropometric characteristics on changes in maximal exercise ventilation and breathing pattern during growth in boys. *Eur. J. Appl. Physiol.*, **63**, 235–241.

Metropolitan Life (1983). *Nutrition and Athletic Performance: Values for Ideal Weight*. New York: Metropolitan Life Insurance.

Miall, W.E., Ashcroft, M.T., Lovell, H.G. & Moore, F.A. (1967). Longitudinal study of the decline of adult height in two Welsh communities. *Hum. Biol.*, **39**, 445–454.

Middaugh, J.P. (1981). Serogroup A meningococcal meningitis in Alaska. In *Circumpolar Health* 81, ed. B. Harvald & J.P. Hart-Hansen, pp. 425–428. Oulu: Nordic Council for Arctic Medical Research, Report Series 33.

(1990). Cardiovascular deaths among Alaskan natives 1980–86. *Am. J. Publ. Health.*, **80**, 282–285.

Middaugh, J.P. & Ritter, D. (1981). A comprehensive rabies control program in Alaska. In *Circumpolar Health* 81, ed. B. Harvald & J.P. Hart-Hansen, pp. 393–397. Oulu: Nordic Council for Arctic Medical Research, Report Series 33.

Middaugh, J.P., Gilchrist, I., Misfeldt, J., From, E. & Chaika, N.A. (1990). Other sexually transmitted diseases. *Arct. Med. Res.*, **49**, Suppl. 3, 17–25.

Middaugh, J.P., Talbot, J. & Roche, J. (1991). Diabetes prevalence in Alaska, 1984–1986. *Arct. Med. Res.*, **50**, 107–119.

Mikkelsen, F. (1976). Infectious hepatitis in Greenland, 1970–72. In *Circumpolar*

Health, ed. R.J. Shephard & S. Itoh, p. 317. Toronto: University of Toronto Press.

Milan, F.A. (1970). The demography of an Alaskan village. *Arct. Anthropol.*, **7**, 26–43.

—— (1980). The demography of selected circumpolar populations. In *The Human Biology of Circumpolar Populations*, ed. F.A. Milan, pp. 13–36. London: Cambridge University Press.

Milan, F.A. & Evonuk, E. (1966). Oxygen consumption and body temperatures of Eskimos during sleep. *J. Appl. Physiol.*, **22**, 565–567.

Milan, F.A., Hannon, J.P. & Evonuk, E. (1963). Temperature regulation of Eskimos, Indians and Caucasians in a bath calorimeter. *J. Appl. Physiol.*, **18**, 378–382.

Millar, W.J. (1990). Smoking prevalence in the Canadian arctic. *Arct. Med. Res.*, **49**, Suppl. 2, 23–28.

Miller, L.K. & Irving, L. (1962). Local reactions to air cooling in an eskimo population. *J. Appl. Physiol.*, **17**, 449–455.

Mills, C.A. (1939). Eskimo sexual functions. *Science*, **89**, 11–12.

Minuk, G.Y., Nicolle, L.E., Postl, B., Waggoner, J.G. & Hoofnagle, J.H. (1981). Hepatitis A and B virus infection in an isolated Canadian Inuit settlement. In *Circumpolar Health* 81, ed. B. Harvald & J.P. Hart-Hansen, pp. 407–409. Oulu: Nordic Council for Arctic Medical Research, Report Series 33.

Minuk, G.Y., Ling, N., Postl, B., Waggoner, J.G., Nicolle, L.E. & Hoofnagle, J.H. (1985). The changing epidemiology of hepatitis B virus infection in the Canadian north. *Am. J. Epidemiol.*, **121**, 598–604.

Mirwald, R.L. & Bailey, D.A. (1986). *Maximal Aerobic Power*. London, Ontario: Sports Dynamics Publishers.

Misfeldt, J. (1990). Trends in tobacco consumption in Greenland and a suggestion for a plan of action. *Arct. Med. Res.*, **49**, Suppl. 2, 29–31.

Misfeldt, J. (1991). Sexually transmitted diseases in Greenland: a three year review. In *Circumpolar Health* 90, ed. B.D. Postl *et al.*, pp. 348–352. Winnipeg: Canadian Society for Circumpolar Health.

Misfeldt, J., Jorgensen, B.B. & Larsen, S.O. (1988). A serological mass examination for syphilis in Greenland in 1987. *Arct. Med. Res.*, **47**, 173–178.

Mitrevski, P.J. (1969). Incomplete right bundle branch block (Lead V1) in athletes. *Med. Sci. Sports*, **1**, 152–153.

Moffatt, M.E.K., Kato, C. & Watters, G.V. (1985). Length, weight and head circumference in Québec Cree children. In *Circumpolar Health* 84, ed. R. Fortuine, pp. 170–172. Seattle: University of Washington Press.

Møller, B.R., Krømann, N., Cordtz, T., Vestergaard-Andersen, T. & From, E. (1981). Occurrence of Chlammydiae and mycoplasmas in Nuuk/Godthab, Greenland. In *Circumpolar Health* 81, ed. B. Harvald & J.P. Hart-Hansen, pp. 451–453. Oulu: Nordic Council for Arctic Research, Report Series 33.

Montoye, H.J. (1975). *Physical activity and Health. An Epidemiological Study of an Entire Community*. Englewood Cliffs, NJ: Prentice Hall.

Moran, E.F. (1981). Human adaptation to Arctic zones. *Ann. Rev. Anthropol.*, **10**, 1025–1043.

Morrell, R.E., Marks, M.I., Champlin, R. & Spence, L. (1975). An outbreak of severe pneumonia due to respiratory syncytial virus in isolated arctic populations. *Am. J. Epidemiol.*, **101**, 231–237.

Mourant, A., Domaniewska-Sobczak, K. & Tills, D. (1975). *Sunshine and the*

Geographical Distribution of the Gc Alleles. Abstract of communication, 18th April, 1975. London: Society for the Study of Human Biology.

Mouratoff, G.J. & Scott, E.M. (1973). Diabetes mellitus in Eskimos after a decade. *J. Am. Med. Assoc.*, **266**, 1345–1346.

Mouratoff, G.J., Carroll, N.V. & Scott, E.M. (1969). Diabetes mellitus in Athabaskan Indians in Alaska. *Diabetes*, **18**, 29–32.

Muri, D. & Blackwood, L. (1981). Gonorrhea reinfection rate in an Alaskan native population. In *Circumpolar Health* 81, ed. B. Harvald & J.P. Hart-Hansen, pp. 446–450. Oulu: Nordic Council for Arctic Medical Research, Report Series 33.

Murphy, N.J., Bulkow, L.R., Schraer, C.D. & Lanier, A.P. (1991). Prevalence of diabetes among Yup'ik Eskimos and Alaska coastal Indians. In *Circumpolar Health* 90, ed. B.D. Postl *et al.*, pp. 423–426. Winnipeg: Canadian Society for Circumpolar Health.

Murphy, N.J., Boyko, E.J., Schraer, C.D., Lanier, A.P. & Bulkow, L.R. (1992). Diabetes Mellitus in Alaskan Yup'ik Eskimos and Athabascan Indians after 25 years. *Diabetes Care*, **15**, 1390–1392.

Nagler, F.P., van Rooyen, C.E. & Sturdy, J.H. (1949). An influenza virus epidemic in Victoria Island, N.W.T., Canada. *Can. J. Publ. Health*, **40**, 457–461.

Namiki, M., Kurihara, H., Fujita, H., Nakanishi, T., Shinzaki, R. & Maezawa, N. (1964). Clinical studies on Ainu. *Nihon Naikagaku Z.* **52**, 1206.

Näyhä, S. (1991). Alcohol-related mortality in Finnish Lapland. In *Circumpolar Health* 90, ed. B.D. Postl *et al.*, pp. 261–263. Winnipeg: Canadian Society for Circumpolar Health.

Nazarova, A.F. (1989). The genetic structure of Chukotka Peninsula Eskimos and Chukchi based on study of 13 loci of serum and erythrocyte proteins and enzymes. *J. Phys. Anthropol.*, **79**, 81–88.

Netsky, G.I. (1976). Problems of zooanthroponoses in the arctic area of Siberia. In *Circumpolar Health*, ed. R.J. Shephard & S. Itoh, pp. 317–318. Toronto: University of Toronto Press.

Newman, M.T. (1960). Adaptations in the physique of American aborigenes to nutritional factors. *Hum. Biol.*, **32**, 288–313.

Nicolle, L.E., Postl, B., Kotelewetz, E., Borgault, A.M., Albritton, W., Harding G.K.M. & Ronald, A. (1981). Nasopharyngeal carriage of Neisseria meningitidis and hemophilus influenzae in a Canadian arctic community. In *Circumpolar Health* 81, ed. B. Harvald & J.P. Hart-Hansen. Oulu: Nordic Council for Arctic Medical Research, Report Series 33.

Nielsen, J.C., Martensson, L., Gürtler, H., Gilberg, Å. & Tingsgaard, P. (1971). Gm types of Greenlandic Eskimos. *Human Hered.*, **21**, 405–409.

Nielsen, N.H. (1986). Cancer Incidence in Greenland. *Arct. Med. Res.*, **43**, 1–168.

Nielsen, N.H. & Hansen, J.P. (1981). Cancer incidence in Greenlanders. In *Circumpolar Health* 81, ed. B. Harvald & J.P. Hart-Hansen, pp. 265–267. Oulu: Nordic Council for Arctic Medical Research, Report Series 33.

(1985). Current trends in cancer incidence in Greenland. In *Circumpolar Health* 84, ed. R. Fortuine, pp. 254–255. Seattle: University of Washington Press.

Nielsen, N.H., Mikkelsen, F. & Hansen, J.P.H. (1978). Carcinomas of the uterine cervix and dysplasia in Greenland. *Acta Pathol. Microbiol. Scand.*, Sect. A., **86**, 36–44.

Nielsen, N.H., Jensen, H. & Hansen, P.K. (1988). Cervical cytology in Greenland

and occurrence of cervical carcinoma, carcinoma in situ, and dysplasia. Extent and impact of uncoordinated screening activity, 1976–1985. *Arct. Med Res.,* **47**, 179–188.

Niinimaa, V., Cole, P., Mintz, S. & Shephard, R.J. (1980). The switching point from nasal to oro-nasal breathing. *Resp. Physiol.,* **42**, 61–71.

Nikitin, Y.P. (1985). Some health problems of man in the Soviet far north. In *Circumpolar Health* 84, ed. R. Fortuine, pp. 8–10. Seattle: University of Washington Press.

Nikitin, Y.P., Astakhova, T.I., Gerasimova, E.N. & Polessky, V.A. (1981). Blood plasma lipids and lipoproteins in native and newcoming populations in Chukotka. In *Circumpolar Health* 81, ed. B. Harvald & J.P. Hart-Hansen, pp. 308–309. Oulu: Nordic Council for Arctic Medical Research, Report Series 33.

Nobmann, E.D. (1991). Dietary intake of Alaska native adults 1987–1988. In *Circumpolar Health* 90, ed. B.D. Postl *et al.,* pp. 735–738. Winnipeg: Canadian Circumpolar Health Society.

Nobmann, E.D., Mamleeva, F.Y. & Klachkova, E.V. (1994). A comparison of the diets of Siberian Chukotka and Alaska native adults and recommendations for improved nutrition, a survey of selected previous studies. *Arct. Med. Res.,* **53**, 123–129.

Norgan, N.G., Ferro-Luzzi, A. & Durnin, J.V.G.A. (1974). The energy and nutrient intake and the energy expenditure of 204 New Guinea adults. *Phil. Trans.,* **268**, 309–348.

Norris, M.J. (1990a). The demography of aboriginal people in populations. In *The Human Biology of Circumpolar Populations,* ed. F.A. Milan, pp. 13–35. London: Cambridge University Press.

(1990b). The demography of aboriginal people in Canada. In *Ethnic Demography: Canadian Immigrant, Racial and Cultural Variations,* ed. S.S. Halli, F. Trovato & L. Driedger, pp. 33–59. Ottawa: Carleton University Press.

Nutrition Canada (1973). *Nutrition: A National Priority.* Ottawa: Queen's Printer.

(1975). *Eskimo Survey Report.* Ottawa: Dept. of National Health & Welfare.

Ogrim, M.E. (1970). The nutrition of Lapps. *Arctic Anthropol.,* **7**, 49–52.

Olsen, J., DeGross, D., Greig, M. & Olsen, O.R. (1990). Sexual behaviour. *Arct. Med. Res.,* **49**, Suppl. 3, 26–29.

Omoto, K. (1973a). Polymorphic traits in peoples of Eastern Asia and the Pacific. *Israeli J. Med. Sci.,* **9**, 1195–1215.

(1973b). The Ainu – a racial isolate. *Israeli J. Med. Sci.,* **9**, 1285–1290.

Omoto, K. & Harada, S. (1972). The distribution of polymorphic traits in the Hidaka Ainu. II. Red cell enzyme and serum protein groups. *J. Faculty of Science, University of Tokyo,* Section V, **4**, 171–211.

Omoto, K. & Misawa, S. (1974). The genetic relations of the Ainu. In *The Biological Origins of the Australians,* ed. R.L. Kirk & A.G. Thorne, p. 299. Canberra: Institute of Aboriginal Studies.

O'Neil, J.D. (1985). Self-determination and Inuit youth. Coping with stress in the Canadian north. In *Circumpolar Health* 84, ed. R. Fortuine, pp. 436–442. Seattle: University of Washington Press.

(1991). Regional health boards and the democratization of health care in the North West Territories. In *Circumpolar Health* 90, ed. B.D. Postl *et al.,* pp. 50–53. Winnipeg: Canadian Society for Circumpolar Health.

Page, J. & Brown, G.M. (1953). Effect of heating and cooling the legs on hand and forearm blood flow in the Eskimo. *J. Appl. Physiol.*, **5**, 753–758.

Page, J., Green, J.D., Hatcher, J.D. & Brown, G.M. (1954). The peripheral circulation of the Eskimo. *Rev. Can. Biol.*, **13**, 79–80.

Palva, I.P. & Finell, B. (1976). Tuberculosis in Finnish Lapland. In *Circumpolar Health*, ed. R.J. Shephard & S. Itoh, pp. 331–334. Toronto: University of Toronto Press.

Pandolf, K.B., Haisman, M.F. & Goldman, R.F. (1976). Metabolic energy expenditure and terrain coefficients for walking on snow. *Ergonomics*, **19**, 683–690.

Paraschak, V. (1982). The heterotransplantation of organized sport: a north-west territories case study. In *Proceedings, 5th Canadian Symposium on the History of Sport and Physical Education, Toronto, Ont.*, ed. B. Kidd & R. Beamish, pp. 424–430. Toronto: University of Toronto School of Physical & Health Education.

Pasquale, P.E. (1991). Family violence in the North: What do we know and where do we go from here? In *Circumpolar Health 90*, ed. B.D. Postl *et al.*, pp. 586–589. Winnipeg: Canadian Society for Circumpolar Health.

Peart, A.F., & Nagler, F.P. (1954). Measles in the Canadian Arctic, 1952. *Can. J. Publ. Health*, **45**, 146–157.

Pelto, P.J. (1962). *Individualism in Skolt Lapp Society*. Helsinki: Finnish Antiquities Society. Cited by Forsius (1980).

Persson, I., Rivat, L., Rousseau, P.Y. & Ropartz, C. (1972). Ten Gm factors and the Inv system in Eskimos in Greenland. *Hum. Hered.*, **22**, 519–528.

Petersen, K.M. & Brant, L.J. (1984). Growth and hematological changes in the Eskimo children of Wainwright, Alaska: 1968 to 1977. *Am. J. Clin. Nutr.*, **39**, 460–465.

Petersen, N.J., Barrett, D.H., Bond, W.W., Berquist, K.R., Favero, M.S., Bender, T.R. & Maynard, J.E. (1976). Hepatitis B surface antigen in saliva, impetigionous lesions, and the environment in two remote Alaskan villages. *Appl. Envir. Microbiol.*, **32**, 572–574.

Péwé, T.L. (1966). *Permafrost and its Effects on Life in the North*. Corvallis, Oregon: Oregon State University Press.

Pflaum, J., Morris, K., Young, D., Sanders, N., Predeger, B. & Littell, S. (1991). Providing baccalaureate nursing education to remote populations via telecommunications: problems and solutions. In *Circumpolar Health 90*, ed. B.D. Postl *et al.*, pp. 143–144. Winnipeg: Canadian Society for Circumpolar Health.

Philip, R.N., Reinhard, K.R. & Lackman, D.B. (1959a). Observations on a mumps epidemic in a 'virgin' population. *Am. J. Hygiene*, **69**, 91–111.

Philip, R.N., Weeks, W.T., Reinhard, K.R., Lackman, D.B. & French, C. (1959b). Observations on Asian influenza on two Alaskan islands. *Publ. Health Rep.*, **74**, 737–745.

Pickering, J., Lavallée, C. & Hanley, J. (1989). Cigarette smoking in Cree Indian school children of the James Bay region. *Arct. Med. Res.*, **48**, 6–11.

Poirier, S., Oshima, H., de Thé, G., Huberty, A., Bourgade, M.C. & Bartch, H. (1987). Volatile nitrosamine levels in common foods from Tunisia, China and Greenland, high risk areas for nasopharyngeal carcinoma (NPC). *Int. J. Cancer*, **39**, 293–296.

Pollitt, E. & Amante, P. (1984). *Energy Intake and Activity*. New York: Alan Liss.

Postl, B. (1981). Six year follow-up of Keewatin Inuit children born 1973–74. In

Circumpolar Health 81, ed. B. Harvald & J.P. Hart-Hansen, pp. 175–179. Oulu: Nordic Society for Arctic Medical Research, Report Series 33.

(1985). Northwest Territories perinatal and infant morbidity and mortality study: follow-up 1982. 1. Utilization, morbidity and mortality. In *Circumpolar Health* 84, ed. R. Fortuine, pp. 125–128. Seattle: University of Washington Press.

Rabinowitch, I.M. & Smith, F.C. (1936). Metabolic studies of Eskimos in the Canadian Eastern Arctic. *J. Nutr.*, **12**, 337–356.

Rausch, R.L. (1974). Tropical problems in the Arctic. Infectious and parasitic diseases: a common denominator. In *Industry and Tropical Health*, Vol. VIII, pp. 63–70. Boston: Harvard School of Public Health.

Rausch, R. L. & Wilson, J.F. (1985). The current status of alveolar hydatid disease in northern regions. In *Circumpolar Health* 84, ed. R. Fortuine, pp. 245–248. Seattle: University of Washington Press.

Read, S.C. & Watts, J.E. (1991). Indigenous control of health services. In *Circumpolar Health* 90, ed. B.D. Postl *et al.*, pp. 54–58. Winnipeg: Canadian Society for Circumpolar Health.

Reece, E.R. & Brotton, T.S. (1981). In vitro studies of cell-mediated immunity in Inuit children. In *Circumpolar Health* 81, ed. B. Harvald & J.P. Hart-Hansen, pp. 385–388. Oulu: Nordic Council for Arctic Medical Research, Report Series 33.

Reijula, K., Larmi, E., Hassi, J. & Hannuksela, M. (1990). Respiratory symptoms and ventilatory function among Finnish reindeer herders. *Arct. Med. Res.*, **49**, 74–80.

Renaud, L. & Dumont, C. (1985). Tuberculosis in the James Bay Cree Indian population 1980–1983. In *Circumpolar Health* 84, ed. R. Fortuine, pp. 217–221. Seattle: University of Washington Press.

Renbourn, E.T. (1972). *Materials and Clothing in Health and Disease*. London: H.K. Lewis.

Rennie, D.W. (1988). Tissue heat transfer in water: lessons from the Korean divers. *Med. Sci. Sports Exerc.*, **20**, Suppl., S177–S184.

Rennie, D.W., Covino, B.G., Blair, M.R. & Rodahl, K. (1962). Physical regulation of temperature in Eskimos. *J. Appl. Physiol.*, **17**, 326–332.

Rennie, D.W., di Prampero, P., Fitts, R.W. & Sinclair, L. (1970). Physical fitness and respiratory function of Eskimos of Wainwright, Alaska. *Arctic Anthropol.*, **7**, 73–82.

Rhodes, A.J. (1949). Poliomyelitis among Eskimos. *Can. J. Publ. Health*, **40**, 440–442.

Ritchie, J.C. & Cwynar, L.C. (1982). The late quaternary vegetation of the North Yukon. In *Paleoecology of Beringia*, ed. D.M. Hopkins, J.V. Matthews, C.E. Schweger & S.B. Young, pp. 113–126. New York: Academic Press.

Roberts, D.F. (1952). Basal metabolism, race and climate. *J. Roy. Anthropol. Inst.*, **82**, 169–183.

(1953). Body weight, race and climate. *Am. J. Phys. Anthropol.*, **11**, 533–558.

Rodahl, K. (1952). Basal metabolism of the Eskimo. *J. Nutr.*, **48**, 359–368.

(1954). Nutritional requirements in cold climates. *J. Nutr.*, **53**, 575–588.

(1958). Physical fitness. *J. Am. Geriatr. Soc.*, **6**, 205–209.

Rode, A. & Shephard, R.J. (1971). Cardiorespiratory fitness of an arctic community. *J. Appl. Physiol.*, **31**, 519–526.

(1973a). Pulmonary function of Canadian Eskimos. *Scand. J. Resp. Dis.*, **54**, 191–205.

(1973b). Fitness of the Canadian Eskimo: the influence of season. *Med. Sci. Sports*, **5**, 170–173.

(1973c). On the mode of exercise appropriate to an arctic community. *Int. Z. Angew. Physiol.*, **31**, 187–196.

(1973d). The cardiac output, blood volume and total haemoglobin of the Canadian Eskimo. *J. Appl. Physiol.*, **34**, 91–96.

(1973e). Growth, development and fitness of the Canadian Eskimo. *Med. Sci. Sports Exerc.*, **5**, 161–169.

(1984a). Ten years of 'civilization'. Fitness of the Canadian Inuit. *J. Appl. Physiol.*, **56**, 1472–1477.

(1984b). Growth, development and acculturation –a ten year comparison of Canadian Inuit children. *Hum. Biol.*, **56**, 217–230.

(1984c). Lung function in Canadian Inuit: a follow-up study. *Can. Med. Assoc. J.*, **131**, 741–744.

(1985). Lung function in a cold environment: a current perspective. In *Circumpolar Health* 84, ed. R. Fortuine, pp. 60–63. Seattle: University of Washington Press.

(1992a). *Fitness and Health of an Inuit Community: 20 Years of Cultural Change.* Ottawa: Circumpolar and Scientific Affairs.

(1992b). Cold, fitness and the exercise electrocardiogram. A 20-year longitudinal study of Canadian Inuit. *Int. J. Sorts Med.*, **13**, S176–S178.

(1993). Acculturation and loss of fitness in the Inuit: the preventive role of active leisure. *Arct. Med. Res.*, **52**, 107–112.

(1994a). The physiological consequences of acculturation: a 20-year study in an Inuit community. *Eur. J. Appl. Physiol.*, **69**, 516–524.

(1994b). *A Comparison of Fitness Between Inuit and an Indigenous Siberian Population.* Ottawa: Circumpolar and Scientific Affairs.

(1994c). Growth and fitness of Canadian Inuit: secular trends, 1970–1990. *Am. J. Hum. Biol.*, **6**, 525–542.

(1994d). Prediction of body fat content in an Inuit community. *Am. J. Hum. Biol.*, **6**, 249–254.

(1994e). Acculturation and the growth of lung function: three cross-sectional surveys of an Inuit community. *Respiration*, **61**, 187–194.

(1994f). The ageing of lung function: cross-sectional and longitudinal studies of an Inuit community. *Eur. Respir. J.*, **7**, 1653–1659.

(1994g). *Health and Fitness: A Comparative Study of Canadian Inuit and Siberian nGanasan*, Toronto: University of Toronto School of Physical and Health Education.

(1994h). Secular and age trends in the height of adults among a Canadian Inuit community. *Arct. Med. Res.*, **53**, 18–24.

(1995). Growth patterns of Canadian Inuit children: a longitudinal study. *Arct. Med. Res.*, in press.

Rode, A., Shephard, R.J., Vloshinsky, P.E. & Kuksis, A. (1995). Plasma fatty acid profiles of Canadian Inuit and Siberian nGanasan. *Arct. Med. Res.*, **54**, 422–425.

Rodgers, D.D. (1991). Community crisis intervention in suicide epidemics. In *Circumpolar Health* 90, ed. B.D. Postl *et al.*, pp. 276–280. Winnipeg: Canadian Society for Circumpolar Health.

Romet, T.T., Shephard, R.J. & Frim, J. (1986). The metabolic cost of exercise in cold air. *Arct. Med. Res.*, **44**, 29–36.

Roychoudhury, A.K. & Nei, M. (1988). *Human Polymorphic Genes: World Distribution*. New York: Oxford University Press.

Ryan, C. (1981). Trichinosis from frozen grizzly bear meat. In *Circumpolar Health 81*, ed. B. Harvald & J.P. Hart-Hansen, pp. 442–445. Oulu: Nordic Council for Arctic Medical Research, Report Series 33.

Rychkov, Y.G. & Sheremet'eva, A. (1980). The genetics of circumpolar populations of Eurasia related to the problem of human adaptation. In *The Human Biology of Circumpolar Populations*, ed. F.A. Milan, pp. 37–80. London: Cambridge University Press.

Sabry, J.H., Evers, F.T. & Wein, E.E. (1989). *Food Habits and Nutrient Intakes of Northern Native Canadians near Wood Buffalo National Park, with Emphasis on the Role of Country Foods*. Report to Health & Welfare Canada, 1989.

Sagild, U., Littauer, J., Jespersen, S.C. & Andersen, S. (1966). Epidemiologic studies in Greenland 1962–64. I. Diabetes mellitus in Eskimos. *Acta Med. Scand.*, **179**, 29–39.

Sahi, T., Eriksson, A.W., Isokoski, M. & Kirjarinta, M. (1976). Isolated adult-type lactose malabsorption in Finnish Lapps. In *Circumpolar Health*, ed. R.J. Shephard & S. Itoh, pp. 145–146. Toronto: University of Toronto Press.

Sampath, H.M. (1976). Modernity, social structure, and mental health of Eskimos in the Canadian eastern arctic. In *Circumpolar Health*, ed. R.J. Shephard & S. Itoh, pp. 479–489. Toronto: University of Toronto Press.

Sauberlich, H.E., Goad, W., Herman, Y.F., Milan, F.A. & Jamison, P. (1970). Preliminary report on the nutrition survey conducted among the Eskimos of Wainwright, Alaska, 21–27 January, 1969. *Arctic Anthropol.*, **7**, 122–124.

Sayed, J.E., Hildes, J.A. & Schaefer, O. (1976a). Biochemical indices of nutrition of the Iglooligmiut. In *Circumpolar Health*, ed. R.J. Shephard & S. Itoh, pp. 130–134. Toronto: University of Toronto Press.

Sayed, H., Sayed, J., Hildes, J.A. & Schaefer, O. (1976b). Deficiency of secretory IgA in Eskimo saliva. In *Circumpolar Health*, ed. R.J. Shephard & S. Itoh, p. 221. Toronto: University of Toronto Press.

Schaefer, O. (1968a). Glycosuria and diabetes mellitus in Canadian Eskimos. *Can. Med. Assoc. J.*, **99**, 201–206.

(1968b). Glucose tolerance testing in Canadian Eskimos: a preliminary report and a hypothesis. *Can. Med. Assoc. J.*, **99**, 252–262.

(1970). Pre-and post-natal growth acceleration and increased sugar consumption in Canadian Eskimos. *Can. Med. Ass. J.*, **103**, 1059–1068.

(1971). Otitis media and bottle feeding. An epidemiological study of infant feeding habits and incidence of recurrent and chronic middle ear disease in Canadian Eskimos. *Can. J. Publ. Health*, **62**, 478–489.

(1977). Are Eskimos more or less obese than other Canadians? A comparison of skinfold thickness and ponderal index in Canadian Eskimos. *Am. J. Clin. Nutr.*, **30**, 1623–1628.

(1981). Ethnology, demography and medicine in the arctic. In *Circumpolar Health 81*, ed. B. Harvald & J.P. Hart-Hansen, pp. 187–193. Oulu: Nordic Council for Arctic Medical Research, Report Series 33.

Schaefer, O., Crockford, P.M. & Romanowski, B. (1972). Normalization effect of preceding protein meals on 'diabetic' oral glucose tolerance in Eskimos. *Can.*

Med. Assoc. J., **107**, 733–738.

Schaefer, O., Hildes, J.A., Greidanus, P. & Leung, D. (1974). Regional sweating in Eskimos compared to Caucasians. *Can. J. Physiol. Pharmacol.*, **52**, 960–965.

Schaefer, O., Timmermans, J.F., Eaton, R.D. & Mathews, A.R. (1980a). General and nutritional health in two Eskimo populations at different stages of acculturation. *Can. J. Publ. Health*, **71**, 397–405.

Schaefer, O., Eaton, R.D.P., Timmermans, F.J.W. & Hildes, J.A. (1980b). Respiratory function impairment and cardio-pulmonary consequences in long-term residents of the Canadian arctic. *Can. Med. Ass. J.*, **123**, 997–1004.

Schanfield, M.S., Crawford, M.H., Dossetor, J.B. & Gershowitz, H. (1990). Immunoglobulin allotypes in several North American Eskimo populations. *Hum. Biol.*, **62**, 773–789.

Schraer, C.D., Lanier, A.P., Boyko, E.J., Gohndes, D. & Murphy, N.J. (1988). Prevalence of diabetes mellitus in Alaskan Eskimos, Indians and Aleuts. *Diabetes Care*, **11**, 693–700.

Schreeder, M.T., Bender, T.R., McMahon, B.J., Moser, M.R., Murphy, B.L., Sheller, M.J., Heyward, W.L., Hall, D.B. & Maynard, J.A. (1983). Prevalence of hepatitis B in selected Alaskan Eskimo villages. *Am. J. Epidemiol.*, **118**, 543–549.

Schreider, E. (1967). Body height and in-breeding in France. *Am. J. Phys. Anthropol.*, **26**, 1–3.

Schrire, C. & Steiger, W.L. (1974). A matter of life and death: an investigation into the practice of female infanticide in the Arctic. *Man*, **9**, 161–184.

Schultz, V., Theno, S., Kappes, B. & Morse, C. (1991). Socioeconomic and environmental characteristics of women of domestic violence in Alaska. In *Circumpolar Health 90*, ed. B.D. Postl *et al.*, pp. 590–595. Winnipeg: Canadian Society for Circumpolar Health.

Scott, E.M. Griffith, I.V., Hoskins, D.D. & Whaley, R.D. (1958). Serum cholesterol levels and blood pressure of Alaskan Eskimo men. *Lancet*, **ii**, 667–668.

Scott, E.M. & Griffith, I.V. (1957). Diabetes mellitus in Eskimos. *Metabolism*, **6**, 320–325.

Seckler, D. (1980). 'Malnutrition': an intellectual Odyssey. *West. J. Ag. Econ.*, **5**, 219.

Sedov, K.R. (1991). Social and economic progress of the north and health of northern native population in the USSR. In *Circumpolar Health 90*, ed. B. Postl *et al.*, pp. 210–212. Winnipeg: Canadian Society for Circumpolar Health.

Segal, B. (1985). Drugs in Alaska: Patterns of use by adults and school age youth. In *Circumpolar Health 84*, ed. R. Fortuine, pp. 344–347. Seattle: Washington University Press.

Seitamo, L. (1976). Psychological adaptation of Skolt Lapp children to cultural change. In *Circumpolar Health*, ed. R.J. Shephard & S. Itoh, pp. 497–507. Toronto: University of Toronto Press.

    (1981). Development of the role of Skolt women. In *Circumpolar Health 81*, ed. B. Harvald & J.P. Hart-Hansen, pp. 464–466. Oulu: Nordic Council for Arctic Medical Research, Report Series 33.

Sellers, F.J., Wood, W.J. & Hildes, J.A. (1959). The incidence of anemia in infancy and early childhood among central arctic Eskimos. *Can. Med. Assoc. J.*, **81**, 656–657.

Shephard, R.J. (1956). *The Physiological Sequealae of Segmental Resection and other forms of Thoracic Treatment in Flying Personnel*. RAF Institute of Aviation

Medicine, Farnborough, Hants, Report FPRC 1007, 1–91.

(1974a). Sport for youth: the Eskimo approach. *Br. J. Sports Med.*, **7**, 315–316.

(1974b). Work physiology and activity patterns of circumpolar Eskimos and Ainu. A synthesis of IBP data. *Hum. Biol.*, **46**, 263–294.

(1977). *Endurance Fitness*, 2nd edn. Toronto: University of Toronto Press.

(1978). *Human Physiological Work Capacity.* London: Cambridge University Press.

(1980). Work physiology and activity patterns. In *Human Biology of Circumpolar Populations,* ed. F.A. Milan, pp. 305–338. London: Cambridge University Press.

(1982). *Physiology and Biochemistry of Exercise.* New York: Praeger Publishing, 672 pp.

(1983). *Biochemistry of Physical Activity.* Springfield, IL: C.C. Thomas, 391 pp.

(1985). Adaptation to exercise in the cold. *Sports Med.*, **2**, 59–71.

(1986). *Fitness of a Nation: Lessons from the Canada Fitness Survey.* Basel: Karger Publications, 186 pp.

(1987). *Physical Activity and Aging*, 2nd edn. London: Croom Helm Publishing, 354 pp.

(1988). Work physiology and activity patterns. In *The Biology of Circumpolar Populations*, ed. F.A. Milan, p. 306. London: Cambridge University Press.

(1992). Respiratory irritation from environmental tobacco smoke. *Arch. Env. Health*, **47**, 123–130.

(1993a). Metabolic adaptations to exercise in the cold. *Sports Med.*, **16**, 266–289.

(1993b). *Aerobic Fitness and Health.* Champaign, IL: Human Kinetics Publishers, 358 pp.

(1993c). Aging, lung function and exercise. *J. Phys. Activ. Aging*, **1**, 59–83.

Shephard, R.J. & Lavallée, H. (1993). Enhanced physical education and body fat in the primary school child. *Am. J. Hum. Biol.*, **5**, 697–704.

(1995). Effects of enhanced physical education, gender and environment on lung volumes of primary school children. *Respiration*, in press.

Shephard, R.J. & Rode, A. (1973). Cardio-respiratory status of the Canadian Eskimo. In *Polar Human Biology.* ed. O.G. Edholm & E.K.E. Gunderson, pp. 216–239. London: Heinemann Medical Books.

Shephard, R.J. & Bouchard, C. (1994). Population evaluations of health-related fitness from perceptions of physical activity and fitness. *Can. J. Appl. Physiol.*, **19**, 151–173.

Shephard, R.J., Allen, C., Benade, S., Davies, C.T.M., DiPrampero, P.E., Hedman, R., Merriman, J.E., Myhre, K. & Simmons, R. (1968). The maximum oxygen intake – an international reference standard of cardiorespiratory fitness. *Bull. WHO*, **38**, 757–764.

Shephard, R.J., Hatcher, J. & Rode, A. (1973). On the body composition of the Eskimo. *Eur. J. Appl. Physiol.*, **30**, 1–13.

Shephard, R.J., Lavallée, H., Beaucage, C., Pérusse, M., Rajic, M., Brisson, G., Jéquier, J-C., Larivière, G. & LaBarre, R. (1975). La capacité physique des enfants canadiens; une comparaison entre les enfants canadiens francais, canadiens anglais et equimaux. III. Psychologie et sociologie des enfants canadiens francais. *Union Méd.*, **104**, 1131–1136.

Shephard, R.J. Lavallée, H. & Larivière, G. (1978a). Competitive selection among age-class ice-hockey players. *Br. J. Sports Med.*, **12**, 11–13.

Shephard, R.J., Lavallée, H., Rajic, M., Jéquier, J-C., Brisson, G. & Beaucage, C.

(1978b). Radiographic age in the interpretation of physiological and anthropological data. In *Pediatric Work Physiology*, ed. J. Borms & M. Hebbelinck, pp. 124–133. Basel: Karger Publishing.

Shephard, R.J., Lavallée, H., Jéquier, J-C., Rajic, M., & LaBarre, R. (1980). On the basis of data standardization in prepubescent children. In *Kinanthropometry II*, ed. M. Ostyn, G. Beunen & J. Simons, pp. 306–316. Baltimore: University Park Press.

Shephard, R.J., Goodman, J., Rode, A. & Schaefer, O. (1984). Snowmobile use and decrease of stature among the Inuit. *Arct. Med. Res.*, **38**, 32–36.

Sherif, C. & Rattray, C.D. (1976). Psychological development and activity in middle childhood (5–12 years). In *Child in Sport and Physical Activity*, ed. J. G. Albinson & G.M. Andrew, pp. 97–132. Baltimore: University Park Press.

Shields, G.F., Hecker, K., Voevoda, M.I. & Reed, K. (1992). Absence of the Asian-specific Region V mitochondrial marker in native Beringians. *Am. J. Hum. Genet.*, **50**, 758–765.

Siebke, J.C., Jorgensen, J.B., Laughlin, W.S. & Reinicke, V. (1986). Prevalence of hepatitis A and B markers in the Eskimos of Kodiak Island, Alaska. *Arct. Med. Res.*, **44**, 48–52.

Siemssen, O.J., Siemssen, S.J., Nielsen, N.H., Hart-Hansen, J.P. & Pedersen, E. (1981). Volatile nitrosamines in food and beverages in southern Greenland. In *Circumpolar Health 81*, ed. B. Harvald & J.P. Hart-Hansen, pp. 289–290. Oulu: Nordic Council for Arctic Medical Research, Report Series 33.

Simons, M.J., Wee, G.B., Day, N.E., Morris, P.J., Shanmugaratnam, K. & de Thé, G.B. (1974). Immuno-genetic aspects of nasopharyngeal carcinoma. (i). Differences in HLA antigen profile between patients and comparison groups. *Int. J. Cancer*, **13**, 122–134.

Simopoulos, A.P. (1991). Omega-3 fatty acids in health and disease and in growth and development. *Am. J. Clin. Nutr.*, **54**, 438–463.

Simpson, N.E. (1981). The load of genetic disease and genetic predisposition to disease in the Canadian north. In *Circumpolar Health 81*, ed. B. Harvald & J.P. Hart-Hansen, pp. 145–153. Oulu: Nordic Council for Arctic Medical Research, Report Series 33.

Sjoflot, L. (1982). The tractor as a workplace. *Ergonomics*, **25**, 11–18.

Skinhøj, P. (1991). Epidemiology of viral hepatitis in circumpolar populations. *Arct. Med. Res.*, **50**, 177–179.

Skinhøj, P., Mikkelsen, F. & Hollinger, F.B. (1977). Hepatitis A in Greenland: importance of specific antibody testing in epidemiologic surveillance. *Am. J. Epidemiol.*, **105**, 140–147.

Skrobak-Kaczynski, J. & Lewin, T. (1976). Secular changes Lapps of Northern Finland. In *Circumpolar Health*, ed. R.J. Shephard, & S. Itoh, pp. 239–247. Toronto: University of Toronto Press.

Skrobak-Kaczynski, J., Lewin, T. & Karlberg, J. (1974). Secular changes in body dimensions in a homogenous population. *Arct. Med. Res. Rept.*, **8**, 17–46.

Smith, M.C. (1976). Changing health hazards in infancy and childhood in northern Canada. In *Circumpolar Health*, ed. R.J. Shephard & S. Itoh, pp. 448–452. Toronto: University of Toronto Press.

Society of Actuaries (1959). *Build and Blood Pressure Study*. Chicago: Society of Actuaries.

Soininen, L. (1991a). Infectious diseases in Lapland in the years 1987–1989. In

Circumpolar Health 90, ed. B.D. Postl *et al.*, pp. 336–338. Winnipeg: Canadian Society for Circumpolar Health.

(1991b). Toullmos, an old healing method of the Skoltsami. In *Circumpolar Health* 90, ed. B.D. Postl *et al.*, pp. 221–222. Winnipeg: Canadian Society for Circumpolar Health.

Soininen, L. & Åkerblom, H. (1981). Infant mortality in the Lapp areas of Finnish Lapland. In *Circumpolar Health* 81, ed. B. Harvald & J.P. Hart-Hansen, pp. 169–171. Oulu: Nordic Council for Arctic Medical Research, Report Series 33.

Sollas, W.J. (1924). *Ancient Hunters and their Modern Representatives*, 3rd edn. London: MacMillan.

Spitzer, W., Baxter, D., Barrows, H., Thomas, D.C., Tamblyn, R., Wolfson, C.M., Dinsdale, H.B., Dauphinee, W.D., Anderson, D.P., Roberts, R.C., Palmer, W.H., Hollomby, D., Reiher, J., Alleyne, B.C. & Helliwell, B.E. (1988). Methyl mercury and the health of Autochtons in Northwestern Quebec. *Clin. Invest. Med.*, **11**, 71–97.

Spuhler, J.N. (1979). Genetic distances, trees and maps of North American Indians. In *The First Americans: Origins, Affinities and Adaptations*, ed. W.S. Laughlin & A.B. Harper, pp. 135–183. New York: Gustav Fischer.

Spurr, G.B., Prentice, A.M., Murgatroyd, P.R., Goldberg, G.R., Reina, J.C. & Christman, N.T. (1988). Energy expenditure from minute by minute heart rate recording: comparison with indirect calorimetry. *Am. J. Clin. Nutr.*, **48**, 552–559.

Stager, J.K. & McSkimming, R.J. (1984). Physical environment. In *Arctic*, ed. D. Damas, pp. 27–35. Washington, DC: Smithsonian Institute.

Statistics Canada (1989). *A Data Book on Canada's Aboriginal Population from the 1986 Census of Canada*. Ottawa: Statistics Canada.

Steegman, A.T. (1975). Human adaptation to cold. In *Physical Anthropology*, ed. A. Damon, pp. 130–166. New York: Oxford University Press.

Steegman, A.T., Hurlich, M.G. & Winterhalder, B. (1983). Coping with cold and other challenges of the boreal forest: an overview. In *Boreal Forest Adaptations: The Northern Algonkians*, ed. T.A. Steegman, pp. 317–351. New York: Plenum Press.

Stepanova, E.G. & Shubnikov, E.W. (1991). Diabetes, glucose tolerance and some risk factors of diabetes mellitus in natives and newcomers of Chukotka. In *Circumpolar Health* 90, ed. B.D. Postl *et al.*, pp. 413–414. Winnipeg: Canadian Society for Circumpolar Health.

Stephens, M.C. (1991). The special premedical studies program – Review of ten years of experience. In *Circumpolar Health* 90, ed. B.D. Postl *et al.*, pp. 134–137. Winnipeg: Canadian Society for Circumpolar Health.

Stephens, T. (1989). Fitness and activity measurements in the 1981 Canada Fitness Survey. In *Assessing Physical Fitness and Physical Activity*, ed. T. Drury, pp. 401–432. Hyattsville, MD: US Dept. of Health & Human Services.

Stephens, T. & Craig, C. (1990). *The Well-being of Canadians: The 1988 Campbell's Survey*. Ottawa: Canadian Fitness and Lifestyle Research Institute.

Stewart, T.D. (1939). Anthropometric observations on the Eskimos and Indians of Labrador. *Field Museum of Natural History, Anthropology Series*, **31**, (1), 3–163. (cited by Zammit *et al.*, 1993).

Stich, H.F. & Hornby, A.P. (1985). Endogenous and exogenous sources of nitrate/nitrite and nitrosamines in Canadian Inuit with a traditional, or a western lifestyle. In *Circumpolar Health 84*, ed. R. Fortuine, pp. 92–95. Seattle:

University of Washington Press.

Storm, H.H., Nielsen, N.H., Prener, A. & Jensen, O.M. (1991). A comparison of cancer in Greenland and Denmark: a study based on routinely collected incidence data 1973–1985, using the Danish population as baseline. In *Circumpolar Health* 90, ed. B.D. Postl *et al.*, pp. 470–471. Winnipeg: Canadian Society for Circumpolar Health.

Sundberg, S., Luukka, P., Andersen, K.L.A., Eriksson, A.W. & Siltanen, P. (1975). Blood pressure in adult Lapps and Skolts. *Ann. Clin. Res.*, **7**, 17–22.

Sundström, I., Zetterqvist, H. & Björnstig, U. (1994). Snowmobile injuries in Kiruna, Northern Sweden. *Arct. Med. Res.*, **53**, 189–195.

Suzuki, S. (1970). Experimental studies on factors on growth. *Monogr. Soc. Res. Child Dev.*, **35**, 6–11.

Svejgaard, A., Platz, P., Ridjer, L.P., Nielsen, L.S. & Thomsen, M. (1975). HL-A and disease associations – a survey. *Transplant. Rev.*, **22**, 3–43.

Szabo, E.L. (1991). Mortality related to alcohol use among the status Indian population of Saskatchewan. In *Circumpolar Health* 90, ed. B.D. Postl *et al.*, pp. 267–275. Winnipeg: Canadian Society for Circumpolar Health.

Szathmary, E.J. (1981). Genetic markers in Siberia and northern North American populations. *Yearb. Phys. Anthropol.*, **24**, 37–73.

Szathmary, E.J. & Holt, N. (1983). Hyperglycemia in Dogrib Indians of the North West Territories, Canada: association with age and a centripetal distribution of body fat. *Hum. Biol.*, **55**, 493–515.

Szathmary, E.J. & Ossenberg, N.S. (1978). Are the biological differences between North American Indians and Eskimos truly profound? *Curr. Anthropol.*, **19**, 673–701.

Szathmary, E.J., Ritenbaugh, C. & McDonald, C.S. (1987). Dietary change and plasma glucose levels in an Amerindian population undergoing cultural transition. *Soc. Sci. Med.*, **24**, 791–804.

Tarp, U. & Hansen, J.C. (1991). Influence of diet on blood cadmium levels in Greenlanders. In *Circumpolar Health* 90, ed. B.D. Postl *et al.*, pp. 768–769. Winnipeg: Canadian Society for Circumpolar Health.

Taylor, W. (1963). Hypothesis on the origin of the Canadian Thule culture. *American Antiquity*, **28**, 456–464.

Thomas, G.W. & Williams, J.H. (1976). Cancer in native populations in Labrador. In *Circumpolar Health*, ed. R.J. Shephard & S. Itoh, pp. 283–289. Toronto: University of Toronto Press.

Thomlinson, E., Gregory, D. & Larsen, J. (1991). A northern Bachelor of Nursing program: one solution to problems in health care provision. In *Circumpolar Health* 90, ed. B.D. Postl *et al.*, pp. 145–148. Winnipeg: Canadian Society for Circumpolar Health.

Thomsen, M., Platz, P., Andersen, O., Christy, M., Lynsgoe, J., Nerup, J., Rasmussen, K., Ryder, L.P., Nielsen, L.S. & Svejgaard, A. (1975). MLC typing in juvenile diabetes mellitus and idopathic Addison's disease. *Transplant. Rev.*, **22**, 125–147.

Thorslund, J. (1991). Suicide among Inuit youth in Greenland. In *Circumpolar Health* 90, ed. B.D. Postl *et al.*, pp. 299–302. Winnipeg: Canadian Society for Circumpolar Health.

Thouez, J.P., Rannmou, A. & Foggin, P. (1989). The other face of development: native population health status and indicators of malnutrition – the case of the

Cree and Inuit of Northern Québec. *Soc. Sci. Med.*, **29**, 965–974.

Thouez, J-P., Ekoe, J.M., Foggin, P.M., Verdy, M., Nadeau, M., Laroche, P., Rannou, A. & Ghadirian, P. (1990). Obesity, hypertension, hyperuricemia and diabetes mellitus among the Cree and Inuit of northern Québec. *Arct. Med. Res.*, **49**, 180–188.

Tkachev, A.V., Bojko, J.V. & Ramenskaya, E.B. (1991). Endocrine status and plasma lipids in the inhabitants of the northern European part of the USSR. *Arct. Med. Res.*, **50**, Suppl. 6, 148–151.

Torrey, E.F., Reiff, F.W. & Noble, G.R. (1979). Hypertension among Aleuts. *Am. J. Epidemiol.*, **110**, 7–14.

Tumnaettuvge, O. (1991). Medico-social aspects of Chukotka natives health improvement: materials and spiritual problems. In *Circumpolar Health* 90, ed. B.D. Postl *et al.*, pp. 37–38. Winnipeg: Canadian Society for Circumpolar Health.

Turner, G.G. (1989). Teeth and prehistory in Asia. *Sci. Am.*, **260**, 70–77.

U.S. Congress, Office of Technology Assessment (1986). *Indian Health Care*. Washington, DC: U.S. Government Printing Office OTA-H-290.

Vahtkin, N. (1992). *Native Peoples of the Russian Far North*. London: Minority Rights Group.

Valle, F.G. (1968). Stresses of change and mental health among the Canadian Eskimos. *Arch. Environ. Health*, **17**, 578–585.

Vallerand, A. & Jacobs, I. (1989). Rates of energy substrate utilization during cold exposure. *Eur. J. Appl. Physiol.*, **58**, 873–878.

(1990). Influence of cold exposure on plasma triglyceride clearance in humans. *Metabolism*, **39**, 1211–1218.

Vas, S.I., Parry, T.L., Watson, C., Young, H. & Bollaerts, S. (1977). Study of Canadian Inuit sera for HBsAg and HBsAb. *Canada Diseases Weekly Rep.*, **3**, 57–58.

Virokannas, H., Hassi, J., Anttonen, H., Järvenpää, I. & Pääkatyössä, R. (1984). Terveysvaarat moottorikelkkatyössä (Health hazards in the use of snowmobiles). *Työterveyslaitoksen tutmuksia*, **2**, 103–117.

Virokannas, H., Anttonen, H. & Niskanen, J. (1988). Effects of multiexposures on the hands in snowmobile driving. In *Recent Advances in Researches on the Combined Effects of Environmental Factors*, ed. O. Manninen, 591-604. Tampere, Institute of Occupational Health.

Voevoda, M.I., Astakhova, T.I. & Nikitin, Y.P. (1991). Estimation of the relative contribution of genetic and environmental factors to population variability of blood serum lipid and arterial blood pressure in the native population of Chukotka. In *Circumpolar Health* 90, ed. B.D. Postl *et al.*, pp. 517–518. Winnipeg: Canadian Society for Circumpolar Health.

Vollmer, W.M., Johnson, L.R., McCamant, L.E. & Buist, A.S. (1988). Longitudinal versus cross-sectional estimation of lung function decline: further insights. *Stats. in Med.*, **7**, 685–696.

von Bonsdorff, C., Fellman, J. & Lewin, T. (1974). Demographic studies on the Inari Lapps in Finland, with special reference to genealogy. *Arct. Med. Res. Rep.*, **6**, 4–10.

von Döbeln, W. (1966). Kroppstorlek, Energie omsättning och Kondition. In *Handbok i Ergonomi*, ed. U. Aberg & N. Lundgrem. Stockholm: Almqvist & Wiksell.

Ward, J.I., Lum, M.K., Hall, D.B., Silimperi, D.R. & Bender, T.R. (1986). Invasive

Haemophilus influenzae type B disease in Alaska: background epidemiology for a vaccine efficacy trial. *J. Infect. Dis.*, **153**, 17–26.

Ward, J.I., Brenneman, G., Lepow, M., Lum, M., Burkhart, K. & Chiu, C.Y. (1988). Hemophilus influenzae type B anticapsular antibody responses to PRP-pertussis and PRP-D vaccines in Alaska Native infants. *J. Infect. Dis.*, **158**, 719–723.

Ware, J.H., Dockery, D.W., Louis, T.A., Xu, X., Ferris, B.G. & Speizer, F.E. (1990). Longitudinal and cross-sectional estimates of pulmonary function decline in never-smoking adults. *Am. J. Epidemiol.*, **132**, 685–700.

Waterlow, J.C. (1986). Global nutritional status. *Bull. Wld. Health. Org.*, **64**, 929–941.

Weiner, J.S. (1964). *Proposals for International Research: Human Adaptability Project Document 5.* London: Royal Anthropological Institute.

Weiner, J.S. & Lourie, J.A. (1969). *Human Biology: A Guide to Field Methods.* Oxford: Blackwell Scientific.

(1981). *Practical Human Biology.* London: Academic Press.

Weiss, K.M. (1988). In search of times past: gene flow and invasion in the generation of human diversity. In *Biological Aspects of Human Migration*, ed. C.G.N. Mascie-Taylor & G.W. Lasker, pp. 130–166. London: Cambridge University Press.

Wenzel, G. (1991). *Animal Rights, Human Rights: Ecology, Economy and Ideology in the Canadian Arctic.* Toronto: University of Toronto Press.

Westlund, K. (1981). Circumpolar non-infectious epidemiology. In *Circumpolar Health* 81, ed. B. Harvald & J.P. Hart-Hansen, pp. 215–228. Oulu: Nordic Council for Arctic Medical Research, Report Series 33.

Weyer, E.M. (1932). *The Eskimos: Their Environment and Folkways.* Hamden, CT: Archon Books.

Wheatley, M.A. (1991). Developing an integrated traditional/clinical health system in the Yukon. In *Circumpolar Health* 90, ed. B.D. Postl *et al.*, pp. 217–220. Winnipeg: Canadian Society for Circumpolar Health.

Wheatley, M.A. & Wheatley, B. (1981). The effect of eating habits on mercury levels among Inuit residents of Sugluk, P.Q. *Etudes Inuit*, **5**, 27–43.

Wherrett, G.J. (1969). A study of tuberculosis in the Eastern Arctic. *Can. J. Publ. Health*, **60**, 7–14.

Wika, M. (1971). Skin circulation of the cheeks of Arctic populations and adaptation to cold. *Acta Physiol. Scand.*, **82**, 39A–40A.

Wilber C.G. & Levine, V.E. (1950). Fat metabolism in Alaskan Eskimos. *Exp. Med. Surg.*, **8**, 422–425.

Williams, R.C., Steinberg, A.G., Gershowitz, H., Bennett, P.H., Knowler, W.C., Pettitt, D.J., Butler, W., Baird, R., Dowder-Rea, L., Burch, T.A., Morse, H.G. & Smith, C.G. (1985). Gm allotypes in Native Americans: evidence for three distinct migrations across the Bering land bridge. *Am. J. Phys. Anthropol.*, **66**, 1–19.

Wilson, C.R., Casson, R.I., Wherrett, B. & Fraser, N. (1991). Toxigenic diphtheria in two isolated northern communities. In *Circumpolar Health* 90, ed. B.D. Postl *et al.*, pp. 346–347. Winnipeg: Canadian Society for Circumpolar Health.

Wintrob, R.M., Sindell, P.S., Berry, J.W. & Mawhinney, T. (1981). The psychosocial impact of cultural change on Cree Indian women: 1966–1980. In *Circumpolar Health* 81, ed. B. Harvald & J.P. Hart-Hansen, pp. 467–480. Oulu: Nordic Council for Arctic Medical Research, Report Series 33.

Wo, C.K.W. & Draper, H.H. (1975). Vitamin E status of Alaskan Eskimos. *Am. J.*

Clin. Nutr., **28**, 808–813.

World Health Organization (1948). *Official Records #2*. Geneva: World Health Organization.

Worthington, E.B. (1978). *Evolution of the IBP*. London: Cambridge University Press.

Wotton, K.A. (1985). Mortality of Labrador Innu and Inuit, 1971–1982. In *Circumpolar Health* 84, ed. R. Fortuine, pp. 139–142. Seattle: University of Washington Press.

Wotton, K.A., Stiver, H.G. & Hildes, J.A. (1981). Meningitis in the central arctic: a 4-year experience. *Can. Med. Assoc. J.*, **124**, 887–890.

Young, T.K. (1991). Prevalence and correlates of hypertension in a subarctic Indian population. In *Circumpolar Health* 90, ed. B.D. Postl *et al.*, p. 410. Winnipeg: Canadian Society for Circumpolar Health.

(1994). Human obesity and arctic adaptation. Epidemiological patterns, metabolic effects and evolutionary implications. PhD dissertation. Oxford: Linacre College, Oxford University.

Young, T.K. & Frank, J.W. (1983). Cancer surveillance in a remote Indian population in northwestern Ontario. *Am. J. Publ. Health*, **73**, 515–520.

Young, T.K. & Krahn, J. (1988). Comparison of screening methods in a diabetes prevalence survey among Northern Indians. *Clin. Invest. Med.*, **11**, 380–385.

Young, T.K. & McIntyre, L.L. (1985). Prevalence of diabetes mellitus among the Cree-Ojibwa of North-western Ontario. In *Circumpolar Health* 84, ed. R. Fortuine, pp. 276–281. Seattle: University of Washington Press.

Young, T.K. & Sevenhuysen, G. (1989). Obesity in Northern Canadian Indians: patterns, determinants, and consequences. *Am. J. Clin. Nutr.*, **49**, 786–793.

Young, T.K., McIntyre, L.L., Dooley, J. & Rodriguez, J. (1985). Epidemiologic features of diabetes mellitus among Indians in northwestern Ontario and northeastern Manitoba. *Can. Med. Ass. J.*, **132**, 793–797.

Young, T.K., Szathmary, E.J.E., Evers, S. & Wheatley, B. (1990). Geographical distribution of diabetes among the native population of Canada: a national survey. *Soc. Sci. Med.*, **31**, 129–139.

Young, T.K., Schraer, C.D., Shubnikoff, E.V., Szathmary, E.J. & Nikitin, Y.P. (1992). Prevalence of diagnosed diabetes in circumpolar indigenous populations. *Int. J. Epidemiol.*, **21**, 730–736.

Young, T.K., Moffatt, M.E.K. & O'Neil, J.D. (1993). Cardiovascular diseases in an arctic population: an epidemiological perspective. *Am. J. Publ. Health*, **83**, 881–887.

Zachau-Christiansen, B. (1976). Height and weight in West Greenland children. *Arct. Med. Res.*, **14**, 40–46.

Zammit, M.P. (1993). Growth patterns of Labrador Inuit youth: I. Height and weight. *Arct. Med. Res.*, **52**, 153–160.

Zammit, M.P., Kalra, V., Nelson, S., Broadbent, B.H. & Hans, M.G. (1994). Growth patterns of Labrador Inuit youth: II. Skeletal age. *Arct. Med. Res.*, **53**, 176–183.

Zemel, B.S. & Johnston, F.E. (1994). Application of the Preece–Baines growth model to cross-sectional data: problems of validity and interpretation. *Am. J. Hum. Biol.*, **6**, 563–570.

Zimmet, P. (1979). Epidemiology of diabetes and its microvascular complications in Pacific populations. The medical effect of social progress. *Diabetes Care*, **2**, 144–153.

Index

302